Introductory Readings

in the

Philosophy of Science

Introductory Readings in the Philosophy of Science

Edited by

E. D. Klemke
Robert Hollinger
A. David Kline

Iowa State University
Ames, Iowa

Prometheus Books
1203 Kensington Avenue
Buffalo, New York 14215

Published 1980 by Prometheus Books

1203 Kensington Avenue, Buffalo, New York 14215

Library of Congress Catalog Number: 80-659-40

ISBN: 0-87975-134-7

Printed in the United States of America

CONTENTS

PREFACE

It is a truism that in our time science has become a highly respected and venerated enterprise. It is widely deemed to be a Good Thing. There are many who hold it to be the most successful pursuit of knowledge in the history of the world. Others consider it to be the noblest intellectual achievement of mankind. It is also true that (in recent years, especially) many have found science to be undeserving of such high praise. Not only has science been attacked by proponents of ESP, astrology, and other pseudosciences, and by critics of modern technology; it has even been found wanting by some scientists and ex-scientists and others who know and write about science. (See Paul Feyerabend's essay in Part 1 of this volume.) Who is right, or more nearly correct, in this controversy?

It is our hope that the reader of this volume will be able to arrive at his or her own answer to this question. However, in our view, one cannot attempt to answer such a question without a clear understanding of what science is and what it is not. One of our major goals has been to provide that understanding and to provide it in such a manner that it is accessible to the beginning student and the layman.

Hence we would like to stress that this is an *introductory* work in the philosophy of science. Our principles of selection for the readings included have been these: The selection should be intrinsically interesting. It should be comprehensible to a beginning student. It should serve to provoke discussion and criticism. We have also tried to stimulate a kind of dialogue among the authors by selecting works which present varying and even conflicting points of view.

We have presented the topics in the order we have found to be most desirable. However, there may be one exception. Part 1 could be used either first or last — or, perhaps best of all, both. Otherwise we believe that the order of presentation should be followed. (However, it is not necessary that every reading within each part be used.)

A few other comments are in order. First, although all three of us worked closely together on all of the material included, the general introduction and the introductions to the six parts were written individually, not collectively. (The initials at the end of each indicate primary authorship.) Second, the Study Questions at the end of each of the six parts were composed by the author of the introduction to that part. Third, the bibliographies — one is given at the end of each part (and at the end of the general introduction) — are not exhaustive. They are intended to provide further sources which deal with some of the major issues discussed in the volume. Fourth, although we have spent much time in revising our format and the selection of readings, we may have overlooked some items which ought to have been included. If so, we shall be grateful to hear from instructors who use the volume and to receive their suggestions with regard to this — or any other — issue.

We would like to express our gratitude to all those who helped in various ways with regard to the preparation of our initial proposal and the final manuscript. We are especially grateful to: Rowena Wright, Bernice Power, Annette Van Cleave, Richard Kniseley, David Hauser, and Steven Isaacson. We would also like to express our appreciation to the members of the Philosophy Department of Iowa State University for their constant help and encouragement.

ACKNOWLEDGMENTS

The editors gratefully acknowledge the kind permission of the authors, editors, and publishers that has enabled us to print the essays included in this book.

PART ONE

Sir Karl Popper, "Science: Conjectures and Refutations." From *Conjectures and Refutations,* pp. 33–41, 52–59. New York: Harper and Row, 1963. Also published by Basic Books, New York, and Routledge and Kegan Paul, London. © 1963, by Karl Popper.

John Ziman, "What Is Science?" *Public Knowledge,* pp. 5–27. New York: Cambridge University Press, 1968. © 1968, by Cambridge University Press.

Paul Feyerabend, "How to Defend Society Against Science." *Radical Philosophy,* No. 11 (1975), pp. 3–8. © 1975, by Paul Feyerabend.

Paul R. Thagard, "Why Astrology Is a Pseudoscience." *Proceedings of Philosophy of Science Association,* 1978, Volume One, pp. 223–224. Edited by P. D. Asquith and I. Hacking. East Lansing: Philosophy of Science Association, 1978. © 1978, by Philosophy of Science Association.

PART TWO

John Hospers, "What is Explanation?" A. Flew (ed.), *Essays in Conceptual Analysis,* pp. 94–119. London: Macmillan Publishers, Ltd., 1956. © 1956, by Macmillan Publishers, Ltd.

John Hospers, *An Introduction to Philosophical Analysis,* 2nd Ed., © 1967, pp. 229–236. Reprinted by permission of Prentice-Hall, Inc., Englewood Cliffs, New Jersey. ["Law"]

Baruch Brody, "Towards an Aristotelian Theory of Scientific Explanation." *Philosophy of Science,* 39 (1972), pp. 20–31. © 1972, by Philosophy of Science Association.

William Dray, "Historical Understanding as Re-thinking." *University of Toronto Quarterly* (1958), pp. 200–215. © 1958 by *University of Toronto Quarterly.*

PART THREE

John Hospers, *An Introduction to Philosophical Analysis,* 2nd Ed., © 1967, pp. 236–239. Reprinted by permission of Prentice-Hall, Inc., Englewood Cliffs, New Jersey. ["Theory"]

N. R. Hanson, "Observation." From "Observation," *Patterns of Discovery,* New York: Cambridge University Press, pp. 4–11, 15–19. © 1958, by Cambridge University Press.

Carl R. Kordig, "The Theory-Ladenness of Observation." *The Review of Metaphysics,* XXIV (1971), pp. 450–455. © 1971 by *The Review of Metaphysics.* Reprinted by permission.

W. T. Stace, "Science and the Physical World." Reprinted from *Man Against Darkness* by W. T. Stace by permission of the University of Pittsburgh Press. © 1967 by the University of Pittsburgh Press.

Grover Maxwell, "The Ontological Status of Theoretical Entities." *Minnesota Studies in the Philosophy of Science,* Vol. III, pp. 3–14. Edited by H. Feigl and G. Maxwell. Minneapolis: University of Minnesota Press, 1962. © 1964, by University of Minnesota Press.

PART FOUR

W. V. Quine and J. S. Ullian, "Hypothesis." From *The Web of Belief,* Second Edition, by W. V. Quine and J. S. Ullian, pp. 64–82. Copyright © 1970, 1978 by Random House, Inc. Reprinted by permission of Random House, Inc.

T. S. Kuhn, "Theory-Choice." *Criticism and the Growth of Knowledge.* Edited by I. Lakatos and A. Musgrave, pp. 260–262. New York: Cambridge University Press, 1970. © 1970, by Cambridge University Press.

Philipp G. Frank, "The Variety of Reasons for the Acceptance of Scientific Theories." *Scientific Monthly,* Vol. 79, September 1954, pp. 139–145. © 1954, by the American Association for the Advancement of Science.

PART FIVE

Richard Rudner, "The Scientist *Qua* Scientist Makes Value Judgments." *Philosophy of Science,* XX (1953), pp. 1-6. © 1953 by Williams and Wilkins Co., Baltimore, Maryland.

Nicholas Rescher, "The Ethical Dimension of Scientific Research." In *Beyond the Edge of Certainty,* Robert G. Colodny, editor, © 1965, pp. 261-276. Reprinted by permission of Prentice-Hall, Inc., Englewood Cliffs, New Jersey.

Carl G. Hempel, "Science and Human Values." *Social Control in a Free Society,* ed. by R. E. Spiller, pp. 39-64. Philadelphia: University of Pennsylvania Press, 1960. © 1960, by University of Pennsylvania Press.

Michael Scriven, "The Exact Role of Value Judgments in Science." *Proceedings of the 1972 Biennial Meeting of the Philosophy of Science Association,* edited by R. S. Cohen and K. Schaffner, pp. 219-247. Dordrecht: D. Reidel Publishing Co., 1974. © 1974, by D. Reidel Publishing Co. Reprinted by permission of D. Reidel Publishing Co.

PART SIX

Theodore Roszak, "The Monster and the Titan: Science, Knowledge, and Gnosis." *Daedalus,* Summer, 1974, pp. 17-32. Reprinted by permission of the American Academy of Arts and Sciences, Boston, Massachusetts. Summer, 1974, *Science and Its Public: The Changing Relationship.*

Leo Marx, "Reflections on the Neo-Romantic Critique of Science." *Daedalus,* Spring, 1978, pp. 61-74. Reprinted by permission of the American Academy of Arts and Sciences, Boston, Massachusetts. Spring, 1978, *Limits of Scientific Inquiry.*

Kurt Baier, "The Meaning of Life." Inaugural Lecture delivered at the Canberra University College, 1957. © 1957, by Kurt Baier.

Introduction

What Is Philosophy of Science?

Most readers of this volume probably have some familiarity with science — or with one or more of the sciences. But the following question may come to mind: Just what is philosophy of science? How does it differ from science? How is it related to other areas of philosophy? We shall here attempt to provide answers to these and related questions.[1]

I. What Philosophy of Science Is Not

Let us begin with a discussion of what philosophy of science is *not*.

(1) Philosophy of science is not the history of science. The history of science is a valuable pursuit for both scientists and nonscientists. But it must not be confused with the philosophy of science. This is not to deny that the two disciplines may often be interrelated. Indeed, some have held that certain problems within the philosophy of science cannot be adequately dealt with apart from the context of the history of science. Nevertheless, it is generally held that we must distinguish between the two.

(2) Philosophy of science is not cosmology or "philosophy of nature." The latter attempts to provide cosmological or ethical speculations about the origin, nature, and purpose of the universe, or generalizations about the universe as a whole. As examples we may cite the views of Hegel and Marx, that the universe is dialectical in character; or the view of Whitehead, that it is organismic. Such cosmologies are often imaginative, metaphorical, and anthropomorphic constructions. They frequently involve interpreted extrapolations from science. Again, certain problems within the philosophy of science may aid the construction of or involve a consideration of such

cosmological theories. But here, too, there is wide agreement that they must be distinguished.

(3) Philosophy of science is not the psychology or sociology of science. The latter disciplines constitute a study of science as an activity, as one social phenomenon among many. Some of the topics that fall within such an inquiry are: scientists' motives for doing what they do; the behavior and activity of scientists; how (in fact) they make discoveries; what the impact of such discoveries is on society; and the sorts of governmental structures under which science has flourished. Again, certain problems in the philosophy of science may on occasion be related to such issues. But once more, it is reasonable to hold that these inquiries must be distinguished.

For the purposes of our study, the philosophy of science will not primarily mean or apply to any of the above. We will not try to comprehend the history of science. We will not present any grand cosmological speculations. We will not try to understand the scientific enterprise in terms of human or social needs. However, with regard to the latter, it is desirable to make a distinction. It is one thing to present a psychological or sociological account of science. This we will not do. It is another thing to examine philosophically the relationship of science and culture and generally of science and values. The latter parts of this volume will be devoted to these issues.

II. What Philosophy of Science Is

Let us attempt now to see what the philosophy of science *is*. By one widely held conception, philosophy of science is the attempt to understand the meaning, method, and logical structure of science by means of a logical and methodological analysis of the aims, methods, criteria, concepts, laws, and theories of science. Let us accept this as a preliminary characterization.

In order to illustrate or apply this characterization, let us focus on the matter of the concepts of science.

(1) There are numerous concepts that are used in many sciences but not investigated by any particular science. For example, scientists often use such concepts as: causality, law, theory, and explanation. Several questions arise: What is meant by saying that one event is the cause of another? That is, what is the correct analysis of the concept of cause? What is a law of nature? How is it related to other laws? What is the nature of a scientific theory? How are laws related to theories? What are description and explanation in science? How is explanation related to prediction? To answer such questions is to engage in logical and methodological analysis. Such an analysis is what philosophy of science, in part, is (according to this conception).

(2) There are many concepts used in the sciences that differ from the ones mentioned above. Scientists often speak of ordinary things—such as beakers, scales, pointers, tables. Let us call these observables. But they also often speak of unobservables: electrons, ions, genes, psi-functions, and so on. Several

questions then arise: How are these entities (if they are entities) related to things in the everyday world? What does a word such as 'positron' mean in terms of things we can see, hear, and touch? What is the logical justification for introducing these words which (purport to) refer to unobservable entities? To answer such questions by means of logical and methodological analysis constitutes another part or aspect of what philosophy of science is (according to the conception we are considering).

Now, with regard to the kinds of concepts mentioned in (2), one might ask: Why analyze these concepts? Don't scientists know how to use them? Yes, they certainly know how to use terms such as 'electron,' 'friction coefficient,' and so on. And often they pretty much agree about whether statements employing such expressions are true or false. But a philosopher, on the other hand, might be puzzled by such terms. Why? Well no one has ever directly seen a certain sub-atomic particle, or a frictionless body, or an ideal gas. Now we generally agree that we see physical objects and some of their properties — spatial relations, and so on. The philosopher of science asks (among other things) whether it is possible that a term such as 'positron' can be "defined" so that all the terms occurring in the definition (except logical terms, such as 'not,' 'and,' 'all') refer to physical objects and their properties. He attempts to reduce or trace such "theoretical constructs" to a lower level in the realm of the observable. Why? Because unless this is done, the doors all open to arbitrarily postulating entities such as gremlins, vital forces, and whatnot.

As we can see, throughout such conceptual investigations as those mentioned above, the standpoint adopted by the philosopher of science is often a commonsense standpoint. Thus certain questions which may be asked by other divisions of philosophy (such as epistemology) are not asked here. For example, whether a table really exists. If one wants to say that this means that philosophy of science has certain limitations, then we must agree. But not much follows from admitting this, for those other questions can always be raised later when we turn to other kinds of philosophical problems. Hence for the philosophy of science, we do not need to raise them. We may use the standpoint of common sense.

III. Some Main Topics in Philosophy of Science

The characterization of philosophy of science we have given in the preceding section does not adequately cover all of the kinds of issues and problems generally recognized as falling within the scope of philosophy of science. Hence it is perhaps best to resist trying to find a single formula or "definition" of philosophy of science and to turn to a different task.

Let us now briefly consider some of the main specific topics and questions with which philosophy of science is concerned. (In this volume, we will be able to focus on only some of these issues.)

(1) The formal sciences: logic and mathematics. Logic and math are often referred to as sciences. In what sense, if any, are they sciences? How do we know logical and mathematical truths? What, if anything, are they true of? What is the relation of mathematics to empirical science?

(2) Scientific description. What constitutes an adequate scientific description? What is the "logic" of concept formation which enters into such description?

(3) Scientific explanation. What is meant by saying that science explains? What is a scientific explanation? Are there other kinds of explanations? If so, how are they related to those of science?

(4) Prediction. We say that science predicts. What makes this possible? What is the relation of prediction to explanation? What is the relation of testing to both?

(5) Causality and law. We sometimes hear it said that science explains by means of laws. What are scientific laws? How do they serve to explain? Further, we sometimes speak of explaining laws. How can that be? Many laws are known as causal laws. What does that mean? Are there noncausal laws? If so, what are they?

(6) Theories, models, and scientific systems. We also hear it said that science explains by means of theories. What are theories? How are they related to laws? How do they function in explanation? What is meant by a "model" in science? What role do models play in science?

(7) Determinism. Discussions of lawfulness lead to the question of determinism. What is meant by determinism in science? Is the deterministic thesis (if it is a thesis) true? Or what reason, if any, do we have for thinking it to be true?

(8) Philosophical problems of physical science. The physical sciences have, in recent years, provided a number of philosophical problems. For example, some have held that relativity theory introduces a subjective component into science. Is this true? Others have said that quantum physics denies or refutes determinism. Is this true or false?

(9) Philosophical problems of biology and psychology. First, are these sciences genuinely distinct? If so, why? If not, why not? Further, are these sciences ultimately reducible to physics, or perhaps to physics and chemistry? This gets us into the old "vitalism/mechanism" controversy.

(10) The social sciences. There are some who deny that the social sciences are genuine sciences. Why? Are they right or wrong? Is there any fundamental difference between the natural sciences and the social sciences?

(11) History. Is history a science? We often speak of historical laws. Are there really any such laws? Or are there only general trends? Or neither?

(12) Reduction and the unity of science. We have already briefly referred to this issue. The question here is whether it is possible to reduce one science to another and whether all of the sciences are ultimately reducible to a single science or a combination of fundamental sciences (such as physics and chemistry).

(13) Extensions of science. Sometimes scientists turn into metaphysicians. They make "radical" statements about the universe—e.g., about the ultimate

heat-death, or that it is imbued with moral progress. Is there any validity in these claims?

(14) Science and values. Does science have anything to say with regard to values? Or is it value-neutral?

(15) Science and religion. Do the findings and conclusions of science have any implications for traditional religious or theological commitments? If so, what are they?

(16) Science and culture. Both religion and the domain of values may be considered to be parts or aspects of culture. But surely the term *culture* also refers to other activities and practices. What is the relationship of science to these?

(17) The limits of science. Are there limits of science? If so, what are they? By what criteria, if any, can we establish that such limits are genuine?

IV. Philosophy of Science and Science

We hope that by now the reader has a fair grasp of what philosophy of science is. In order to provide further understanding, let us examine one way by which one might contrast science with the philosophy of science. We may best do this by focusing on the activities and concerns of scientists and of philosophers of science. There are many ways in which these differ. Let us look at just a couple of them. According to one widely held view:

(1) Scientists (among other things, and not necessarily in this order): (a) observe what happens in the world and note regularities; (b) experiment— i.e., manipulate (some) things so that they can be observed under special circumstances; (c) discover (or postulate) laws of nature which are intended to explain regularities; (d) combine laws of nature into theories or subsume those laws under theories. Philosophers of science do none of the above things. Rather, they ask questions such as: What is a law of nature? What is a scientific (vs. a nonscientific or unscientific) theory? What are the criteria (if any) by which to distinguish or demarcate those theories which are genuinely scientific from those which are not? Furthermore, according to this view:

(2) Scientists, like almost everyone else, make deductions. For example, they often construct a certain theory from various laws and observations and then from it deduce other theories or laws, or even certain specific occurrences which serve to test a theory. Philosophers of science do not do that. Rather they clarify the nature of deduction (and how it differs from other inferences or reasoning), and they describe the role deduction plays in science. For example, they ask how deduction is involved in the testing of theories.

From these examples of (some of) the activities of scientists and philosophers of science, we may see that (according to the view we are considering): Whereas science is largely empirical, synthetic, and experimental, philosophy of science is largely verbal, analytic, and reflective. To be sure, in the works of some scientists—especially those who are in the more "theoretical" sciences—verbal, analytic, and reflective features may be found.

But the converse is not generally true. The activities of philosophers of science are, for the most part, not empirical or experimental, and they do not add to our store of factual knowledge of the actual world. And even in those cases where the more "philosophical" activities are found in science, they are usually not pursued with the same rigor or toward the same ends as they are by philosophers of science.

We may roughly see the difference by examining the following table:

Philosophy of science	— — —	is comprised of	— — —	a meta-language
Science	— — —	is comprised of	— — —	an object-language
Reality (or the world)	— — —	is comprised of	— — —	objects, processes, etc.

a meta-language which refers to *an object-language* which refers to *objects, processes, etc.*

Thus we may see that, whereas science uses (an object-) language to talk about the objects of the world, philosophy of science (or at least a large "part" of it) uses (a meta-) language to talk about the language of science. In short, as a slight oversimplification, we may say: Science is talk about the world (a certain *kind* of talk, of course). Philosophy is talk about language (again, a certain kind of talk, of course).

To summarize the view we have considered: (1) The sciences consist of such things as listings of data, generalizations from them, the formulation of laws or trends, theoretical interpretations of data or laws, and arguments and evidence in favor of them. (2) Philosophy of science, to a large extent, consists of remarks about the language of science: the analysis of concepts, methods, and arguments of the various sciences; and also the analysis of the principles underlying science.

It is hoped now that our earlier characterization of the philosophy of science may be more readily understood and appreciated. Once again, according to that characterization, philosophy of science is the attempt to understand the meaning, method, and logical structure of science by means of a logical and methodological analysis of the aims, methods, criteria, concepts, laws, and theories of science.

One might reasonably object: But this view of philosophy of science does not do justice or apply to the list of topics in the philosophy of science provided in the preceding section. We are sympathetic to such an objection. Whereas our initial characterization does apply to many of the problems and concerns found in that list, it does not apply to others—for example, the topics of science and religion, or science and culture. Hence we propose that our initial characterization be modified in order to take such matters into account. We propose the following as an amended characterization of the philosophy of science. Philosophy of science is the attempt (a) to understand the method, foundations, and logical structure of science and (b) to examine the relations and interfaces of science and other human concerns, institutions, and quests, by means of (c) a logical and methodological analysis both of the aims, methods, and criteria of science and of the aims, methods, and concerns of various cultural phenomena in their relations to science.

V. The Scope of This Book

As we have mentioned, we cannot within a single volume do justice to all of the topics which fall within the domain of philosophy of science. We have therefore chosen six topics which (a) are crucial ones in philosophy of science, (b) are intrinsically interesting to the layperson as well as to the scientist or philosopher, and (c) are accessible to the beginning student. Similarly, the readings we have selected reflect those features. The topics are:

1. Science and Nonscience
2. Explanation and Law
3. Theory and Observation
4. Confirmation and Acceptance
5. Science and Values
6. Science and Culture

Since we have provided discussions of these topics in the introductions to the parts of the book, we shall not make further comments about them at this point.

We truly hope that the readers of this volume will derive as much enjoyment from the book as we have had in our production of it. We urge that the Study Questions at the end of each part be utilized. For further reading we have provided selected bibliographies.

E.D.K.

NOTES

1. Many of the views regarding science and the philosophy of science presented in this introduction and in the introduction to Part 1 stem from the lectures and writings of Herbert Feigl, May Brodbeck, John Hospers, and Sir Karl Popper.

SELECTED BIBLIOGRAPHY

[1] Brody, B., ed. *Readings in the Philosophy of Science.* Englewood Cliffs, N.J.: Prentice-Hall, 1970.

[2] Carnap, R. *An Introduction to the Philosophy of Science.* New York: Basic Books, 1966.

[3] Danto, A. and Morgenbesser, S., eds. *Philosophy of Science.* New York: Meridian, 1960.

[4] Hempel, C. *Philosophy of Natural Science.* Englewood Cliffs, N.J.: Prentice-Hall, 1966.

[5] Michalos, A., ed. *Philosophical Problems of Science and Technology.* Rockleigh, N.J.: Allyn & Bacon, 1974.

[6] Morgenbesser, S., ed. *Philosophy of Science Today.* New York: Basic Books, 1967.

[7] Nidditch, P. H., ed. *Philosophy of Science.* New York: Oxford University Press, 1968.

[8] Quine, W. V., and Ullian, J. *The Web of Belief,* 2nd ed. New York: Random House, 1978.

[9] Shapere, D, ed. *Philosophical Problems of Natural Science.* New York: Macmillan, 1964.

[10] Toulmin, Stephen. *The Philosophy of Science.* New York: Harper & Row, 1960.

Part One

Science and Nonscience

Part 1: Science and Nonscience

Introduction

The major topics we shall discuss in this essay are: the aims of science; the criteria of science, or the criteria for distinguishing that which is scientific from that which is nonscientific; the question 'What is science?'; and the central issues of the readings which follow. But, first, let us begin by making some distinctions.

I. Some Distinctions

Before turning to the topics above it will be helpful to consider some ways of classifying the various sciences. Among these, the following should be noted.

(1) Pure sciences versus applied sciences. It is widely held that we must distinguish: (A) science as a field of knowledge (or set of cognitive disciplines) from (B) the applications of science. It is common to refer to these as the pure and applied sciences. (A) Among the pure sciences we may distinguish: (a) the formal sciences, logic and mathematics; and (b) the factual or empirical sciences. Among the latter we may also distinguish: (b1) the natural sciences, which include the physical sciences, physics, chemistry, and so on, and the life and behavioral sciences, such as biology and psychology; and (b2) the social sciences, such as sociology and economics. (B) The applied sciences include the technological sciences—such as engineering and aeronautics—medicine, agriculture, and so on.

It should be noted that there are at least two levels of application among the various sciences. There is, first, the application of the formal sciences to the pure, factual sciences. Since the factual sciences must have logical form and usually utilize some mathematics, such application is often held to be essential for the development of the pure factual sciences. Different from this is, second, the application of the factual sciences to the applied sciences. Here the findings of the pure, empirical sciences are applied (in a different sense of 'applied') to disciplines which fulfill various social, human purposes, such as building houses or roads and health care.

(2) Law-finding sciences versus fact-finding sciences. We recognize that such sciences as chemistry and physics attempt to discover universal laws which are applicable everywhere at all times, whereas such sciences as geography, history (if it is a science), and perhaps economics are concerned with local events. It is often said that the subject matter of the latter consists of particular facts, not general laws. As a result, there are some who wish to limit the term 'science' to the law-finding sciences. Upon the basis of the criteria of science (such as those which will be presented later, or others), we believe that we may say that both the law-finding disciplines and the fact-finding disciplines are capable of being sciences if those (or other) criteria are met. Furthermore, one might argue that there are no purely fact-finding sciences. If so, to speak of law-finding *versus* fact-finding may, in many cases, indicate an artificial disjunction.

(3) Natural sciences versus social sciences. Related to (2), we find that some would limit the giving of scientific status to the natural sciences alone. Sometimes the reason given is the distinction referred to above—that the natural sciences are primarily law-finding, whereas the social sciences are predominantly fact-finding. But sometimes the distinction is based on subject matter. Hence it is held by some that natural phenomena constitute the field of science but cultural phenomena constitute the field of scholarship and require understanding, *verstehen,* and empathy. But there are points at which the classification does not hold up. First, there are some predominantly fact-finding natural sciences, such as geography, geology, and palaeontology. And there are some law-finding social sciences, such as sociology and linguistics. Second, the distinction according to subject matter is not a clear-cut one. Hence we shall take a "liberal" view of science and allow the use of the term 'science' to apply to both the natural and the social sciences—with the recognition that there are some differences.

It is widely held that distinctions (2) and (3) do not hold up but that (1) is an acceptable distinction. However, as we shall see in the readings which follow, some have even raised doubts about the significance of (1). Here, as always, we urge the reader to reflect upon these matters.

II. The Aims of Science

Let us now turn to the question 'What are the aims of science?' Using the above distinction between pure (empirical) and applied science, we may then cite the following as some of the aims of science.

(1) The aims of applied science include: control, planning, technological progress; the utilization of the forces of nature for practical purposes. Obvious examples are: flood control, the construction of sturdier bridges, and the improvement of agriculture. Since this is all fairly obvious, no further elaboration is needed.

(2) The aims of the pure, factual sciences may be considered from two standpoints. (a) Psychologically considered, the aims of the pure, empirical sciences are: the pursuit of knowledge; the attainment of truth (or the closest possible realization of truth); the satisfaction of using our intellectual powers to explain and predict accurately. Scientists, of course, derive enjoyment from rewards, prestige, and competing with others. But they often achieve a genuine inner gratification which goes with the search for truth. In some ways this is similar in quality to artistic satisfaction. It is seen, for example, in the enjoyment one derives from the solution of a difficult problem.

(b) Logically considered, the aims of the pure, factual sciences are often held to be: description, explanation, and prediction. (b1) Description includes giving an account of what we observe in certain contexts, the formulation of propositions which apply to (or correspond to) facts in the world. (b2) Explanation consists of accounting for the facts and regularities we observe. It involves asking and answering 'Why?' or 'How come?' This may be done by subsuming facts under laws and theories. (b3) Prediction is closely related to explanation. It consists in deriving propositions which refer to events which have not yet happened, the deducing of propositions from laws and theories and then seeing if they are true, and hence provide a testing of those laws and theories. (b4) We might also mention post- or retro-diction, the reconstruction of past events. This process is also inferential in character. Since these issues will be discussed in subsequent parts of this volume, we shall not elaborate upon them at this time. (See some of the readings in Part 1 and those in Parts 2, 3 and 4.) However, we might mention that, here again, there is not unanimity with regard to the aims characterized above. Once more we urge the reader to think about these (and other) issues.

III. The Criteria of Science

In this section we shall state and discuss one view with regard to what are the essential criteria of science, that is, those criteria which may be used for at least two purposes: first, to distinguish science from commonsense knowledge (without claiming that the two are radically disjunctive—in some cases they may differ only in degree, not in kind); second, to distinguish that which is scientific, on the one hand, from that which is either nonscientific or unscientific, on the other—for example, to distinguish between theories which are genuinely scientific and those which are not. It has been maintained that any enterprise, discipline, or theory is scientific if it is characterized by or meets those criteria.

Before turning to the view which we have selected for consideration, let us consider an example. It is quite likely that most scientists and others who reflect upon science would hold that (say) Newton's theory of gravitation is scientific (even if it had to be modified), whereas (say) astrology is not scientific. Perhaps the reader would agree. But just what is it that allows us to rule in Newton's theory and to rule out astrology? In order to stimulate the reader's reflection, we shall consider one view of what the criteria for making such distinctions are. These criteria have been stated by Professor Herbert Feigl in various lectures. Our discussion of them corresponds fairly closely, but not entirely, to the discussion given by Professor Feigl.

The five criteria are:

(1) Intersubjective testability. This refers to the possibility of being, in principle, capable of corroboration or "check-up" by anyone. Hence: *inter*subjective. (Hence, private intuitions and so forth must be excluded.)

(2) Reliability. This refers to that which, when put to a test, turns out to be true, or at least to be that which we can most reasonably believe to be true. Testing is not enough. We want theories which, when tested, are found to be true.

(3) Definiteness and precision. This refers to the removal of vagueness and ambiguity. We seek, for example, concepts which are definite and delimited. We are often helped here by measurement techniques and so forth.

(4) Coherence or systematic character. This refers to the organizational aspect of a theory. A set of disconnected statements is not as fruitful as one which has systematic character. It also refers to the removal of, or being free from, contradictoriness.

(5) Comprehensiveness or scope. This refers to our effort to attain a continual increase in the completeness of our knowledge and also to our seeking theories which have maximum explanatory power—for example, to account for things which other theories do not account for.

Let us consider these criteria in greater detail.

(1) Intersubjective testability. (a) Testability. We have noted that in science we encounter various kinds of statements: descriptions, laws, theoretical explanations, and so on. These are put forth as knowledge-claims. We must (if possible) be able to tell whether evidence speaks for or against such knowledge-claims. If the propositions which express those claims are not capable of tests, we cannot call those propositions true or false or even know how to go about establishing their truth or falsity. It should be noted that the criterion is one of *testability*, not *tested*. For example, at a given point in time, 'There are mountains on the far side of the moon' was testable though not tested.

(b) Intersubjective. 'Intersubjectivity' is often employed as a synonym for 'objectivity.' And the latter term has various meanings. Some of these are: (i) A view or belief is said to be objective if it is not based on illusions, hallucinations, deceptions, and so on. (ii) Something is referred to as objective if it is not merely a state of mind but is really "out there" in the external world. (iii) We often use 'objective' to indicate the absence of bias and the presence of disinterestedness and dispassionateness. (iv) 'Objectivity' also refers to the possibility of verification by others, and hence excludes beliefs, which stem from private, unique, unrepeatable experiences. Science strives for objectivity in all of these senses. Hence Feigl takes 'intersubjective' to include all of them.

(c) Intersubjective testability. It is often held that (according to the view we are considering) in order for a proposition or theory to be judged scientific it must meet this first requirement. Indeed, many of the other criteria presuppose intersubjective testability. We cannot even begin to talk of reliability or precision unless this first criterion has been met.

(2) Reliability. Science is not merely interested in hypotheses which are intersubjectively testable. It is also interested in those which are true or at least have the greatest verisimilitude or likelihood of being true. Hence the need arises for the criterion of reliability. Whereas the first criterion stressed the possibility of finding assertions which are true *or* false, the second stresses the end result of that process. We judge a claim or body of knowlege to be reliable if it contains not merely propositions which are capable of being true *or* false but rather those which *are true* or which have the greatest verisimilitude. We find such propositions to be true (or false) by means of confirmation. Complete verification, and hence complete certainty, cannot be achieved in the factual sciences.

It should be noted that, first, the reliability of scientific assertions make them useful for prediction; second, although the assertions of many enterprises are testable (for example, those of astrology as much as those of astronomy), only some of them are reliable. And we reject some of them precisely because they are unreliable. The evidence is against them; we do not attain truth by means of them.

(3) Definiteness and precision. The terms 'definiteness' and 'precision' may be used in at least two related senses. First, they refer to the delimitation of our *concepts* and to the removal of ambiguity or vagueness. Second, they refer to a more rigid or exact formulation of *laws*. For example, 'It is more probable than not that X causes disease Y' is less desirable than 'The probability that X causes Y is 9.1.'

(4) Coherence or systematic character. In the sciences, we seek not merely disorganized or loosely related facts but a well-connected account of the facts. It has been held by many that we achieve this via what has been called the hypothetico-deductive procedure of science. This procedure includes: (a) our beginning with a problem (which pertains to some realm of phenomena); (b) the formulation of hypotheses, laws, and theories by which to account for those phenomena or by which to resolve the problem; (c) the deriving (from (b)) of statements which refer to observable facts; (d) the testing of those deduced assertions to see if they hold up. Thus we seek an integrated, unified network, not merely a congeries of true statements.

But, of course, we also seek theories which are consistent, which are free from self-contradictions. The reason for insisting upon such coherence is obvious; hence there is no need for elaboration.

(5) Comprehensiveness or scope. The terms 'comprehensiveness' and 'scope' are also used in two senses, both of which are essential in science. First, a theory is said to be comprehensive if it possesses maximum explanatory power. Thus Newton's theory of gravitation was ranked high partly because it accounted not only for the laws of falling bodies but also for the revolution of the heavenly bodies and for the laws of tides. Second, by 'comprehesive' we often refer to the completeness of our knowledge. This of course does not mean finality. We do not think of the hypotheses of the empirical sciences as being certain for all time. Rather we must be ready to modify them or even, on occasion, to abandon them.

To summarize: According to the view we have presented, we judge a law, hypothesis, theory, or enterprise to be scientific if it meets all five of the above criteria. If it fails to meet all five, it is judged to be unscientific or at least non-scientific. To return to our earlier example, it seems clear that Newton's theory thus passes the test. Astrological theory or Greek mythology does not.

It should be noted that, in presenting Professor Feigl's criteria for the reader's consideration, we do not claim that they are correct or free from defects. Indeed, as we shall see in the readings which follow, many writers have rejected some (or all) or those criteria. The reader should once again attempt to seek an acceptable criterion or set of criteria, if such can be found.

IV. What is Science?

A common characterization of science (or sometimes of scientific method) runs as follows. Science is knowledge obtained by: (1) making observations as accurate and definite as possible; (2) recording these intelligibly; (3) classifying them according to the subject matter being studied; (4) extracting from them, by induction, general statements (laws) which assert regularities; (5) deducing other statements from these; (6) verifying those statements by further observation; and (7) propounding theories which connect and so account for the largest possible number of laws. It is further maintained that this process runs from (1) through (7) *in that order.*

The conception of science has been challenged in recent years. Its most severe critic is Sir Karl Popper. (See the selection in Part 1 of this volume.) We shall not repeat Popper's criticisms. Instead we offer a characterization of science which some believe to be more adequate than the one mentioned above and which they deem to be free from the defects it possesses.

According to this view, the following is at least a minimal characterization of (factual) science (or of a science).

Science is a body of knowledge which consists of the following, coherently organized in a systematic way:

(a) Statements which record and classify observations which are relevant for the solution of a problem in as accurate and definite a way as possible.

(b) General statements—laws or hypotheses—which assert regularities among certain classes of observed or observable phenomena.

(c) Theoretical statements which connect and account for the largest possible number of laws.

(d) Other general or specific statements which are deducible from the initial descriptions and from laws and theories and which are confirmed by further observation and testing.

At least two things should be noted about this characterization. First, it indicates the role of the formal sciences in the empirical sciences. Mathematics is important for (a); logic is important for (d). Second, nothing is said in this characterization about the *method* of obtaining knowledge or of obtaining laws. It may be induction, but it may also be a guess, intuition, hunch, or whatever.

Since a number of the readings in Part 1 deal with the question "What is science?" we shall not attempt to provide a "final" answer. Instead, we encourage the student to come up with the best answer possible, based on his or her reading and reflection.

VI. The Readings in Part 1

Since the essays contained in Part 1 are clearly written and since they are accessible to the beginning student or ordinary reader, no detailed summaries will be presented here. We urge the student to prepare his or her own summaries and to make use of the Study Questions at the end. However, a few brief remarks may be helpful.

Throughout many of his works, Sir Karl Popper has been concerned with the problem of how to distinguish between science and pseudoscience (or non-science). He claims to have solved that problem by having provided a criterion of demarcation, a criterion by which to distinguish theories which are genuinely scientific from those which are not. By means of this criterion—of falsifiability or refutability—he attempts to show that Einstein's theory of gravitation satisfies the criterion (and hence is scientific) whereas astrology, the Marxist theory of history, and various psychoanalytic theories—for varying reasons—are not scientific. He also wishes to separate the problem of demarcation from the problem of meaning and maintains that the latter is a pseudoproblem. (The reader should reflect upon why he holds that it is a pseudoproblem and whether he has succeeded in showing that it is.)

In the middle sections of Popper's essay, he claims that the problem of demarcation has provided a key for solving a number of philosophical problems, especially the problem of induction. Since this issue does not pertain to the main topics of Part 1, most of those sections of the essay have been deleted here. The problem of induction is: How, if at all, can we justify our knowledge-claims concerning matters of fact which we have not yet experienced or are not now experiencing? In the eighteenth century, David Hume maintained that we cannot provide any rational justification. Popper agrees with Hume's logical refutation of induction but disagrees with his psychological explanation of induction (in terms of custom or habit). In the concluding sections of the essay (reprinted here), Popper returns to the problem of demarcation and relates it to the problem of induction.

The selection by John Ziman consists of extracts from his book on science. In the first part he discusses and rejects various definitions of science which have been held. And he attempts to formulate a more accurate and tenable characterization based on what he takes to be the goal or objective of science, namely, consensus of rational opinion "over the widest possible field." In the second part, he provides his answer to the question "What distinguishes science from nonscience?" The reader should attempt to decide whether his "criterion of demarcation" is an improvement over Popper's and, if so, why. Since this selection is unusually clear and readable, no further comments are required.

Feyerabend's essay is, no doubt, one of the most controversial ones in this volume. Feyerabend claims that he wishes to defend society and its inhabitants from *all* ideologies, including science. He likens them (again, including science) to fairytales "which have lots of interesting things to say but which also contain wicked lies." He goes on to consider an argument designed to defend the exceptional status which science has in society today. According to this argument "(1) science has finally found the correct *method* for achieving results and (2) . . . there are many *results* to prove the excellence of the method." In the next sections he argues against both (1) and (2). He concludes his essay with a provocative discussion of education and myth. We urge the reader to reflect seriously upon Feyerabend's somewhat unorthodox views and to ask whether Feyerabend has adequately defended them.

Paul R. Thagard's essay constitutes both a further discussion of some of the above-mentioned topics (such as the criterion of demarcation) and an example of the application of them. Most scientists and philosophers agree that astrology is a pseudoscience. Thagard attempts to show *why* it is. After presenting a brief description of astrology, he attempts to show that the major objections which have been provided do not show *that* it is a pseudoscience. Thagard then proposes his principle of demarcation and, upon the basis of it, claims to show that and why astrology is unscientific.

There is a kind of dialogue which runs through the essays in this part. We urge the reader to critically evaluate the various positions presented and attempt to come to his or her own conclusion with regard to the questions "What is science?," "By what criteria can we distinguish science from non-science or pseudoscience?" and so on. The Study Questions should provide assistance in gauging the reader's understanding of the selections and in grappling with these and related questions.

E.D.K.

1

Science: Conjectures and Refutations

Sir Karl Popper

> Mr. Turnbull had predicted evil consequences, . . .
> and was now doing the best in his power to bring
> about the verification of his own prophecies.
>
> ANTHONY TROLLOPE

I

When I received the list of participants in this course and realized that I had been asked to speak to philosophical colleagues I thought, after some hesitation and consultation, that you would probably prefer me to speak about those problems which interest me most, and about those developments with which I am most intimately acquainted. I therefore decided to do what I have never done before: to give you a report on my own work in the philosophy of science, since the autumn of 1919 when I first began to grapple with the problem, *"When should a theory be ranked as scientific?"* or *"Is there a criterion for the scientific character or status of a theory?"*

The problem which troubled me at the time was neither, "When is a theory true?" nor, "When is a theory acceptable?" My problem was different. I *wished to distinguish between science and pseudo-science;* knowing very well that science often errs, and that pseudo-science may happen to stumble on the truth.

I knew, of course, the most widely accepted answer to my problem: that science is distinguished from pseudo-science — or from "metaphysics" — by its *empirical method,* which is essentially *inductive,* proceeding from observation or experiment. But this did not satisfy me. On the contrary, I often formulated my problem as one of distinguishing between a genuinely empirical method

A lecture given at Peterhouse, Cambridge, in Summer 1953, as part of a course on developments and trends in contemporary British philosophy, organized by the British Council; originally published under the title "Philosophy of Science: a Personal Report" in British Philosophy in Mid-Century, *ed. C. A. Mace, 1957. [Portions have been deleted by the editors for this publication.]*

and a non-empirical or even a pseudo-empirical method—that is to say, a method which, although it appeals to observation and experiment, nevertheless does not come up to scientific standards. The latter method may be exemplified by astrology with its stupendous mass of empirical evidence based on observation—on horoscopes and on biographies.

But as it was not the example of astrology which led me to my problem I should perhaps briefly describe the atmosphere in which my problem arose and the examples by which it was stimulated. After the collapse of the Austrian Empire there had been a revolution in Austria: the air was full of revolutionary slogans and ideas, and new and often wild theories. Among the theories which interested me Einstein's theory of relativity was no doubt by far the most important. Three others were Marx's theory of history, Freud's psycho-analysis, and Alfred Adler's so-called "individual psychology."

There was a lot of popular nonsense talked about these theories, and especially about relativity (as still happens even today), but I was fortunate in those who introduced me to the study of this theory. We all—the small circle of students to which I belonged—were thrilled with the result of Eddington's eclipse observations which in 1919 brought the first important confirmation of Einstein's theory of gravitation. It was a great experience for us, and one which had a lasting influence on my intellectual development.

The three other theories I have mentioned were also widely discussed among students at that time. I myself happened to come into personal contact with Alfred Adler, and even to co-operate with him in his social work among the children and young people in the working-class districts of Vienna where he had established social guidance clinics.

It was during the summer of 1919 that I began to feel more and more dissatisfied with these three theories—the Marxist theory of history, psychoanalysis, and individual psychology; and I began to feel dubious about their claims to scientific status. My problem perhaps first took the simple form, "What is wrong with Marxism, psycho-analysis, and individual psychology? Why are they so different from physical theories, from Newton's theory, and especially from the theory of relativity?"

To make this contrast clear I should explain that few of us at the time would have said that we believed in the *truth* of Einstein's theory of gravitation. This shows that it was not my doubting the *truth* of those other three theories which bothered me, but something else. Yet neither was it that I merely felt mathematical physics to be more *exact* than the sociological or psychological type of theory. Thus what worried me was neither the problem of truth, at that stage at least, nor the problem of exactness or measurability. It was rather that I felt that these other three theories, though posing as sciences, had in fact more in common with primitive myths than with science; that they resembled astrology rather than astronomy.

I found that those of my friends who were admirers of Marx, Freud, and Adler, were impressed by a number of points common to these theories, and especially by their apparent *explanatory power*. These theories appeared to be able to explain practically everything that happened within the fields to which they referred. The study of any of them seemed to have the effect of an intellectual conversion or revelation, opening your eyes to a new truth hidden from those not yet initiated. Once your eyes were thus opened you saw confirming instances everywhere: the world was full of *verifications* of the theory. Whatever happened always confirmed it. Thus its truth appeared manifest; and unbelievers were clearly people who did not want to see the manifest truth; who refused to see it, either because it was against their class interest, or because of their repressions which were still "un-analysed" and crying aloud for treatment.

The most characteristic element in this situation seemed to me the incessant stream of confirmations, of observations which "verified" the theories in question; and this point was constantly emphasized by their adherents. A Marxist could not open a newspaper without finding on every page confirming evidence for his interpretation of history; not only in the news, but also in its presentation—which revealed the class bias of the paper—and especially of course in what the paper *did not* say. The Freudian analysts emphasized that their theories were constantly verified by their "clinical observations." As for Adler, I was much impressed by a personal experience. Once, in 1919, I reported to him a case which to me did not seem particularly Adlerian, but which he found no difficulty in analysing in terms of his theory of inferiority feelings, although he had not even seen the child. Slightly shocked, I asked him how he could be so sure. "Because of my thousandfold experience," he replied; whereupon I could not help saying: "And with this new case, I suppose, your experience has become thousand-and-one-fold."

What I had in mind was that his previous observations may not have been much sounder than this new one; that each in its turn had been interpreted in the light of "previous experience," and at the same time counted as additional confirmation. What, I asked myself, did it confirm? No more than that a case could be interpreted in the light of the theory. But this meant very little, I reflected, since every conceivable case could be interpreted in the light of Adler's theory, or equally of Freud's. I may illustrate this by two very different examples of human behaviour: that of a man who pushes a child into the water with the intention of drowning it; and that of a man who sacrifices his life in an attempt to save the child. Each of these two cases can be explained with equal ease in Freudian and in Adlerian terms. According to Freud the first man suffered from repression (say, of some component of his Oedipus complex), while the second man had achieved sublimation. According to Adler the first man suffered from feelings of inferiority (producing perhaps the need to prove to himself that he dared to commit some crime), and so did the second man (whose need was to prove to himself that he dared to rescue the child). I could not think of any human behaviour which could not be interpreted in terms of either theory. It was precisely this fact—that they always fitted, that they were always confirmed—which in the eyes of their admirers constituted the strongest argument in favour of these theories. It began to dawn on me that this apparent strength was in fact their weakness.

With Einstein's theory the situation was strikingly different. Take one typical instance—Einstein's prediction, just then confirmed by the findings of Eddington's expedition. Einstein's gravitational theory had led to the result that light must be attracted by heavy bodies (such as the sun), precisely as material bodies were attracted. As a consequence it could be calculated that light from a distant fixed star whose apparent position was close to the sun would reach the earth from such a direction that the star would seem to be slightly shifted away from the sun; or, in other words, that stars close to the sun would look as if they had moved a little away from the sun, and from one another. This is a thing which cannot normally be observed since such stars are rendered invisible in daytime by the sun's overwhelming brightness; but during an eclipse it is possible to take photographs of them. If the same constellation is photographed at night one can measure the distances on the two photographs, and check the predicted effect.

Now the impressive thing about this case is the *risk* involved in a prediction of this kind. If observation shows that the predicted effect is definitely absent, then the theory is simply refuted. The theory is *incompatible with certain possible results of observation*—in fact with results which everybody before Einstein would have expected.[1] This is quite different from the situation I have previously described, when it turned out that the theories in question were compatible with the most divergent human behaviour, so that it was practically impossible to describe any human behaviour that might not be claimed to be a verification of these theories.

These considerations led me in the winter of 1919–20 to conclusions which I may now reformulate as follows.

(1) It is easy to obtain confirmations, or verifications, for nearly every theory—if we look for confirmations.

(2) Confirmations should count only if they are the result of *risky predictions*; that is to say, if, unenlightened by the theory in question, we should have expected an event which was incompatible with the theory—an event which would have refuted the theory.

(3) Every "good" scientific theory is a prohibition: it forbids certain things to happen. The more a theory forbids, the better it is.

(4) A theory which is not refutable by any conceivable event is nonscientific. Irrefutability is not a virtue of theory (as people often think) but a vice.

(5) Every genuine *test* of a theory is an attempt to falsify it, or to refute it. Testability is falsifiability; but there are degrees of testability; some theories are more testable, more exposed to refutation, than others; they take, as it were, greater risks.

(6) Confirming evidence should not count *except when it is the result of a genuine test of the theory*; and this means that it can be presented as a serious but unsuccessful attempt to falsify the theory. (I now speak in such cases of "corroborating evidence.")

(7) Some genuinely testable theories, when found to be false, are still upheld by their admirers—for example by introducing *ad hoc* some auxiliary assumption,

or by re-interpreting the theory *ad hoc* in such a way that it escapes refutation. Such a procedure is always possible, but it rescues the theory from refutation only at the price of destroying, or at least lowering, its scientific status. (I later described such a rescuing operation as a *"conventionalist twist"* or a *"conventionalist stratagem."*)

One can sum up all this by saying that the *criterion of the scientific status of a theory is its falsifiability, or refutability, or testability.*

II

I may perhaps exemplify this with the help of the various theories so far mentioned. Einstein's theory of gravitation clearly satisfied the criterion of falsifiability. Even if our measuring instruments at the time did not allow us to pronounce on the results of the tests with complete assurance, there was clearly a possibility of refuting the theory.

Astrology did not pass the test. Astrologers were greatly impressed, and misled, by what they believed to be confirming evidence—so much so that they were quite unimpressed by any unfavourable evidence. Moreover, by making their interpretations and prophecies sufficiently vague they were able to explain away anything that might have been a refutation of the theory had the theory and the prophecies been more precise. In order to escape falsification they destroyed the testability of their theory. It is a typical soothsayer's trick to predict things so vaguely that the predictions can hardly fail: that they become irrefutable.

The Marxist theory of history, in spite of the serious efforts of some of its founders and followers, ultimately adopted this soothsaying practice. In some of its earlier formulations (for example in Marx's analysis of the character of the "coming social revolution") their predictions were testable, and in fact falsified.[2] Yet instead of accepting the refutations the followers of Marx reinterpreted both the theory and the evidence in order to make them agree. In this way they rescued the theory from refutation; but they did so at the price of adopting a device which made it irrefutable. They thus gave a "conventionalist twist" to the theory; and by this stratagem they destroyed its much advertised claim to scientific status.

The two psycho-analytic theories were in a different class. They were simply non-testable, irrefutable. There was no conceivable human behaviour which could contradict them. This does not mean that Freud and Adler were not seeing certain things correctly: I personally do not doubt that much of what they say is of considerable importance, and may well play its part one day in a psychological science which is testable. But it does mean that those "clinical observations" which analysts naively believe confirm their theory cannot do this any more than the daily confirmations which astrologers find in their practice.[3] And as for Freud's epic of the Ego, the Super-ego, and the Id, no substantially stronger claim to scientific status can be made for it than for Homer's collected stories from Olympus. These theories describe some facts, but in the manner of myths. They contain most interesting psychological suggestions, but not in a testable form.

At the same time I realized that such myths may be developed, and become testable; that historically speaking all—or very nearly all—scientific theories originate from myths, and that a myth may contain important anticipations of scientific theories. Examples are Empedocles' theory of evolution by trial and error, or Parmenides' myth of the unchanging block universe in which nothing ever happens and which, if we add another dimension, becomes Einstein's block universe (in which, too, nothing ever happens, since everything is, four-dimensionally speaking, determined and laid down from the beginning). I thus felt that if a theory is found to be non-scientific, or "metaphysical" (as we might say), it is not thereby found to be unimportant, or insignificant, or "meaningless," or "nonsensical."[4] But it cannot claim to be backed by empirical evidence in the scientific sense—although it may easily be, in some genetic sense, the "result of observation."

(There were a great many other theories of this pre-scientific or pseudo-scientific character, some of them, unfortunately, as influential as the Marxist interpretation of history; for example, the racialist interpretation of history—another of those impressive and all-explanatory theories which act upon weak minds like revelations.)

Thus the problem which I tried to solve by proposing the criterion of falsifiability was neither a problem of meaningfulness or significance, nor a problem of truth or acceptability. It was the problem of drawing a line (as well as this can be done) between the statements, or systems of statements, of the empirical sciences, and all other statements—whether they are of a religious or of a metaphysical character, or simply pseudo-scientific. Years later—it must have been in 1928 or 1929—I called this first problem of mine the *"problem of demarcation."* The criterion of falsifiability is a solution to this problem of demarcation, for it says that statements or systems of statements, in order to be ranked as scientific, must be capable of conflicting with possible, or conceivable, observations.

III

Today I know, of course, that this *criterion of demarcation*—the criterion of testability, or falsifiability, or refutability—is far from obvious; for even now its significance is seldom realized. At that time, in 1920, it seemed to me almost trivial, although it solved for me an intellectual problem which had worried me deeply, and one which also had obvious practical consequences (for example, political ones). But I did not yet realize its full implications, or its philosophical significance. When I explained it to a fellow student of the Mathematics Department (now a distinguished mathematician in Great Britain), he suggested that I should publish it. At the time I thought this absurd; for I was convinced that my problem, since it was so important for me, must have agitated many scientists and philosophers who would surely have reached my rather obvious solution. That this was not the case I learnt from Wittgenstein's work, and from its reception; and so I published my results thirteen years later in the form of a criticism of Wittgenstein's *criterion of meaningfulness.*

Wittgenstein, as you all know, tried to show in the *Tractatus* (see for example his propositions 6.53; 6.54; and 5) that all so-called philosophical or metaphysical propositions were actually non-propositions or pseudo-propositions: that they were senseless or meaningless. All genuine (or meaningful) propositions were truth functions of the elementary or atomic propositions which described "atomic facts," i.e. — facts which can in principle be ascertained by observation. In other words, meaningful propositions were fully reducible to elementary or atomic propositions which were simple statements describing possible states of affairs, and which could in principle be established or rejected by observation. If we call a statement an "observation statement" not only if it states an actual observation but also if it states anything that *may* be observed, we shall have to say (according to the *Tractatus,* 5 and 4.52) that every genuine proposition must be a truth-function of, and therefore deducible from, observation statements. All other apparent propositions will be meaningless pseudo-propositions; in fact they will be nothing but nonsensical gibberish.

This idea was used by Wittgenstein for a characterization of science, as opposed to philosophy. We read (for example in 4.11, where natural science is taken to stand in opposition to philosophy): "The totality of true propositions is the total natural science (or the totality of the natural sciences)." This means that the propositions which belong to science are those deducible from *true* observation statements; they are those propositions which can be *verified* by true observation statements. Could we know all true observation statements, we should also know all that may be asserted by natural science.

This amounts to a crude verifiability criterion of demarcation. To make it slightly less crude, it could be amended thus: "The statements which may possibly fall within the province of science are those which may possibly be verified by observation statements; and these statements, again, coincide with the class of *all* genuine or meaningful statements." For this approach, then, *verifiability, meaningfulness, and scientific character all coincide.*

I personally was never interested in the so-called problem of meaning; on the contrary, it appeared to me a verbal problem, a typical pseudo-problem. I was interested only in the problem of demarcation, i.e. in finding a criterion of the scientific character of theories. It was just this interest which made me see at once that Wittgenstein's verifiability criterion of meaning was intended to play the part of a criterion of demarcation as well; and which made me see that, as such, it was totally inadequate, even if all misgivings about the dubious concept of meaning were set aside. For Wittgenstein's criterion of demarcation — to use my own terminology in this context — is verifiability, or deducibility from observation statements. But this criterion is too narrow (*and* too wide): it excludes from science practically everything that is, in fact, characteristic of it (while failing in effect to exclude astrology). No scientific theory can ever be deduced from observation statements, or be described as a truth-function of observation statements.

All this I pointed out on various occasions to Wittgensteinians and members of the Vienna Circle. In 1931–2 I summarized my ideas in a largish book (read by several members of the Circle but never published; although part of it was incorporated in my *Logic of Scientific Discovery*); and in 1933 I published a letter to the Editor of *Erkenntnis* in which I tried to compress into two pages my ideas on the problems of demarcation and induction.[5] In this letter and elsewhere I described the problem of meaning as a pseudo-problem, in contrast to the problem of demarcation. But my contribution was classified by members of the Circle as a proposal to replace the verifiability criterion of *meaning* by a falsifiability criterion of *meaning*—which effectively made nonsense of my views.[6] My protests that I was trying to solve, not their pseudo-problem of meaning, but the problem of demarcation, were of no avail.

My attacks upon verification had some effect, however. They soon led to complete confusion in the camp of the verificationist philosophers of sense and nonsense. The original proposal of verifiability as the criterion of meaning was at least clear, simple, and forceful. The modifications and shifts which were now introduced were the very opposite.[7] This, I should say, is now seen even by the participants. But since I am usually quoted as one of them I wish to repeat that although I created this confusion I never participated in it. Neither falsifiability nor testability were proposed by me as criteria of meaning; and although I may plead guilty to having introduced both terms into the discussion, it was not I who introduced them into the theory of meaning.

Criticism of my alleged views was widespread and highly successful. I have yet to meet a criticism of my views.[8] Meanwhile, testability is being widely accepted as a criterion of demarcation. . . .

IV

Let us now turn from our logical criticism of the *psychology of experience* to our real problem—the problem of *the logic of science*. Although some of the things I have said may help us here, in so far as they may have eliminated certain psychological prejudices in favour of induction, my treatment of the *logical problem of induction* is completely independent of this criticism, and of all psychological considerations. Provided you do not dogmatically believe in the alleged psychological fact that we make inductions, you may now forget my whole story with the exception of two logical points: my logical remarks on testability or falsifiability as the criterion of demarcation; and Hume's logical criticism of induction.

From what I have said it is obvious that there was a close link between the two problems which interested me at that time: demarcation, and induction or scientific method. It was easy to see that the method of science is criticism, i.e. attempted falsifications. Yet it took me a few years to notice that the two problems—of demarcation and of induction—were in a sense one.

Why, I asked, do so many scientists believe in induction? I found they did so because they believed natural science to be characterized by the inductive method — by a method starting from, and relying upon, long sequences of observations and experiments. They believed that the difference between genuine science and metaphysical or pseudo-scientific speculation depended solely upon whether or not the inductive method was employed. They believed (to put it in my own terminology) that only the inductive method could provide a satisfactory *criterion of demarcation*.

I recently came across an interesting formulation of this belief in a remarkable philosophical book by a great physicist — Max Born's *Natural Philosophy of Cause and Chance*.[9] He writes: "Induction allows us to generalize a number of observations into a general rule: that night follows day and day follows night . . . But while everyday life has no definite criterion for the validity of an induction, . . . science has worked out a code, or rule of craft, for its application." Born nowhere reveals the contents of this inductive code (which, as his wording shows, contains a "definite criterion for the validity of an induction"); but he stresses that "there is no logical argument" for its acceptance: "it is a question of faith"; and he is therefore "willing to call induction a metaphysical principle." But why does he believe that such a code of valid inductive rules must exist? This becomes clear when he speaks of the "vast communities of people ignorant of, or rejecting, the rule of science, among them the members of anti-vaccination societies and believers in astrology. It is useless to argue with them; I cannot compel them to accept the same criteria of valid induction in which I believe: the code of scientific rules." This makes it quite clear that *"valid induction" was here meant to serve as a criterion of demarcation between science and pseudo-science.*

But it is obvious that this rule or craft of "valid induction" is not even metaphysical: it simply does not exist. No rule can ever guarantee that a generalization inferred from true observations, however often repeated, is true. (Born himself does not believe in the truth of Newtonian physics, in spite of its success, although he believes that it is based on induction.) And the success of science is not based upon rules of induction, but depends upon luck, ingenuity, and the purely deductive rules of critical argument.

I may summarize some of my conclusions as follows:

(1) Induction, i.e. inference based on many observations, is a myth. It is neither a psychological fact, nor a fact of ordinary life, nor one of scientific procedure.

(2) The actual procedure of science is to operate with conjectures: to jump to conclusions — often after one single observation (as noticed for example by Hume and Born).

(3) Repeated observations and experiments function in science as *tests* of our conjectures or hypotheses, i.e. as attempted refutations.

(4) The mistaken belief in induction is fortified by the need for a criterion of demarcation which, it is traditionally but wrongly believed, only the inductive method can provide.

(5) The conception of such an inductive method, like the criterion of verifiability, implies a faulty demarcation.

(6) None of this is altered in the least if we say that induction makes theories only probable rather than certain.

V

If, as I have suggested, the problem of induction is only an instance or facet of the problem of demarcation, then the solution to the problem of demarcation must provide us with a solution to the problem of induction. This is indeed the case, I believe, although it is perhaps not immediately obvious.

For a brief formulation of the problem of induction we can turn again to Born, who writes: ". . . no observation or experiment, however extended, can give more than a finite number of repetitions"; therefore, "the statement of a law — B depends on A — always transcends experience. Yet this kind of statement is made everywhere and all the time, and sometimes from scanty material."[10]

In other words, the logical problem of induction arises from *(a)* Hume's discovery (so well expressed by Born) that it is impossible to justify a law by observation or experiment, since it "transcends experience"; *(b)* the fact that science proposes and uses laws "everywhere and all the time." (Like Hume, Born is struck by the "scanty material," i.e. the few observed instances upon which the law may be based.) To this we have to add *(c) the principle of empiricism* which asserts that in science, only observation and experiment may decide upon the *acceptance or rejection* of scientific statements, including laws and theories.

These three principles, *(a), (b),* and *(c),* appear at first sight to clash; and this apparent clash constitutes the *logical problem of induction*.

Faced with this clash, Born gives up *(c),* the principle of empiricism (as Kant and many others, including Bertrand Russell, have done before him), in favour of what he calls a "metaphysical principle"; a metaphysical principle which he does not even attempt to formulate; which he vaguely describes as a "code or rule of craft"; and of which I have never seen any formulation which even looked promising and was not clearly untenable.

But in fact the principles *(a)* to *(c)* do not clash. We can see this the moment we realize that the acceptance by science of a law or of a theory is *tentative only*; which is to say that all laws and theories are conjectures, or tentative *hypotheses* (a position which I have sometimes called "hypotheticism"); and that we may reject a law or theory on the basis of new evidence, without necessarily discarding the old evidence which originally led us to accept it.[11]

The principles of empiricism *(c)* can be fully preserved, since the fate of a theory, its acceptance or rejection, is decided by observation and experiment — by the result of tests. So long as a theory stands up to the severest tests we can design, it is accepted; if it does not, it is rejected. But it is never inferred, in any sense, from the empirical evidence. There is neither a psychological nor a logical induction. *Only the falsity of the theory can be inferred from empirical evidence, and this inference is a purely deductive one.*

Hume showed that it is not possible to infer a theory from observation statements; but this does not affect the possibility of refuting a theory by observation statements. The full appreciation of this possibility makes the relation between theories and observations perfectly clear.

This solves the problem of the alleged clash between the principles *(a)*, *(b)*, and *(c)*, and with it Hume's problem of induction.

VI

Thus the problem of induction is solved. But nothing seems less wanted than a simple solution to an age-old philosophical problem. Wittgenstein and his school hold that genuine philosophical problems do not exist;[12] from which it clearly follows that they cannot be solved. Others among my contemporaries do believe that there are philosophical problems, and respect them; but they seem to respect them too much; they seem to believe that they are insoluble, if not taboo; and they are shocked and horrified by the claim that there is a simple, neat, and lucid, solution to any of them. If there is a solution it must be deep, they feel, or at least complicated.

However this may be, I am still waiting for a simple, neat and lucid criticism of the solution which I published first in 1933 in my letter to the Editor of *Erkenntnis,*[13] and later in *The Logic of Scientific Discovery.*

Of course, one can invent new problems of induction, different from the one I have formulated and solved. (Its formulation was half its solution.) But I have yet to see any reformulation of the problem whose solution cannot be easily obtained from my old solution. I am now going to discuss some of these re-formulations.

One question which may be asked is this: how do we really jump from an observation statement to a theory?

Although this question appears to be psychological rather than philosophical, one can say something positive about it without invoking psychology. One can say first that the jump is not from an observation state-ment, but from a problem-situation, and that the theory must allow *to explain* the observations which created the problem (that is, *to deduce* them from the theory strengthened by other accepted theories and by other observation statements, the so-called initial conditions). This leaves, of course, an immense number of possible theories, good and bad; and it thus appears that our ques-tion has not been answered.

But this makes it fairly clear that when we asked our question we had more in mind than, "How do we jump from an observation statement to a theory?" The question we had in mind was, it now appears, "How do we jump from an obser-vation statement to a *good* theory?" But to this the answer is: by jumping first to *any* theory and then testing it, to find whether it is good or not; i.e. by repeatedly applying the critical method, eliminating many bad theories, and inventing many new ones. Not everybody is able to do this; but there is no other way.

Other questions have sometimes been asked. The original problem of induction, it was said, is the problem of *justifying* induction, i.e. of justifying inductive inference. If you answer this problem by saying that what is called an "inductive inference" is always invalid and therefore clearly not justifiable, the following new problem must arise: how do you justify your method of trial and error? Reply: the method of trial and error is a *method of eliminating false theories* by observation statements; and the justification for this is the purely logical relationship of deducibility which allows us to assert the falsity of universal statements if we accept the truth of singular ones.

Another question sometimes asked is this: why is it reasonable to prefer non-falsified statements to falsified ones? To this question some involved answers have been produced, for example pragmatic answers. But from a pragmatic point of view the question does not arise, since false theories often serve well enough: most formulae used in engineering or navigation are known to be false, although they may be excellent approximations and easy to handle; and they are used with confidence by people who know them to be false.

The only correct answer is the straightforward one: because we search for truth (even though we can never be sure we have found it), and because the falsified theories are known or believed to be false, while the non-falsified theories may still be true. Besides, we do not prefer *every* non-falsified theory—only one which, in the light of criticism, appears to be better than its competitors: which solves our problems, which is well tested, and of which we think, or rather conjecture or hope (considering other provisionally accepted theories), that it will stand up to further tests.

It has also been said that the problem of induction is, "Why is it *reasonable* to believe that the future will be like the past?," and that a satisfactory answer to this question should make it plain that such a belief is, in fact, reasonable. My reply is that it is reasonable to believe that the future will be very different from the past in many vitally important respects. Admittedly it is perfectly reasonable to *act* on the assumption that it will, in many respects, be like the past, and that well-tested laws will continue to hold (since we can have no better assumption to act upon); but it is also reasonable to believe that such a course of action will lead us at times into severe trouble, since some of the laws upon which we now heavily rely may easily prove unreliable. (Remember the midnight sun!) One might even say that to judge from past experience, and from our general scientific knowledge, the future will *not* be like the past, in perhaps most of the ways which those have in mind who say that it will. Water will sometimes not quench thirst, and air will choke those who breathe it. An apparent way out is to say that the future will be like the past *in the sense that the laws of nature will not change,* but this is begging the question. We speak of a "law of nature" only if we think that we have before us a regularity which does not change; and if we find that it changes then we shall not continue to call it a "law of nature." Of course our search for natural laws indicates that we hope to find them, and that we believe that there are natural laws; but our belief in any particular natural law cannot have a safer basis than our unsuccessful critical attempts to refute it.

I think that those who put the problem of induction in terms of the *reasonableness* of our beliefs are perfectly right if they are dissatisfied with a Humean, or post-Humean, skeptical despair of reason. We must indeed reject the view that a belief in science is as irrational as a belief in primitive magical practices—that both are a matter of accepting a "total ideology," a convention or a tradition based on faith. But we must be cautious if we formulate our problem, with Hume, as one of the reasonableness of our *beliefs*. We should split this problem into three—our old problem of demarcation, or of how to *distinguish* between science and primitive magic; the problem of the rationality of the scientific or critical *procedure,* and of the role of observation within it; and lastly the problem of the rationality of our *acceptance* of theories for scientific and for practical purposes. To all these three problems solutions have been offered here.

One should also be careful not to confuse the problem of the reasonableness of the scientific procedure and the (tentative) acceptance of the results of this procedure—i.e. the scientific theories—with the problem of the rationality or otherwise *of the belief that this procedure will succeed.* In practice, in practical scientific research, this belief is no doubt unavoidable and reasonable, there being no better alternative. But the belief is certainly unjustifiable in a theoretical sense. Moreover, if we could show, on general logical grounds, that the scientific quest is likely to succeed, one could not understand why anything like success has been so rare in the long history of human endeavours to know more about our world.

Yet another way of putting the problem of induction is in terms of probability. Let *t* be the theory and *e* the evidence: we can ask for *P(t,e),* that is to say, the probability of *t,* given *e*. The problem of induction, it is often believed, can then be put thus: construct a *calculus of probability* which allows us to work out for any theory *t* what its probability is, relative to any given empirical evidence *e*; and show that *P(t,e)* increases with the accumulation of supporting evidence, and reaches high values—at any rate values greater than ½.

In *The Logic of Scientific Discovery* I explained why I think that this approach to the problem is fundamentally mistaken.[14] To make this clear, I introduced there the distinction between *probability* and *degree of corroboration or confirmation.* (The term "confirmation" has lately been so much used and misused that I have decided to surrender it to the verificationists and to use for my own purposes "corroboration" only. The term "probability" is best used in some of the many senses which satisfy the well-known calculus of probability, axiomatized, for example, by Keynes, Jeffreys, and myself; but nothing of course depends on the choice of words, as long as we do not *assume,* uncritically, that degree of corroboration must also be a probability—that is to say, that it must satisfy the calculus of probability.)

I explained in my book why we are interested in theories with a *high degree of corroboration.* And I explained why it is a mistake to conclude from this that we are interested in *highly probable* theories. I pointed out that the probability of a statement (or set of statements) is always the greater the less the statement says: it is inverse to the content or the deductive power of the statement, and thus to its explanatory power. Accordingly every interesting and powerful statement must have a low probability; and *vice versa*: a statement with a high probability will be scientifically uninteresting, because it says little and has no explanatory power. Although we seek theories with a high degree of corroboration, *as scientists we do not seek highly probable theories but explanations; that is to say, powerful and improbable theories.*[15] The opposite view—that science aims at high probability—is a characteristic development of verificationism: if you find that you cannot verify a theory, or make it certain by induction, you may turn to probability as a kind of *"Ersatz"* for certainty, in the hope that induction may yield at least that much. . . .

NOTES

1. This is a slight oversimplification, for about half of the Einstein effect may be derived from the classical theory, provided we assume a ballistic theory of light.

2. See, for example, my *Open Society and Its Enemies,* ch. 15, section iii, and notes 13–14.

3. "Clinical observations," like all other observations, are *interpretations in the light of theories*; and for this reason alone they are apt to seem to support those theories in the light of which they were interpreted. But real support can be obtained only from observations undertaken as tests (by "attempted refutations"); and for this purpose *criteria of refutation* have to be laid down beforehand; it must be agreed which observable situations, if actually observed, mean that the theory is refuted. But what kind of clinical responses would refute to the satisfaction of the analyst not merely a particular analytic diagnosis but psycho-analysis itself? And have such criteria ever been discussed or agreed upon by analysts? Is there not, on the contrary, a whole family of analytic concepts, such as "ambivalence" (I do not suggest that there is no such thing as ambivalence), which would make it difficult, if not impossible, to agree upon such criteria? Moreover, how much headway has been made in investigating the question of the extent to which the (conscious or unconscious) expectations and theories held by the analyst influence the "clinical responses" of the patient? (To say nothing about the conscious attempts to influence the patient by proposing interpretations to him, etc.) Years ago I introduced the term *"Oedipus effect"* to describe the influence of a theory or expectation or prediction *upon the event which it predicts* or describes: it will be remembered that the causal chain leading to Oedipus' parricide was started by the oracle's prediction of this event. This is a characteristic and recurrent theme of such myths, but one which seems to have failed to attract the interest of the analysts, perhaps not accidentally. (The problem of confirmatory dreams suggested by the analyst is discussed by Freud, for example in *Gesammelte Schriften,* III, 1925, where he says on p. 314: "If anybody asserts that most of the dreams which can be utilized in an analysis . . . owe their origin to [the analyst's] suggestion, then no objection can be made from the point of view of analytic theory. Yet there is nothing in this fact," he surprisingly adds, "which would detract from the reliability of our results.")

4. The case of astrology, nowadays a typical pseudo-science, may illustrate this point. It was attacked, by Aristotelians and other rationalists, down to Newton's day, for the wrong reason—for its now accepted assertion that the planets had an "influence" upon terrestrial ("sublunar") events. In fact Newton's theory of gravity, and especially the lunar theory of the tides, was historically speaking an offspring of astrological lore. Newton, it seems, was most reluctant to adopt a theory which came from the same stable as for example the theory that "influenza" epidemics are due to an astral "influence." And Galileo, no doubt for the same

reason, actually rejected the lunar theory of the tides; and his misgivings about Kepler may easily be explained by his misgivings about astrology.

5. My *Logic of Scientific Discovery* (1959, 1960, 1961), here usually referred to as *L.Sc.D.,* is the translation of *Logik der Forschung* (1934), with a number of additional notes and appendices, including (on pp. 312-14) the letter to the Editor of *Erkenntnis* mentioned here in the text which was first published in *Erkenntnis*, **3**, 1933, pp. 426 f.

Concerning my never published book mentioned here in the text, see R. Carnap's paper *"Ueber Protokollstäze"* (On Protocol-Sentences), *Erkenntnis*, **3**, 1932, pp. 215-28 where he gives an outline of my theory on pp. 223-8, and accepts it. He calls my theory "procedure B," and says (p. 224, top): "Starting from a point of view different from Neurath's" (who developed what Carnap calls on p. 223 "procedure A"), "Popper developed procedure B as part of his system." And after describing in detail my theory of tests, Carnap sums up his views as follows (p. 228): "After weighing the various arguments here discussed, it appears to me that the second language form with procedure B—that is in the form here described—is the most adequate among the forms of scientific language at present advocated . . . in the . . . theory of knowledge." This paper of Carnap's contained the first published report of my theory of critical testing. (See also my critical remarks in *L.Sc.D.,* note 1 to section 29, p. 104, where the date "1933" should read "1932"; and ch. 11, below, text to note 39.)

6. Wittgenstein's example of a nonsensical pseudo-proposition is: 'Socrates is identical'. Obviously, 'Socrates is not identical' must also be nonsense. Thus the negation of any nonsense will be nonsense, and that of a meaningful statement will be meaningful. *But the negation of a testable (or falsifiable) statement need not be testable,* as was pointed out, first in my *L.Sc.D.,* (e.g. pp. 38 f) and later by my critics. The confusion caused by taking testability as a criterion of *meaning* rather than of *demarcation* can easily be imagined.

7. The most recent example of the way in which the history of this problem is misunderstood is A. R. White's "Note on Meaning and Verification," *Mind*, **63**, 1954, pp. 66 ff. J. L. Evans's article, *Mind*, **62**, 1953, pp. 1 ff., which Mr. White criticizes, is excellent in my opinion, and unusually perceptive. Understandably enough, neither of the authors can quite reconstruct the story. (Some hints may be found in my *Open Society*, notes 46, 51 and 52 to ch. 11; and a fuller analysis in ch. 11 of *Conjectures and Refutations* (1963).

8. In *L.Sc.D.* I discussed, and replied to, some likely objections which afterwards were indeed raised, without reference to my replies. One of them is the contention that the falsification of a natural law is just as impossible as its verification. The answer is that this objection mixes two entirely different levels of analysis (like the objection that mathematical demonstrations are impossible since checking, no matter how often repeated, can never make it quite certain that we have not overlooked a mistake). On the first level, there is a logical asymmetry: one singular statement—say about the perihelion of Mercury—can formally falsify Kepler's laws; but these cannot be formally verified by any number of singular statements. The attempt to minimize this asymmetry can only lead to confusion. On another level, we may hesitate to accept any statement, even the simplest observation statement; and we may point out that every statement involves *interpretation in the light of theories,* and that it is therefore uncertain. This does not affect the fundamental asymmetry, but it is important: most dissectors of the heart before Harvey observed the wrong things—those, which they expected to see. There can never be anything like a completely safe observation, free from the dangers of misinterpretation. (This is one of the reasons why the theory of induction does not work.) The "empirical basis" consists largely of a mixture of *theories* of lower degree of universality (of "reproducible effects"). But the fact remains that, relative to whatever basis the investigator may accept (at his peril), he can test his theory only by trying to refute it.

9. Max Born, *Natural Philosophy of Cause and Chance,* Oxford, 1949, p. 7.

10. *Natural Philosophy of Cause and Chance,* p. 6.

11. I do not doubt that Born and many others would agree that theories are accepted only tentatively. But the widespread belief in induction shows that the far-reaching implications of this view are rarely seen.

12. Wittgenstein still held this belief in 1946.

13. See note 5 above.

14. *L.Sc.D.* (see note 5 above), ch. x, especially sections 80 to 83, also section 34 ff. See also my note "A Set of Independent Axioms for Probability," *Mind,* N.S. **47**, 1938, p. 275. (This note has since been reprinted, with corrections, in the new appendix ii of *L.Sc.D.*)

15. A definition, in terms of probabilities, of *C(t,e),* i.e. of the degree of corroboration (of a theory *t* relative to the evidence *e*) satisfying the demands indicated in my *L.Sc.D.,* sections 82 to 83, is the following:

$$C(t,e) = E(t,e) (1 + P(t)P(t,e)),$$

where $E(t,e) = (P(e,t) - P(e))/(P(e,t) + P(e))$ is a (non-additive) measure of the explanatory power of *t* with respect to *e*. Note that $C(t,e)$ is not a probability: it may have values between -1 (refutation of *t* by *e*) and $C(t,t) \leqslant + 1$. Statements *t* which are lawlike and thus non-verifiable cannot even reach $C(t,e) = C(t,t)$ upon empirical evidence *e*. $C(t,t)$ is the *degree of corroborability* of *t,* and is equal to the *degree of testability* of *t,* or to the *content* of *t*. Because the demands implied in point (6) at the end of section I above, I do not think, however, that it is possible to give a complete formalization of the idea of corroboration (or, as I previously used to say, of confirmation).

(Added 1955 to the first proofs of this paper:)

See also my note "Degree of Confirmation," *British Journal for the Philosophy of Science,* 5, 1954, pp. 143 ff. (See also **5**, pp. 334.) I have since simplified this definition as follows (*B.J.P.S.,* 1955, **5**, p. 359):

$$C(t,e) = (P(e,t) - P(e))/(P(e,t) - P(e,t) + P(e))$$

For a further improvement, see *B.J.P.S.* **6**, 1955, p. 56.

2

What Is Science?

John Ziman

To answer the question "What is Science?" is almost as presumptuous as to try to state the meaning of Life itself. Science has become a major part of the stock of our minds; its products are the furniture of our surroundings. We must accept it, as the good lady of the fable is said to have agreed to accept the Universe.

Yet the question is puzzling rather than mysterious. Science is very clearly a conscious artifact of mankind, with well-documented historical origins, with a definable scope and content, and with recognizable professional practitioners and exponents. The task of defining Poetry, say, whose subject matter is by common consent ineffable, must be self-defeating. Poetry has no rules, no method, no graduate schools, no logic: the bards are self-anointed and their spirit bloweth where it listeth. Science, by contrast, is rigorous, methodical, academic, logical and practical. The very facility that it gives us, of clear understanding, of seeing things sharply in focus, makes us feel that the instrument itself is very real and hard and definite. Surely we can state, in a few words, its essential nature.

It is not difficult to state the order of being to which Science belongs. It is one of the categories of the intellectual commentary that Man makes on his World. Amongst its kith and kin we would put Religion, Art, Poetry, Law, Philosophy, Technology, etc.—the familiar divisions or "Faculties" of the Academy or the Multiversity.

At this stage I do not mean to analyse the precise relationship that exists between Science and each of these cognate modes of thought; I am merely asserting that they are on all fours with one another. It makes some sort of sense (though it may not always be stating a truth) to substitute these words for one another, in phrases like "Science teaches us . . ." or "The Spirit of *Law* is . . ." or "*Technology* benefits mankind by . . ." or "He is a student of

Philosophy.'' The famous "conflict between Science and Religion" was truly a battle between combatants of the same species—between David and Goliath if you will—and not, say, between the Philistine army and a Dryad, or between a point of order and a postage stamp.

Science is obviously like Religion, Law, Philosophy, etc. in being a more or less coherent set of ideas. In its own technical language, Science is information; it does not act directly on the body; it speaks to the mind. Religion and Poetry, we may concede, speak also to the emotions, and the statements of Art can seldom be written or expressed verbally—but they all belong in the non-material realm.

But in what ways are these forms of knowledge *unlike* one another? What are the special attributes of Science? What is the criterion for drawing lines of demarcation about it, to distinguish it from Philosophy, or from Technology, or from Poetry?

This question has long been debated. Famous books have been devoted to it. It has been the theme of whole schools of philosophy. To give an account of all the answers, with all their variations, would require a history of Western thought. It is a daunting subject. Nevertheless, the types of definition with which we are familiar can be stated crudely.

Science Is the Mastery of Man's Environment. This is, I think, the vulgar conception. It identifies Science with its products. It points to penicillin or to an artificial satellite and tells us of all the wonderful further powers that man will soon acquire by the same agency.

This definition enshrines two separate errors. In the first place it confounds Science with Technology. It puts all its emphasis on the applications of scientific knowledge and gives no hint as to the intellectual procedures by which that knowledge may be successfully obtained. It does not really discriminate between Science and Magic, and gives us no reason for studies such as Cosmology and Pure Mathematics, which seem entirely remote from practical use.

It also confuses ideas with things. Penicillin is not Science, any more than a cathedral is Religion or a witness box is Law. The material manifestations and powers of Science, however beneficial, awe-inspiring, monstrous, or beautiful, are not even symbolic; they belong in a different logical realm, just as a building is not equivalent to or symbolic of the architect's blueprints. A meal is not the same as a recipe

Science Is the Study of the Material World. This sort of definition is also very familiar in popular thought. It derives, I guess, from the great debate between Science and Religion, whose outcome was a treaty of partition in which Religion was left with the realm of the Spirit whilst Science was allowed full sway in the territory of Matter.

Now it is true that one of the aims of Science is to provide us with a Philosophy of Nature, and it is also true that many questions of a moral or spiritual kind cannot be answered at all within a scientific framework. But the dichotomy between Matter and Spirit is an obsolete philosophical notion which does not stand up very well to careful critical analysis. If we stick to this definition we may end up in a circular argument in which Matter is only recognizable as the subject matter of Science. Even then, we shall have stretched the meaning of words a long way in order to accommodate Psychology, or Sociology, within the Scientific stable.

This definition would also exclude Pure Mathematics. Surely this is wrong. Mathematical thinking is so deeply entangled with the physical sciences that one cannot draw a line between them. Modern mathematicians think of themselves as exploring the logical consequences (the "theorems") of different sets of hypotheses or "axioms," and do not claim absolute truth, in a material sense, for their results. Theoretical physicists and applied mathematicians try to confine their explorations to systems of hypotheses that they believe to reflect properties of the "real" world, but they often have no license for this belief. It would be absurd to have to say that Newton's *Principia,* and all the work that was built upon it, was not now Science, just because we now suppose that the inverse square law of gravitation is not perfectly true in an Einsteinian universe. I suspect that the exclusion of the "Queen of the Sciences" from her throne is a relic of some ancient academic arrangement, such as the combination of classical literary studies with mathematics in the Cambridge Tripos, and has no better justification than that Euclid and Archimedes wrote in Greek.

Science Is the Experimental Method. The recognition of the importance of experiment was the key event in the history of Science. The Baconian thesis was sound; we can often do no better today than to follow it.

Yet this definition is incomplete in several respects. It arbitrarily excludes Pure Mathematics, and needs to be supplemented to take cognizance of those perfectly respectable sciences such as Astronomy or Geology where we can only observe the consequences of events and circumstances over which we have no control. It also fails to give due credit to the strong theoretical and logical sinews that are needed to hold the results of experiments and observations together and give them force. Scientists do not in fact work in the way that operationalists suggest; they tend to look for, and find, in Nature little more than they believe to be there, and yet they construct airier theoretical systems than their actual observations warrant. Experiment distinguishes Science from the older, more speculative ways to knowledge but it does not fully characterize the scientific method.

Science Arrives at Truth by Logical Inferences from Empirical Observations
This is the standard type of definition favoured by most serious philosophers. It is usually based upon the principle of induction—that what has been seen to happen a great many times is almost sure to happen invariably and may be treated as a basic fact or Law upon which a firm structure of theory can be erected.

There is no doubt that this is the official philosophy by which most practical scientists work. From it one can deduce a number of practical procedures, such as the testing of theory by "predictions" of the results of future observations, and their subsequent confirmation. The importance of speculative thinking is recognized, provided that it is curbed by conformity to facts. There is no restriction of a metaphysical kind upon the subject matter of Science, except that it must be amenable to observations and inference.

But the attempt to make these principles logically watertight does not seem to have succeeded. What may be called the positivist programme, which would assign the label "True" to statements that satisfy these criteria, is plausible but not finally compelling. Many philosophers have now sadly come to the conclusion that there is no ultimate procedure which will wring the last drops of uncertainty from what scientists call their knowledge.

And although working scientists would probably state that this is the Rule of their Order, and the only safe principle upon which their discoveries may be based, they do not always obey it in practice. We often find complex theories—quite good theories—that really depend on very few observations. It is extraordinary, for example, how long and complicated the chains of inference are in the physics of elementary particles; a few clicks per month in an enormous assembly of glass tubes, magnet fields, scintillator fluids and electronic circuits becomes a new "particle," which in its turn provokes a flurry of theoretical papers and ingenious interpretations. I do not mean to say that the physicists are not correct; but no one can say that all the possible alternative schemes of explanation are carefully checked by innumerable experiments before the discovery is acclaimed and becomes part of the scientific canon. There is far more faith, and reliance upon personal experience and intellectual authority, than the official doctrine will allow.

A simple way of putting it is that the logico-inductive scheme does not leave enough room for genuine scientific error. It is too black and white. Our experience, both as individual scientists and historically, is that we only arrive at partial and incomplete truths; we never achieve the precision and finality that seem required by the definition. Thus, nothing we do in the laboratory or study is "really" scientific, however honestly we may aspire to the ideal. Surely, it is going too far to have to say, for example, that it was "unscientific" to continue to believe in Newtonian dynamics as soon as it had been observed and calculated that the rotation of the perihelion of Mercury did not conform to its predictions.

This summary of the various conceptions of science obviously fails to do justice to the vast and subtle literature on the subject. If I have emphasized the objections to each point of view, this is merely to indicate that none of the definitions is entirely satisfactory. Most practicing scientists, and most people generally, take up one or other of the attitudes that I have sketched, according to the degree of their intellectual sophistication—but without fervour. One can be zealous for Science, and a splendidly successful research worker, without pretending to a clear and certain notion of what Science really is. In practice it does not seem to matter.

Perhaps this is healthy. A deep interest in theology is not welcome in the average churchgoer, and the ordinary taxpayer should not really concern himself about the nature of sovereignty or the merits of bicameral legislatures. Even though Church and State depend, in the end, upon such abstract matters, we may reasonably leave them to the experts if all goes smoothly. The average scientist will say that he knows from experience and common sense what he is doing, and so long as he is not striking too deeply into the foundations of knowledge he is content to leave the highly technical discussion of the nature of Science to those self-appointed authorities the Philosophers of Science. A rough and ready conventional wisdom will see him through.

Yet in a way this neglect of—even scorn for—the Philosophy of Science by professional scientists is strange. They are, after all, engaged in a very difficult, rather abstract, highly intellectual activity and need all the guidance they can get from general theory. We may agree that the general principles may not in practice be very helpful, but we might have thought that at least they would be taught to young scientists in training, just as medical students are taught Physiology and budding administrators were once encouraged to acquaint themselves with Plato's *Republic*. When the student graduates and goes into a laboratory, how will he know what to do to make scientific discoveries if he has not been taught the distinction between a scientific theory and a non-scientific one? Making all allowances for the initial prejudice of scientists against speculative philosophy, and for the outmoded assumption that certain general ideas would communicate themselves to the educated and cultured man without specific instruction, I find this an odd and significant phenomenon.

The fact is that scientific investigation, as distinct from the theoretical *content* of any given branch of science, is a practical art. It is not learnt out of books, but by imitation and experience. Research workers are trained by apprenticeship, by working for their Ph.D.'s under the supervision of more experienced scholars, not by attending courses in the metaphysics of physics. The graduate student is given his "problem": "You might have a look at the effect of pressure on the band structure of the III–V compounds; I don't think it has been done yet, and it would be interesting to see whether it fits into the pseudopotential theory." Then, with considerable help, encouragement and criticism, he sets up his apparatus, makes his measurements, performs his calculations, etc. and in due course writes a thesis and is accounted a qualified professional. But notice that he will not at any time have been made to study formal logic, nor will he be expected to defend his thesis in a step by step deductive procedure. His examiners may ask him why he had made some particular assertion in the course of his argument, or they may enquire as to the reliability of some particular measurement. They may even ask him to assess the value of the "contribution" he has made to the subject as a whole. But they will not ask him to give any opinion as to whether Physics is ultimately *true,* or whether he is justified now in believing in an external world, or in what sense a theory is verified by the observation of favourable instances. The examiners will assume that the candidate shares with them the common language and principles of their discipline. No scientist really doubts that theories are verified by observation, any more than a Common Law judge hesitates to rule that hearsay evidence is inadmissible.

What one finds in practice is that scientific argument, written or spoken, is not very complex or logically precise. The terms and concepts that are used may be extremely subtle and technical, but they are put together in quite simple logical forms, with expressed or implied relations as the machinery of deduction. It is very seldom that one uses the more sophisticated types of proof used in Mathematics, such as asserting a proposition by proving that its negation implies a contradiction. Of course actual mathematical or numerical analysis of data may carry the deduction through many steps, but the symbolic machinery of algebra and the electronic circuits of the computer are then relied on to keep the argument straight.[1] In my own experience, one more often detects elementary *non sequiturs* in the verbal reasoning than actual mathematical mistakes in the calculations that accompany them. This is not said to disparage the intellectual powers of scientists; I mean simply that the reasoning used in scientific papers is not very different from what we should use in an everyday careful discussion of an everyday problem.

. . . [This point] is made to emphasize the inadequacy of the "logico-inductive" metaphysic of Science. How can this be correct, when few scientists are interested in or understand it, and none ever uses it explicitly in his work? But then if Science is distinguished from other intellectual disciplines neither by a particular style or argument nor by a definable subject matter, what is it?

The answer proposed in this essay is suggested by its title: *Science Is Public Knowledge.* This is, of course, a very cryptic definition, with almost the suggestion of a play upon words. What I mean is something along the following lines. Science is not merely *published* knowledge or information. Anyone may make an observation, or conceive a hypothesis, and, if he has the financial means, get it printed and distributed for other persons to read. Scientific knowledge is more than this. Its facts and theories must survive a period of critical study and testing by other competent and disinterested individuals, and must have been found so persuasive that they are almost universally accepted. The objective of Science is not just to acquire information nor to utter all non-contradictory notions; its goal is a *consensus* of rational opinion over the widest possible field.

In a sense, this is so obvious and well-known that it scarcely needs saying. Most educated and informed people agree that Science is true, and therefore impossible to gainsay. But I assert my definition much more positively; this is the basic principle upon which Science is founded. It is not a subsidiary consequence of the "Scientific Method"; it *is* the scientific method itself.

The defect of the conventional philosophical approach to Science is that it considers only two terms in the equation. The scientist is seen as an individual, pursuing a somewhat one-sided dialogue with taciturn Nature. *He* observes phenomena, notices regularities, arrives at generalizations, deduces consequences, etc. and eventually, Hey Presto! a law of Nature springs into being. But it is not like that at all. The scientific enterprise is corporate. It is not merely, in Newton's incomparable phrase, that one stands on the shoulders of giants, and hence can see a little farther. Every scientist sees through his own eyes — and also through the eyes of his predecessors and colleagues. It is never one individual that goes through all the steps in the logico-inductive chain; it is a group of individuals, dividing their labour but continuously and jealously checking each other's contributions. The cliché of scientific prose betrays itself "Hence we arrive at the conclusion that . . ." The audience to which scientific publications are addressed is not passive; by its cheering or booing, its bouquets or brickbats, it actively controls the substance of the communications that it receives.

In other words, scientific research is a social activity. Technology, Art and Religion are perhaps possible for Robinson Crusoe, but Law and Science are not. To understand the nature of Science, we must look at the way in which scientists behave towards one another, how they are organized and how information passes between them. The young scientist does not study formal logic, but he learns by imitation and experience a number of conventions that embody strong social relationships. In the language of Sociology, he learns to play his *role* in a system by which knowledge is acquired, sifted and eventually made public property.

It has, of course, long been recognized that Science is peculiar in its origins to the civilization of Western Europe. The question of the social basis of Science, and its relations to other organizations and institutions of our way of life, is much debated. Is it a consequence of the "Bourgeois Revolution," or of Protestantism — or what? Does it exist despite the Church and the Universities, or because of them? Why did China, with its immense technological and intellectual resources, not develop the same system? What should be the status of the scientific worker in an advanced society; should he be a paid employee, with a prescribed field of study, or an aristocratic dilettante? How should decisions be taken about expenditure on research? And so on.

These problems, profoundly sociological, historical and political though they may be, are not quite what I have in mind. Only too often the element in the argument that gets the least analysis is the actual institution about which the whole discussion hinges — scientific activity itself. To give a contemporary example, there is much talk nowadays about the importance of creating more effective systems for storing and indexing scientific literature, so that every scientist can very quickly become aware of the relevant work of every other scientist in his field. This recognizes that publication is important, but the discussion usually betrays an absence of careful thought about the part that conventional systems of scientific communication play in sifting and sorting the material that they handle. Or again, the problem of why Greek Science never finally took off from its brilliant taxying runs is discussed in terms of, say, the aristocratic citizen despising the servile labour of practical experiment, when it might have been due to the absence of just such a communications system between scholars as was provided in the Renaissance by alphabetic printing. The internal sociological analysis of Science itself is a necessary preliminary to the study of the Sociology of Knowledge in the secular world.

The present essay cannot pretend to deal with all such questions. The "Science of Science" is a vast topic, with many aspects. The very core of so many difficulties is suggested by my present argument—that Science stands in the region where the intellectual, the psychological and the sociological coordinate axes intersect. It is knowledge, therefore intellectual, conceptual and abstract. It is inevitably created by individual men and women, and therefore has a strong psychological aspect. It is public, and therefore moulded and determined by the social relations between individuals. To keep all these aspects in view simultaneously, and to appreciate their hidden connections, is not at all easy.

It has been put to me that one should in fact distinguish carefully between Science as a body of knowledge, Science as what scientists do, and Science as a social institution. This is precisely the sort of distinction that one must *not* make; in the language of geometry, a solid object cannot be reconstructed from its projections upon the separate cartesian planes. By assigning the intellectual aspects of Science to the professional philosophers we make of it an arid exercise in logic; by allowing the psychologists to take possession of the personal dimension we overemphasize the mysteries of "creativity" at the expense of rationality and the critical power of well-ordered argument; if the social aspects are handed over to the sociologists, we get a description of research as an N-person game, with prestige points for stakes and priority claims as trumps. The problem has been to discover a unifying principle for Science in all its aspects. The recognition that scientific knowledge must be public and *consensible* (to coin a necessary word) allows one to trace out the complex inner relationships between its various facets. Before one can distinguish and discuss separately the philosophical, psychological or sociological dimensions of Science, one must somehow have succeeded in characterizing it as a whole.[2]

In an ordinary work of Science one does well not to dwell too long on the hypothesis that is being tested, trying to define and describe it in advance of reporting the results of the experiments or calculations that are supposed to verify or negate it. The results themselves indicate the nature of the hypothesis, its scope and limitations. The present essay is organized in the same manner. Having sketched a point of view in this chapter, I propose to turn the discussion to a number of particular topics that I think can be better understood when seen from this new angle. To give a semblance of order to the argument, the various subjects have been arranged according to whether they are primarily *intellectual*—as, for example, some attempt to discriminate between scientific and non-scientific disciplines; *psychological*—e.g., the role of education, the significance of scientific creativity; *sociological*—the structure of the scientific community and the institutions by which it maintains scientific standards and procedures. Beyond this classification, the succession of topics is likely to be pretty haphazard; or, as the good lady said, "How do I know what I think until I have heard what I have to say?"

The subject is indeed endless. . . . The present brief essay is meant only as an exposition of a general theory, which will be applied to a variety of more specific instances in a larger work. The topics discussed here are chosen, therefore, solely to exemplify the main argument, and are not meant to comprehend the whole field. In many cases, also, the discussion has been kept abstract and schematic, to avoid great marshlands of detail. The reader is begged, once more, to forgive the inaccuracies and imprecisions inevitable in such an account, and to concentrate his critical attention upon the validity of the general principle and its power of explaining how things really are.

Science and Non-science

In this chapter Science will be considered mainly in its intellectual aspects, as a system of ideas, as a compilation of abstract knowledge. The first question to be answered has already been posed in the introductory chapter: what distinguishes Science from its sister "Faculties" — Law, Philosophy, Technology, etc.? The argument is that Science is unique in striving for, and insisting on, a consensus.

Take Law, for example. We all feel that legal thought is quite different from scientific thought — but what is the basis of this intuition? There are many ways in which legal argument is very close to Science. There is undoubtedly an attempt to make every judgement follow logically on statutes and precedents. Every lawyer seeks to clarify a path of implications through successive stages to validate his case. The judge reasons it out, on the basis of universal principles of equity, in the effort to arrive at a decision that will command the assent of all just and learned men.

The kinship of Law with the mathematical sciences is emphasized by the interesting suggestion that legal decisions might be arrived at automatically by a computer, into which all the conditions and precedents of the case would be fed and a purely mechanical process of logical reduction would produce exactly the correct judgement.[3] Although perhaps the idea is somewhat fanciful, if this procedure were technically feasible it would provide decisions that could not but command the assent of all lawyers — just as a table of values of a mathematical function printed out by a computer commands the assent of all mathematicians. To the extent, therefore, that the Law is strictly logical, it can be made "scientific."

Again, in the concept of "evidence" there is close similarity. This is too primary and basic an idea to be defined readily, but, roughly speaking, it means "any information that is relevant to a disputed hypothesis." In Science, as in Law, we are almost always dealing with theories that are disputable, and that can only be challenged by an appeal to evidence for and against them. It is the duty of scientists, as of lawyers, to bring out this evidence, on both sides, to the full.

In the end, the case may hang upon some very minor item of information—was the man who got off the 3:57 at Little Puddlecome on Monday, 27 May, wearing a black hat? A scientific theory also may be validated by some tiny fact—for example, the almost imperceptible changes in the orbit of the planet Mercury. The question of the *credibility* of evidence can become very important. We may find everyone in full agreement that, if a fact is as stated by a witness, it has vital logical implications for the hypothesis under consideration; yet the court may be completely undecided as to whether this evidence is true or not. The existence of honest error has to be allowed for. This sort of thing happens in Science too, though it does not usually get remembered in the conventional histories. For example, many scientists will recall the interest that was aroused by the publication of evidence for organic compounds in meteorites—probably an erroneous interpretation of a complex observation, but of the most profound significance if it proved to be true. In such cases there may even be questions about the relative reliability, in general, of two different observers—an assessment, perhaps, based upon their scientific standing and expert authority—just as the relative veracity of conflicting witnesses may become the key issue in a legal case.

But, of course, in Science, when the evidence is conflicting, we withhold our assent or dissent, and do the experiment again. This cannot be done in legal disputes, which must be terminated yea or nay. If we are forced to a premature opinion on a scientific question, we are bound to give the Scottish verdict *Not Proven,* or say that the jury have disagreed, and a new trial is needed. In Criminal Law, where the case for the prosecution must be proved up to the hilt, or the accused acquitted, this is well enough; but Civil Law demands a decision, however difficult the case.

The Law is thus unscientific because it *must* decide upon matters which are not at all amenable to a consensus of opinion. Indeed, legal argument is concerned with the conflict between various principles, statutes, precedents, etc.; if there were not an area of uncertainty and contradiction, then there would be no need to go to law about it and get the verdict of the learned judge. In Science, too, we are necessarily interested in those questions that are not automatically resolved by the known "Laws of Nature" (the analogy here with man-made Laws is only of historical interest) but we agree to work and wait until we can arrive at an interpretation or explanation that is satisfactory to all parties.

There are other elements in the Law that are quite outside science—normative principles and moral issues that underlie any notion of justice. As is so often said, Science cannot tell us what *ought* to be done; it can only chart the consequences of what *might* be done.

Normative and moral principles cannot, by definition, be embraced in a consensus; to assert that one *ought* to do so and so is to admit that some people, at least, will not freely recognize the absolute necessity of not doing otherwise. Legal principles and norms are neither eternal nor universal; they are attached to the local, ephemeral situation of this country here and now; their arbitrariness can never be mended by any amount of further logical manipulation. Thus, there are components of legal argument that are necessarily refractory to the achievement of free and general agreement and these quite clearly discriminate between Law and Science as academic Faculties.

To the ordinary Natural Scientist this discussion may perhaps have seemed quite unnecessary—Law, he would say, is a man-made set of social conventions, whilst Science deals with material, objective, eternal verities. But to the Social Scientist this distinction is by no means so clear. He may, for example, find it impossible to disentangle such legal concepts as personal responsibility from his scientific understanding of the power of social determination in a pattern of delinquency. The criterion of consensibility might temper some of the scientific arrogance of the expert witness—"Would *every* criminologist agree with you on this point, Dr. X?"—whilst at the same time throwing the full weight of personal decision and responsibility upon the judge, who should never be allowed to shelter behind the cruel and mechanical absolutes of "Legal Science." The intellectual authority of Science is such that it must not be wielded incautiously or irresponsibly.

At first sight, one would not suppose that much need be said about what distinguishes Science from those disciplines and activities that belong to the Arts and Humanities—Literature, Music, Fine Arts, etc. Our modern view of Poetry, say, is that it is an expression of a private personal opinion. By his skill the poet may strike unsuspected chords of emotion in a vast number of other men, but this is not necessarily his major intention. A poem that is immediately acceptable and agreeable to everyone must be banal in the extreme.

But, of course Arts dons do not write Poetry: they write about it. Literary and artistic critics do sometimes pretend that their judgements are so convincing that it is wilful to oppose them. An imperious temper demands that we accept their every utterance of interpretation and valuation. Fortunately, we have the right of dissent and if our heart and mind carry us along a different path we have no need to be frightened by their shrill cries of contempt.

The point here is that there are genuine differences of taste and feeling, just as there are genuine differences of moral principle. At the back of our definition of Science itself is the assumption that men are free to express their true feelings; without this condition, the notion of a consensus loses meaning. Under a dictatorship we might be constrained to pay lip service to a uniform standard of style or taste, but this is the death of criticism.

There are, of course, periods of "classicism" and "academicism" when some style of technique is overwhelmingly praised and practised, but no one supposes that this is in obedience to the commands of absolute necessity. The attempts of the stupider sort of academic critic to rationalize the taste of his age by rules of "harmony" or "dramatic unity" are invariably by-products of the fashion whose dominance he is seeking to justify, not its determining factors. No sooner are such rules formulated than a great artist cannot resist the temptation to break them, and a new fashion sweeps the land. By their very nature, the Arts are not consensible, and hence are quite distinct from Science as I conceive it.

Science is not immune from fashion—a sure sign of its socio-psychological nature. . . . But what, abstractly, *is* fashion? It means doing what other people do for no better reason than that that is what is done. If everyone were to follow only fashionable lines of thought, there would be a false impression of a consensus; the inhibition of the critical imagination by such a conformist sentiment is the antithesis of the scientific attitude. It is also, of course, another way of death for true Poetry and Art.

But the products and producers of Literature, Art and Music may be studied in more factual aspects than for their emotional or spiritual message. For example, they are the outcome of, or participants in, historical events.

The place of *History* in this analysis is very significant; it seems to be truly one of the borderlands marching between scientific and non-scientific pursuits. Suppose that we are investigating such a problem as the date and place of birth of a writer or statesman. We search in libraries and other collections of material documents for written evidence. From various oblique references we might build up an argument in favour of some particular hypothesis—an argument to persuade our colleagues by its invincible logic that no other interpretation is tenable. This procedure seems quite as scientific as the research of a palaeontologist, who might reconstruct the anatomy of an extinct animal by piecing together fragments of fossil bone. Our aim is the same—to make a thoroughly convincing case which no reasonable person can refute. If, unfortunately, we cannot find sufficient evidence to clinch the case, we do not cling to our hypothesis and abuse our opponents for not accepting it; we quietly concede that the matter is uncertain, and return once more to the search. On such material points, the mood of historical scholarship is perfectly scientific.

The other mood in History is much more akin to Literature or Theology; it is the attempt to understand human history imaginatively and to "explain" it. Having ascertained the "facts," the historian tries to uncover the hidden motives and forces at work, just as the scientist goes behind the phenomena to the laws of their being.

The trouble is that the complex events of history can seldom be explained convincingly in the language of elementary cause and effect. To ascribe the English Civil War, for example, to the "Rise of the Gentry" may be a brilliant and fruitful hypothesis, but it is almost impossible to prove. Even though one may feel that this is the essence of the matter, and though one may marshal factual evidence forcefully in its favour, the case can be no more than circumstantial and hedged with vagueness and provisos. It will go into the canon of interesting historical theories, but experience tells us that it will not, as would a valid scientific theory, be so generally acceptable as to eliminate all competitors.

The rule in Science is not to attempt explanations of such complex phenomena at all, or at least to postpone this enterprise until answers are capable of being agreed upon. Imagination in the search for such problems is essential, but speculation is always kept rigidly under control. Even in such disciplines as Cosmology, where it sometimes seems as if a new theory of the Universe is promulgated each week, the range of discussion is limited quite narrowly to model systems whose mathematical properties are calculable and can be critically assessed by other scholars.

History does not impose such restrictions upon its pronouncements. It is felt, quite naturally, that the larger questions, although more difficult, are very important and must be discussed, even if they cannot be answered with precision. To restrict oneself to decidable propositions would be to miss the lessons that the strange sad story has for mankind. A history of "facts," of dates and kings and queens, although acceptable to the consensus, would be banal and trivial. In other words, History also has to provide other spiritual values, and to satisfy other normative principles, than scientific accuracy.

There are, of course, historians who have claimed universal "scientific" validity for their larger schemes and "Laws." It is not inconceivable that historical events do follow discernible patterns, and that there are, indeed, hidden forces—the class struggle, say, or the Protestant ethic—which largely determine the outcome of human affairs. It would not be necessary for such a theory to be absolute and mathematically rigorous for it to acquire scientific validity, any more than the proof that smoking causes lung cancer requires every smoker to die at the age of 50. It is not inherently absurd to search for historical laws, any more than it was absurd, 200 years ago, to search for the laws governing smallpox. Seemingly haphazard events often turn out to have their pattern, and to be capable of rational explanation.

All I am saying is that no substantial general principles of historical explanation have yet won universal acceptance. There have been fashionable doctrines, and dogmas backed by naked force, but never the sort of consensus of free and well-informed scholars that we ordinarily find in the Natural Sciences. Many historians assert that historical events are the outcome of such a variety of chance causes that they could never be subsumed to simpler, more general laws. Others say that the number of instances of exactly similar situations is always too small to provide sufficient statistical evidence to support an abstract theoretical analysis.

Whatever their reasons, historians do not agree on the general theoretical foundations or methodology of their studies. Instead of establishing, by mutual criticism and tacit cooperation, a limited common basis of acceptable theory, from which to build upwards and outwards, they often feel bound to set up antagonistic "schools" of interpretation, like so many independent walled cities.

They are not to be blamed for such behaviour; it only shows that this is a field where a scientific consensus is not the main objective. If you insisted that historians should work more closely together, they would object that the knowledge that they have in common is too dull, too trivial, too distant from the interesting problems of History, to circumscribe the thought of a serious scholar. To write about the Civil War without asking why the whole extraordinary thing happened is to compose a mere chronicle. For that reason, much of historical scholarship is not essentially scientific.

It would be wrong, on the other hand, to give the approving label "Science" only to the new techniques of historical research derived from the physical and biological sciences and technologies—carbon dating, aerial photography, demographic statistics, chemical analysis of ink and parchment. Such techniques are often powerful, but they are not more "scientific" than the traditional scholarly exercises of editing texts, verifying references and making rational deductions from the written words of documents. There is no reason at all why marks on paper in comprehensible language should be treated as inherently less evidential than the pointer readings of instruments or the print-out from a computer. In German the word *Wissenschaft,* which we translate as *Science,* includes quite generally all the branches of scholarship, including literary and historical studies.

To maintain, therefore, an impassable divide between Science and the Humanities is to perpetrate a gross misunderstanding that springs in the British case solely from a peculiarity of educational curricula. The Story, the Arts, the Poetry of Mankind are worthy both of spiritual contemplation and scholarly study, whether by laymen in general education or by experts as their life career. In many aspects this study is perfectly akin to the scientific study of electrons, molecules, cells, organisms or social systems: consensible knowledge may be acquired whether as isolated facts or as generally valid explanations. But to confine oneself, in education as in scholarship, to such aspects would impoverish the imagination, and even restrict the scope of possible further advance. Without general concepts as a guide—however uncertain, personal and provisional—we simply could not see any larger patterns in the picture. Historical and literary scholarship cannot therefore pretend to be scientific through and through, but that does not prevent their making progress towards a close definition of the truth. In the end, bold speculative generalization and unverifiable psychological insight may go further in establishing a convincing narrative than a rigid insistence on precise minutiae.

It scarcely needs to be said that *Religion,* as we nowadays study and practise it, is also quite distinct from Science. This seems so obvious in our enlightened age that one wonders how there could have been any conflict and confusion between them. But was not Religion primitive Science—the corpus of generally accepted public knowledge? Should we not see Science as growing out of, and eventually severing itself from, this parent body—or perhaps as a process of differentiation and specialization within the unity of the medieval *Summum*? Just because many religious beliefs are now seen to be wrong, it does not follow that they were not seriously, freely and rationally accepted in their time. Conventional science too can be wrong at times.

Let me give an example. In the late eighteenth and early nineteenth centuries, prehistoric remains were found that we now see as pointing to the great antiquity of Man. But many scholars stood out against this interpretation because it did not square with the Biblical chronology of the past. Is it fair to treat this as a conflict between scientific rationality and religious prejudice? Would it not be more just to say that a widely accepted theory was being ousted by a better one as new evidence came to light?

The point is that this debate was open and free. The participants on one side may have been blinkered by their upbringing, but their beliefs were honestly held and rationally maintained. They may often have used poor arguments to defend their case—but they did not call in the secular arm or the secret police. In the end, they lost; and since then the appeal to Divine Scripture has ceased to be an acceptable element in scientific discussion.

What I am arguing is that there is a progressive improvement in the techniques and criteria of such discussion, and that the use of abstract theological principles was once respectable but is now discredited, just as the absolute justification of Euclidean geometry from the Parallelism Axiom is now discredited. The "Scientific Revolution" of the seventeenth century is not a complete break with the past. The idea of presenting a rational non-contradictory account of the universe is perhaps a legacy of Greece, but it is very strong in medieval Philosophy and Theology. It may be that the very existence of a dogmatic system of metaphysics, implying a rational order of things and fiercely debated in detail, was the prerequisite for the development of an alternative system, using some of the same logical techniques but based upon different principles and more extensive evidence.[4] The doctrinaire consensus of the Church may have been prolonged beyond its acceptability to free men by the power of the Holy Office, but it had originally provided an example of a generally agreed picture of the world. These are subtle and deep questions which I am not competent to discuss, but I wonder whether the failure of Science to grow in China and India was due as much to the general doctrinal permissiveness of their religious systems as to any other cause. Toleration of deviation, and the lack of a very sharp tradition of logical debate may have made the very idea of a consensus of opinion on the Philosophy of Nature as absurd to them as the idea of absolute agreement on ethical principles would be to us.

The relationship between Science and *Philosophy* is altogether more complex and confused. In a sense, all of modern science is the Philosophy of Nature, as distinct from, say, Moral or Political Philosophy. But this terminology is somewhat old-fashioned, and we try to make a distinction between Physics and Metaphysics, between the Philosophy *of* Science and Philosophy *as* Science. Some philosophers attempt to limit themselves to statements as precise and verifiable as those of scientists, and confine their arguments to the rigid categories of symbolic logic. The consensus criterion would be acceptable to them, for they would hold that by a continuous process of analysis and criticism they would make progress towards creating a generally agreed upon set of principles governing the use of words and the establishment of valid truths. Others hold that such a hope can never be realized and that by limiting philosophical discourse in this way they would only allow themselves to make trivial statements, however unexceptionable. For this school of Philosophy it is important to be free to comment on grander topics, even though such comments will only reveal the variety and contradictory character of the views of different philosophers.

As with History we can only say that if Philosophy is what academic philosophers write in their books, then some of it is not very different from Science. But generally the motivation is nonscientific, by our definition, and the multiplicity of viewpoints indicates that there is no dominant urge to find maximum regions of agreement. Whatever their claims, the proponents of "scientific" philosophical systems do not convince the majority of their colleagues that theirs is the only way to truth.

Let us now consider *Technology*—Engineering, Medicine, etc. For the multitude, Science is almost synonymous with its applications, whereas scientists themselves are very careful to stress the distinction between "pure" knowledge, studied "for its own sake," and technological knowledge applied to human ends.

The trouble is that this distinction is very difficult to make in practice. Suppose, for example, that we are researching on the phenomenon of "fatigue" in metals. We are almost forced into the position of saying that on Monday, Wednesday and Friday we are just honest seekers after truth, adding to our understanding of the natural world, etc., whilst on Tuesday, Thursday and Saturday we are practical chaps trying to stop aeroplanes from falling to pieces, advancing the material welfare of mankind and so on. Or we may have to make snobbish distinctions between Box, a pure scientist working in a University, and Cox, a technologist, doing the same research but employed by an aircraft manufacturer. There was once a time when Science was academic and useless and Technology was a practical art, but now they are so interfused that one is not surprised that the multitude cannot tell them apart.

Here again, a definition in terms of the scientific consensus can be really effective. The technologist has to fulfil a need; he must provide the means to do a definite job—bridge this river, cure this disease, make better beer. He must do the best he can with the knowledge available. That knowledge is almost always inadequate for him to calculate the ideal solution to his problem—and he cannot wait while all the research is done to obtain it. The bridge must be built this year; the patient must be saved today; the brewery will go bankrupt if its product is not improved.

So there will be a large element of the incalculable, of sheer art, in what he does. A different engineer would come out with quite a different design; a different doctor would prescribe quite different treatment. These might be better or worse, in their results—but nobody quite knows. Each situation is so complex, and has so many unassessable factors, that the only sensible policy for the client is to choose his engineeer or doctor carefully and then rely upon his skill and experience. To look for a solution acceptable to all the professional experts is a familiar recipe for disaster—"Design by Committee."

The technologist's prime responsibility is towards his employer, his customer or his patient, not to his professional peers. His task is to solve the problem in hand, not to address himself to the opinions of the other experts. If his proposed solution is successful, then it may well establish a lead, and eventually add to the "Science" of his Technology; but that should not be in his mind at the outset.

What we find, of course, is that a corpus of generally accepted principles develops in every technical field. Modern Technology is deliberately scientific, in that there is continuous formal study and empirical investigation of aspects of technique, in addition to the mere accumulation of experience from successfully accomplished tasks. The aim of such research is not to solve immediate specific problems, but to acquire knowledge for the use of the experts in their professional work. It is directed, therefore, at the mind of the profession, as a potential contribution to the consensus opinion. This sort of work is thus genuinely scientific, however trivial and limited its scope may be.

The abstract distinction here being made between a "scientific Technology" and "technological Science" has its psychological counterpart. It is a commonplace in the literature on the Management of Industrial Research that applied scientists often suffer from divided loyalties. On the one hand, they owe their living to the company that employs them, and that expects its return in the profitable solution of immediate problems. On the other hand, they give their intellectual allegiance to their scientific profession—to Colloid Chemistry, or Applied Mathematics, or whatever it is—where they look for scholarly recognition. Although the rewards for "technological" work are greater and more direct, they very often prefer to stick to their "scientific" research.

This preference seems almost incomprehensible to management experts, because they fail to see that the scientific loyalty is not just towards a prestigious professional group but to an ideology. The young scientist is trained to make contributions to public knowledge. All the habits and practices of his years of apprenticeship emphasize the importance of making them convincing, and thus making them part of the common pool. Being a successful scientist is not just winning prizes; it is having other scientists cite your work. To give this up is worse than losing caste; it is to give up one's faith and be made to worship foreign idols.

Nevertheless, one must agree that Science and Technology are now so intimately mingled that the distinction can become rather pedantic. Take, for example, a typical Consumers Association report on a motor car. Some of the tests, such as the measure of petrol consumption, may be perfectly scientific in that their validity would be universally acceptable. Other tests, such as whether the springing was comfortable, would not satisfy this criterion, although it would be one of the important skills of the designer to attend to just such "subjective" and "qualitative" features. For this reason, to say that a car has been "scientifically designed" is merely to assert that it has been well designed by competent engineers. Yet an account, by the designer, of the rationale behind various technical features of the model could rank as a serious

contribution to the Science of Automobile Engineering by adding to the body
of agreed principles at the basis of that mysterious art.

All that I can claim is that these distinctions, although subtle and perhaps
pedantic, are not entirely arbitrary or unreal. We do not need to look far
ahead to some conceivable remote application of the knowledge in question,
nor do we need to examine the hidden, perhaps unconscious, motives of those
who produce it. We do not need to decide whether some particular laboratory
is "technological" or "scientific," and then attach the appropriate label to its
products. The criterion is in the work itself, in the form in which it is
presented, and in the audience to which it is addressed.

What are we to make of the so-called "Social Sciences" in the light of this
discussion? It is obvious that such a subject as Politics is very close to History
and to Philosophy in its goals and achievements; to stick to ascertainable
public "facts" is to limit the discourse to the banal. To give this discipline the
name of Political Science is unfortunate; it offers more than it can deliver and
debases its ethical message.

On the other hand, *Economics* is a very technological subject; the experts
are always being asked to diagnose the ills of the nation and to propose specific
cures, long before they have sufficient scientific understanding to make a valid
analysis. Yet the totally quantitative material medium—money—allows of
convincing proofs, statistical or algebraic, of precise hypotheses, so that a
body of agreed principles is gradually emerging. Leading economists may
debate in public, and seem to be at loggerheads, but behind the scenes they
teach much the same things to their students. It is typical of the tacit cooperation between scholars in a scientific subject that American and Soviet
economists respect and learn from each other's work on Input-Output
Analysis, however much they may disagree on more speculative issues of
general social policy.

These are the neighbours of the new discipline of *Sociology* which is an
attempt to escape from the "unscientific" traditions of History and Politics, and
to make the study of social systems, and of man in society, at least as scientific
as Economics.

That is the reason why so much sociological research is by questionnaire and
statistical analysis; the aim is to provide the necessary factual basis for firmly
scientific theories. To the extent that observations of this sort are verifiable by
repetition, and capable of being made quite convincing to a critical public, this
attitude is sound. But the intractability of the subject must be reckoned with.
Vast quantities of information do not add up to much serious knowledge
without theories to give it meaning. Moreover, even to accept "facts" of this
sort may imply the acceptance of dubious hidden theories. Suppose, for
example, that our car-testing organization decided to assess the comfort of
various models by asking a hundred people to give their views, and reported
that car A "rated" 87 per cent and car B only 63 per cent. This is objective
information, which might well be "verifiable." But there is the implication that

it "measures" something—"comfort"—which may not exist at all. This is a crude case, but sociological research is full of more subtle examples of the same difficulty.

Some sociologists have taken quite a different line. They deal in abstract categories, which they manipulate logically into various hypothetical relations in a sort of formal calculus. This approach also strives towards the creation of a consensus, in that the structure of the argument can be purged of contradictions and hence made unexceptionable and theoretically acceptable. But without much more rigorous connection of these abstractions with real systems and actual phenomena it is vacuous knowlege, without the power to persuade us that thus and thus is the world of men.

Nevertheless, Sociology is often genuinely scientific in spirit, although it has turned out to be an exceedingly difficult science whose positive achievements do not always match the effort expended on it. The "methodological problem" has not been surmounted; there is not yet a reliable procedure for building up interesting hypotheses that can be made sufficiently plausible to a sufficient number of other scholars by well-devised observations, experiments or rational deductions. It was the sort of problem facing Physics before Galileo began seriously to apply mathematical reasoning and numerical measurement to the subject. The ideal of a consensus is there, but the intellectual techniques by which it might be created and enlarged seem elusive.

This survey of the Faculties has necessarily been brief and schematic. Why should we even want to decide whether a particular discipline is scientific or not? The answer is, simply, that, *when it is available,* scientific knowledge is more reliable, on the whole, than non-scientific. When there are conflicts of authority, when Sociology tells us to go one way and History another, we need to weigh their respective claims to validity. Our general argument here is that in a discipline where there is a scientific consensus the amount of *certain* knowledge may be limited, but it will be honestly labelled: "Trust your neck to this," or "This ladder was built by a famous scholar, but no one else has been able to climb it."

In the end, the best way to decide whether a particular body of knowledge is scientific or not is often to study the attitudes of its professional practitioners to one another's work. A sure symptom of non-science is personal abuse and intolerance of the views of one scholar by another. The existence of irreconcilable "schools" of thought is familiar in such academic realms as Theology, Philosophy, Literature and History. When we find them in a "scientific" discipline, we should be on our guard.

This is the reason why for example we should be very suspicious of the claims of Psychoanalysis. The history of this subject is a continuous series of bitter conflicts between persons, schools and theories. Freud himself had the most honest and sincere desire to create a thoroughly respectable scientific discipline, but for some reason he failed to understand this key point—the need to move slowly forward, step by step, from a basis of generally accepted ideas. Perhaps the struggle to get anyone to listen at all was too bitter, or perhaps his mind was too active and impatient to endure continuous critical assessment of each new theory or interpretation. Whatever the reason, the mood of Psychoanalysis in its formative period was antagonistic to the covert cooperative spirit of true Science. Its clinical successes were only of technological significance, and did not scientifically validate the theories on which they were said to be based.

I have given this example, not out of prejudice against psychoanalytic ideas (one or other of the contending schools may well be right: we shall see) but to show that the principle of the consensus is a powerful criterion, with something definite to say on this vexed topic. To some people the words "scientific" and "unscientific" have come to mean no more than "true" and "false," or "rational" and "irrational." In this chapter I have tried to show, by reference to other organized bodies of knowledge, that this usage is quite improper, and grossly unfair to those scholars who seek rationality and truth in bolder ways than by microscopic dissection of minutiae.

NOTES

1. This point I owe to Professor Körner.
2. "Hence a true philosophy of science must be a philosophy of scientists and laboratories as well as one of waves, particles and symbols." Patrick Meredith in *Instruments of Communication,* p. 40.
3. I am indebted to Professor Julius Stone for sending me his fascinating critical essay on this subject.
4. This point is made in *Science in the Modern World* by A. N. Whitehead (New York: Macmillan, 1931).

3

How to Defend Society Against Science

Paul Feyerabend

Practitioners of a strange trade, friends, enemies, ladies and gentlemen: Before starting with my talk, let me explain to you, how it came into existence.

About a year ago I was short of funds. So I accepted an invitation to contribute to a book dealing with the relation between science and religion. To make the book sell I thought I should make my contribution a provocative one and the most provocative statement one can make about the relation between science and religion is that science is a religion. Having made the statement the core of my article I discovered that lots of reasons, lots of excellent reasons, could be found for it. I enumerated the reasons, finished my article, and got paid. That was stage one.

Next I was invited to a Conference for the Defence of Culture. I accepted the invitation because it paid for my flight to Europe. I also must admit that I was rather curious. When I arrived in Nice I had no idea what I would say. Then while the conference was taking its course I discovered that everyone thought very highly of science and that everyone was very serious. So I decided to explain how one could defend culture from science. All the reasons collected in my article would apply here as well and there was no need to invent new things. I gave my talk, was rewarded with an outcry about my "dangerous and ill considered ideas," collected by ticket and went on to Vienna. That was stage number two.

Now I am supposed to address you. I have a hunch that in some respect you are very different from my audience in Nice. For one, you look much younger. My audience in Nice was full of professors, businessmen, and television

[This] article is a revised version of a talk given to the Philosophy Society at Sussex University in November 1974.

executives, and the average age was about 58½. Then I am quite sure that most of you are considerably to the left of some of the people in Nice. As a matter of fact, speaking somewhat superficially I might say that you are a left-ist audience while my audience in Nice was a rightist audience. Yet despite all these differences you have some things in common. Both of you, I assume, respect science and knowledge. Science, of course, must be reformed and must be made less authoritarian. But once the reforms are carried out, it is a valuable source of knowledge that must not be contaminated by ideologies of a different kind. Secondly, both of you are serious people. Knowledge is a serious matter, for the Right as well as for the Left, and it must be pursued in a serious spirit. Frivolity is out, dedication and earnest application to the task at hand is in. These similarities are all I need for repeating my Nice talk to you with hardly any change. So, here it is.

Fairytales

I want to defend society and its inhabitants from all ideologies, science included. All ideologies must be seen in perspective. One must not take them too seriously. One must read them like fairytales which have lots of interesting things to say but which also contain wicked lies, or like ethical prescriptions which may be useful rules of thumb but which are deadly when followed to the letter.

Now, is this not a strange and ridiculous attitude? Science, surely, was always in the forefront of the fight against authoritarianism and superstition. It is to science that we owe our increased intellectual freedom vis-à-vis religious beliefs; it is to science that we owe the liberation of mankind from ancient and rigid forms of thought. Today these forms of thought are nothing but bad dreams—and this we learned from science. Science and enlightenment are one and the same thing—even the most radical critics of society believe this. Kropotkin wants to overthrow all traditional institutions and forms of belief, with the exception of science. Ibsen criticises the most intimate ramifications of nineteenth-century bourgeois ideology, but he leaves science untouched. Levi-Strauss has made us realise that Western Thought is not the lonely peak of human achievement it was once believed to be, but he excludes science from his relativization of ideologies. Marx and Engels were convinced that science would aid the workers in their quest for mental and social liberation. Are all these people deceived? Are they all mistaken about the role of science? Are they all the victims of a chimaera?

To these questions my answer is a firm *Yes and No.*

Now, let me explain my answer.

My explanation consists of two parts, one more general, one more specific.

The general explanation is simple. Any ideology that breaks the hold a comprehensive system of thought has on the minds of men contributes to the liberation of man. Any ideology that makes man question inherited beliefs is an aid to enlightenment. A truth that reigns without checks and balances is a tyrant who must be overthrown, and any falsehood that can aid us in the overthrow of this tyrant is to be welcomed. It follows that seventeenth- and eighteenth-century science indeed *was* an instrument of liberation and enlightenment. It does not follow that science is bound to *remain* such an instrument. There is nothing inherent in science or in any other ideology that makes it *essentially* liberating. Ideologies can deteriorate and become stupid religions. Look at Marxism. And that the science of today is very different from the science of 1650 is evident at the most superficial glance.

For example, consider the role science now plays in education. Scientific "facts" are taught at a very early age and in the very same manner in which religious "facts" were taught only a century ago. There is no attempt to waken the critical abilities of the pupil so that he may be able to see things in perspective. At the universities the situation is even worse, for indoctrination is here carried out in a much more systematic manner. Criticism is not entirely absent. Society, for example, and its institutions, are criticised most severely and often most unfairly and this already at the elementary school level. But science is excepted from the criticism. In society at large the judgement of the scientist is received with the same reverence as the judgement of bishops and cardinals was accepted not too long ago. The move towards "demythologization," for example, is largely motivated by the wish to avoid any clash between Christianity and scientific ideas. If such a clash occurs, then science is certainly right and Christianity wrong. Pursue this investigation further and you will see that science has now become as oppressive as the ideologies it had once to fight. Do not be misled by the fact that today hardly anyone gets killed for joining a scientific heresy. This has nothing to do with science. It has something to do with the general quality of our civilization. Heretics in science are still made to suffer from the *most severe* sanctions this relatively tolerant civilization has to offer.

But—is this description not utterly unfair? Have I not presented the matter in a very distorted light by using tendentious and distorting terminology? Must we not describe the situation in a very different way? I have said that science has become *rigid,* that it has ceased to be an instrument of *change* and *liberation,* without adding that it has found the *truth,* or a large part thereof. Considering this additional fact we realise, so the objection goes, that the rigidity of science is not due to human wilfulness. It lies in the nature of things. For once we have discovered the truth—what else can we do but follow it?

This trite reply is anything but original. It is used whenever an ideology wants to reinforce the faith of its followers. "Truth" is such a nicely neutral word. Nobody would deny that it is commendable to speak the truth and wicked to tell lies. Nobody would deny that—and yet nobody knows what such an attitude amounts to. So it is easy to twist matters and to change allegiance to truth in one's everyday affairs into allegiance to the Truth of an ideology which is nothing but the dogmatic defence of that ideology. And it is of course *not* true that we *have* to follow the truth. Human life is guided by many ideas. Truth is one of them. Freedom and mental independence are others. If Truth, as conceived by some ideologists, conflicts with freedom, then we have a *choice.* We may abandon freedom. But we may also abandon Truth. (Alternatively, we may adopt a more sophisticated idea of truth that no longer contradicts freedom; that was Hegel's solution.) My criticism of modern science is that it inhibits freedom of thought. If the reason is that it has found the truth and now follows it, then I would say that there are better things than first finding, and then following such a monster.

This finishes the general part of my explanation.

There exists a more specific argument to defend the exceptional position science has in society today. Put in a nutshell the argument says (1) that science has finally found the correct *method* for achieving results and (2) that there are many *results* to prove the excellence of the method. The argument is mistaken — but most attempts to show this lead into a dead end. Methodology has by now become so crowded with empty sophistication that it is extremely difficult to perceive the simple errors at the basis. It is like fighting the hydra — cut off one ugly head, and eight formalizations take its place. In this situation the only answer is superficiality: when sophistication loses content then the only way of keeping in touch with reality is to be crude and superficial. This is what I intend to be.

Against Method

There is a method, says part (1) of the argument. What is it? How does it work?

One answer which is no longer as popular as it used to be is that science works by collecting facts and inferring theories from them. The answer is unsatisfactory as theories never *follow from* facts in the strict logical sense. To say that they may yet be *supported* from facts assumes a notion of support that (a) does not show this defect and (b) is sufficiently sophisticated to permit us to say to what extent, say, the theory of relativity is supported by the facts. No such notion exists today, nor is it likely that it will ever be found (one of the problems is that we need a notion of support in which grey ravens can be said to support "all ravens are black"). This was realised by conventionalists and transcendental idealists who pointed out that theories *shape* and *order* facts and can therefore be retained come what may. They can be retained because the human mind either consciously or unconsciously carries out its ordering function. The trouble with these views is that they assume for the mind what they want to explain for the world, viz., that it works in a regular fashion. There is only one view which overcomes all these difficulties. It was invented twice in the nineteenth century, by Mill, in his immortal essay *On Liberty,* and by some Darwinists who extended Darwinism to the battle of ideas. This view takes the bull by the horns: theories cannot be justified and their excellence cannot be shown without reference to other theories. We may explain the *success* of a theory by reference to a more comprehensive theory (we may explain the success of Newton's theory by using the general theory of relativity); and we may explain our *preference* for it by comparing it with other theories.

Such a comparison does not establish the intrinsic excellence of the theory we have chosen. As a matter of fact, the theory we have chosen may be pretty lousy. It may contain contradictions, it may conflict with well-known facts, it may be cumbersome, unclear, ad hoc in decisive places, and so on. But it may still be better than any other theory that is available at the time. It may in fact be the best lousy theory there is. Nor are the standards of judgement chosen in an absolute manner. Our sophistication increases with every choice we make, and so do our standards. Standards compete just as theories compete and we choose the standards most appropriate to the historical situation in which the choice occurs. The rejected alternatives (theories; standards; "facts") are not eliminated. They serve as correctives (after all, we may have made the wrong choice) and they also explain the content of the preferred views (we understand relativity better when we understand the structure of its competitors; we know the full meaning of freedom only when we have an idea of life in a totalitarian state, of its advantages—and there are many advantages—as well as of its disadvantages). Knowledge so conceived is an ocean of alternatives channelled and subdivided by an ocean of standards. It forces our mind to make imaginative choices and thus makes it grow. It makes our mind capable of choosing, imagining, criticising.

Today this view is often connected with the name of Karl Popper. But there are some very decisive differences between Popper and Mill. To start with, Popper developed his view to solve a special problem of epistemology—he wanted to solve "Hume's problem." Mill, on the other hand, is interested in conditions favourable to human growth. His epistemology is the result of a certain theory of man, and not the other way around. Also Popper, being influenced by the Vienna Circle, improves on the logical form of a theory before discussing it, while Mill uses every theory in the form in which it occurs in science. Thirdly, Popper's standards of comparison are rigid and fixed, while Mill's standards are permitted to change with the historical situation. Finally, Popper's standards eliminate competitors once and for all: theories that are either not falsifiable or falsifiable and falsified have no place in science. Popper's criteria are clear, unambiguous, precisely formulated; Mill's criteria are not. This would be an advantage if science itself were clear, unambiguous, and precisely formulated. Fortunately, it is not.

To start with, no new and revolutionary scientific theory is ever formulated in a manner that permits us to say under what circumstances we must regard it as endangered: many revolutionary theories are unfalsifiable. Falsifiable versions do exist, but they are hardly ever in agreement with accepted basic statements: every moderately interesting theory is falsified. Moreover, theories have formal flaws, many of them contain contradictions, ad hoc adjustments, and so on and so forth. Applied resolutely, Popperian criteria would eliminate science without replacing it by anything comparable. They are useless as an aid to science. In the past decade this has been realised by various thinkers, Kuhn and Lakatos among them. Kuhn's ideas are interesting but, alas, they are much too vague to give rise to anything but lots of hot air. If you don't believe me, look at the literature. Never before has the literature on the philosophy of science been invaded by so many creeps and incompetents. Kuhn encourages people who have no idea why a stone falls to the ground to talk with assurance about scientific method. Now I have no objection to incompetence but I do object when incompetence is accompanied by boredom and self-righteousness. And this is exactly what happens. We do not get interesting false ideas, we get boring ideas or words connected with no ideas at all. Secondly, wherever one tries to make Kuhn's ideas more definite one finds that they are *false*. Was there ever a period of normal science in the history of thought? No—and I challenge anyone to prove the contrary.

Lakatos is immeasurably more sophisticated than Kuhn. Instead of theories he considers research programmes which are sequences of theories connected by methods of modification, so-called heuristics. Each theory in the sequence may be full of faults. It may be beset by anomalies, contradictions, ambiguities. What counts is not the shape of the single theories, but the tendency exhibited by the sequence. We judge historical developments and achievements over a period of time, rather than the situation at a particular time. History and methodology are combined into a single enterprise. A research programme is said to progress if the sequence of theories leads to novel predictions. It is said to degenerate if it is reduced to absorbing facts that have been discovered without its help. A decisive feature of Lakatos' methodology is that such evaluations are no longer tied to methodological rules which tell the scientist either to retain or to abandon a research programme. Scientists may stick to a degenerating programme; they may even succeed in making the programme overtake its rivals and they therefore proceed rationally whatever they are doing (provided they continue calling degenerating programmes degenerating and progressive programmes progressive). This means that Lakatos offers *words* which *sound* like the elements of a methodology; he does not offer a methodology. There is no method according to the most advanced and sophisticated methodology in existence today. This finishes my reply to part (1) of the specific argument.

Against Results

According to part (2), science deserves a special position because it has produced *results*. This is an argument only if it can be taken for granted that nothing else has ever produced results. Now it may be admitted that almost everyone who discusses the matter makes such an assumption. It may also be admitted that it is not easy to show that the assumption is false. Forms of life different from science either have disappeared or have degenerated to an extent that makes a fair comparison impossible. Still, the situation is not as hopeless as it was only a decade ago. We have become acquainted with methods of medical diagnosis and therapy which are effective (and perhaps even more effective than the corresponding parts of Western medicine) and which are yet based on an ideology that is radically different from the ideology of Western science. We have learned that there are phenomena such as telepathy and telekinesis which are obliterated by a scientific approach and which could be used to do research in an entirely novel way (earlier thinkers such as Agrippa of Nettesheim, John Dee, and even Bacon were aware of these phenomena). And then—is it not the case that the Church saved souls while science often does the very opposite? Of course, nobody now believes in the ontology that underlies this judgement. Why? Because of ideological pressures identical with those which today make us listen to science to the exclusion of everything else. It is also true that phenomena such as telekinesis and acupuncture may eventually be absorbed into the body of science and may therefore be called "scientific." But note that this happens only after a long period of resistance during which a science *not yet* containing the phenomena wants to get the upper hand over forms of life that contain them. And this leads to a further objection against part (2) of the specific argument. The fact that science has results counts in its favour only if these results were achieved by science alone, and without any outside help. A look at history shows that science hardly ever gets its results in this way. When Copernicus introduced a new view of the universe, he did not consult *scientific* predecessors, he consulted a crazy Pythagorean such as Philolaos. He adopted his ideas and he maintained them in the face of all sound rules of scientific method. Mechanics and optics owe a lot to artisans, medicine to midwives and witches. And in our own day we have seen how the interference of the state can advance science: when the Chinese communists refused to be intimidated by the judgement of experts and ordered traditional medicine back into universities and hospitals there was an outcry all over the world that science would now be ruined in China. The very opposite occurred: Chinese science advanced and Western science learned from it. Wherever we look we see that great scientific advances are due to outside interference which is made to prevail in the face of the most basic and most "rational" methodological rules. The lesson is plain: there does not exist a single argument that could be used to support the exceptional role which science today plays in society. Science has done many things, but so have other ideologies. Science often proceeds systematically, but so do other ideologies (just consult the records of the many doctrinal debates that took place in the Church) and, besides, there are no overriding rules which are adhered to under any circumstances; there is no "scientific methodology" that can be used to separate science from the rest. *Science is just one of the many ideologies that propel society and it should be treated as such* (this statement applies even to the most progressive and most dialectical sections of science). What consequences can we draw from this result?

The most important consequence is that there must be a *formal separation between state and science* just as there is now a formal separation between state and church. Science may influence society but only to the extent to which any political or other pressure group is permitted to influence society. Scientists may be consulted on important projects but the final judgement must be left to the democratically elected consulting bodies. These bodies will consist mainly of laymen. Will the laymen be able to come to a correct judgement? Most certainly, for the competence, the complications and the successes of science are vastly exaggerated. One of the most exhilarating experiences is to see how a lawyer, who is a layman, can find holes in the testimony, the technical testimony, of the most advanced expert and thus prepare the jury for its verdict. Science is not a closed book that is understood only after years of training. It is an intellectual discipline that can be examined and criticised by anyone who is interested and that looks difficult and profound only because of a systematic campaign of obfuscation carried out by many scientists (though, I am happy to say, not by all). Organs of the state should never hesitate to reject the judgement of scientists when they have reason for doing so. Such rejection will educate the general public, will make it more confident, and it may even lead to improvement. Considering the sizeable chauvinism of the scientific establishment we can say: the more Lysenko affairs, the better (it is not the *interference* of the state that is objectionable in the case of Lysenko, but the *totalitarian* interference which kills the opponent rather than just neglecting his advice). Three cheers to the fundamentalists in California who succeeded in having a dogmatic formulation of the theory of evolution removed from the text books and an account of Genesis included. (But I know that they would become as chauvinistic and totalitarian as scientists are today when given the chance to run society all by themselves. Ideologies are marvellous when used in the companies of other ideologies. They become boring and doctrinaire as soon as their merits lead to the removal of their opponents.) The most important change, however, will have to occur in the field of *education*.

Education and Myth

The purpose of education, so one would think, is to introduce the young into life, and that means: into the *society* where they are born and into the *physical universe* that surrounds the society. The method of education often consists in the teaching of some *basic myth*. The myth is available in various versions. More advanced versions may be taught by initiation rites which firmly implant them into the mind. Knowing the myth, the grownup can explain almost everything (or else he can turn to experts for more detailed information). He is the master of Nature and of Society. He understands them both and he knows how to interact with them. However, *he is not the master of the myth that guides his understanding.*

Such further mastery was aimed at, and was partly achieved, by the Presocratics. The Presocratics not only tried to understand the *world*. They also tried to understand, and thus to become the masters of, the *means of understanding the world*. Instead of being content with a single myth they developed many and so diminished the power which a well-told story has over the minds of men. The sophists introduced still further methods for reducing the debilitating effect of interesting, coherent, "empirically adequate" etc. etc. tales. The achievement of these thinkers were not appreciated and they certainly are not understood today. When teaching a myth we want to increase the chance that it will be understood (i.e. no puzzlement about any feature of the myth), believed, *and accepted*. This does not do any harm when the myth is counterbalanced by other myths: even the most dedicated (i.e. totalitarian) instructor in a certain version of Christianity cannot prevent his pupils from getting in touch with Buddhists, Jews and other disreputable people. It is very different in the case of science, or of rationalism where the field is almost completely dominated by the believers. In this case it is of paramount importance to strengthen the minds of the young, and "strengthening the minds of the young" means strengthening them *against* any easy acceptance of comprehensive views. What we need here is an education that makes people *contrary, counter-suggestive,* without making them incapable of devoting themselves to the elaboration of any single view. How can this aim be achieved?

It can be achieved by protecting the tremendous imagination which children possess and by developing to the full the spirit of contradiction that exists in them. On the whole children are much more intelligent than their teachers. They succumb, and give up their intelligence because they are bullied, or because their teachers get the better of them by emotional means. Children can learn, understand, and keep separate two to three different languages ("children" and by this I mean three to five year olds, *not* eight year olds who were experimented upon quite recently and did not come out too well; why? because they were already loused up by incompetent teaching at an earlier age). Of course, the languages must be introduced in a more interesting way than is usually done. There are marvellous writers in all languages who have told marvellous stories—let us begin our language teaching with *them* and not with "der Hund hat einen Schwanz" and similar inanities. Using stories we may of course also introduce "scientific" accounts, say, of the origin of the world and thus make the children acquainted with science as well. But science must not be given any special position except for pointing out that there are lots of people who believe in it. Later on the stories which have been told will be supplemented with "reasons," where by reasons I mean further accounts of the kind found in the tradition to which the story belongs. And, of course, there will also be contrary reasons. Both reasons and contrary reasons will be told by the experts in the fields and so the young generation becomes acquainted with all kinds of sermons and all types of wayfarers. It becomes acquainted with them, it becomes acquainted with their stories, and every individual can make up his mind which way to go. By now everyone knows that you can earn a lot of money and respect and perhaps even a Nobel Prize by becoming a scientist, so many will become scientists. They will *become* scientists *without having been taken in by the ideology of science,* they will *be* scientists *because they have made a free choice.* But has not much time been wasted on unscientific subjects and will this not detract from their competence once they have become scientists? Not at all! The progress of science, of good science depends on novel ideas and on intellectual freedom: science has very often been advanced by outsiders (remember that Bohr and Einstein regarded themselves as outsiders). Will not many people make the wrong choice and end up in a dead end? Well, that depends on what you mean by a "dead end." Most scientists today are devoid of ideas, full of fear, intent on producing some paltry result so that they can add to the flood of inane papers that now constitutes "scientific progress" in many areas. And, besides, what is more important? To lead a life which one has chosen with open eyes, or to spend one's time in the nervous attempt of avoiding what some not so intelligent people call "dead ends"? Will not the number of scientists decrease so that in the end there is nobody to run our precious laboratories? I do not think so. Given a choice many people may choose science, for a science that is run by free agents looks much more attractive than the science of today which is run by slaves, slaves of institutions and slaves of "reason." And if there is a temporary shortage of scientists the situation may always be remedied by various kinds of incentives. Of course, scientists will not play any predominant role in the society I envisage. They will be more than balanced by magicians, or priests, or astrologers. Such a situation is unbearable for many people, old and young, right and left. Almost all of you have the firm belief that at least *some* kind of truth has been found, that it must be preserved, and that the method of teaching I advocate and the form of society I defend will dilute it and make it finally disappear. You have this firm belief; many of you may even have reasons. *But what you have to consider is that the absence of good contrary reasons is due to a historical accident*; it does *not* lie in the nature of things. Build up the kind of society I recommend and the views you now despise (without knowing them, to be sure) will return in such splendour that you will have to work hard to maintain your own position and will perhaps be entirely unable to do so. You do not believe me? Then look at history. Scientific astronomy was firmly founded on Ptolemy and Aristotle, two of the greatest minds in the history of Western Thought. Who upset their well-argued, empirically adequate and precisely formulated system? Philolaos the mad and antediluvian Pythagorean. How was it that Philolaos could stage such a comeback? Because he found an able defender: Copernicus. Of course, you may follow your intuitions as I am following mine. But remember that your intuitions are the result of your "scientific" training where by science I also mean the science of Karl Marx. My training, or, rather, my non-training, is that of a journalist who is interested in strange and bizarre events. Finally, is it not utterly irresponsible, in the present world situation, with millions of people starving, others enslaved, downtrodden, in abject misery of body and mind, to think luxurious thoughts such as these? Is not freedom of choice a luxury under such circumstances? Is not the flippancy and the humour I want to see combined with the freedom of choice a luxury under such circumstances? Must we not give up all self-indulgence and *act*? Join together, and *act*? This is the most important objection which today is raised against an approach such as the one recommended by me. It has tremendous appeal, it has the appeal of unselfish dedication. Unselfish dedication—to what? Let us see!

We are supposed to give up our selfish inclinations and dedicate ourselves to the liberation of the oppressed. And selfish inclinations are what? They are our wish for maximum liberty of thought in the society in which we live *now,* maximum liberty not only of an abstract kind, but expressed in appropriate institutions and methods of teaching. This wish for concrete intellectual and physical liberty in our own surroundings is to be put aside, for the time being. This assumes, first, that we do not need this liberty for our task. It assumes that we can carry out our task with a mind that is firmly closed to some alternatives. It assumes that the correct way of liberating others *has always been found* and that all that is needed is to carry it out. I am sorry, I cannot accept such doctrinaire self-assurance in such extremely important matters. Does this mean that we cannot act at all? It does not. But it means that *while acting we have to try to realise as much of the freedom I have recommended so that our actions may be corrected in the light of the ideas we get while increasing our freedom.* This will slow us down, no doubt, but are we supposed to charge ahead simply because some people tell us that they have found an explanation for all the misery and an excellent way out of it? Also we want to liberate people not to make them succumb to a new kind of slavery, *but to make them realise their own wishes,* however different these wishes may be from our own. Self-righteous and narrow-minded liberators cannot do this. As a rule they soon impose a slavery that is worse, because more systematic, than the very sloppy slavery they have removed. And as regards humour and flippancy the answer should be obvious. Why would anyone want to liberate anyone else? Surely not because of some *abstract* advantage of liberty but because liberty is the best way to free development *and thus to happiness.* We want to liberate people so that *they can smile.* Shall we be able to do this if we ourselves have forgotten how to smile and are frowning on those who still remember? Shall we then not spread another disease, comparable to the one we want to remove, the disease of puritanical self-righteousness? Do not object that dedication and humour do not go together—Socrates is an excellent example to the contrary. *The hardest task needs the lightest hand or else its completion will not lead to freedom but to a tyranny much worse than the one it replaces.*

4

Why Astrology Is a Pseudoscience
Paul R. Thagard

Most philosophers and historians of science agree that astrology is a pseudo-science, but there is little agreement on *why* it is a pseudoscience. Answers range from matters of verifiability and falsifiability, to questions of progress and Kuhnian normal science, to the different sort of objections raised by a large panel of scientists recently organized by *The Humanist* magazine. Of course there are also Feyerabendian anarchists and others who say that no demarcation of science from pseudoscience is possible. However, I shall pro-pose a complex criterion for distinguishing disciplines as pseudoscientific; this criterion is unlike verificationist and falsificationist attempts in that it intro-duces social and historical features as well as logical ones.

I begin with a brief description of astrology. It would be most unfair to evaluate astrology by reference to the daily horoscopes found in newspapers and popular magazines. These horoscopes deal only with sun signs, whereas a full horoscope makes reference to the "influences" also of the moon and the planets, while also discussing the ascendant sign and other matters.

Astrology divides the sky into twelve regions, represented by the familiar signs of the Zodiac: Aquarius, Libra and so on. The sun sign represents the part of the sky occupied by the sun at the time of birth. For example, anyone born between September 23 and October 22 is a Libran. The ascendant sign, often assumed to be at least as important as the sun sign, represents the part of the sky rising on the eastern horizon at the time of birth, and therefore changes every two hours. To determine this sign, accurate knowledge of the time and place of birth is essential. The moon and the planets (of which there are five or eight depending on whether Uranus, Neptune and Pluto are taken into account) are also located by means of charts on one of the parts of the Zodiac. Each planet is said to exercise an influence in a special sphere of human activity; for example, Mars governs drive, courage and daring, while Venus governs love and artistic endeavor. The immense number of combinations of sun, ascendant, moon and planetary influences allegedly determines human personality, behavior and fate.

Astrology is an ancient practice, and appears to have its origins in Chaldea, thousands of years B.C. By 700 B.C., the Zodiac was established, and a few centuries later the signs of the Zodiac were very similar to current ones. The conquests of Alexander the Great brought astrology to Greece, and the Romans were exposed in turn. Astrology was very popular during the fall of the Republic, with many notables such as Julius Caesar having their horoscopes cast. However, there was opposition from such men as Lucretius and Cicero.

Astrology underwent a gradual codification culminating in Ptolemy's *Tetrabiblos* [20], written in the second century A.D. This work describes in great detail the powers of the sun, moon and planets, and their significance in people's lives. It is still recognized as a fundamental textbook of astrology. Ptolemy took astrology as seriously as he took his famous work in geography and astronomy; this is evident from the introduction to the *Tetrabiblos,* where he discusses two available means of making predictions based on the heavens. The first and admittedly more effective of these concerns the relative movements of the sun, moon and planets, which Ptolemy had already treated in his celebrated *Almagest* [19]. The secondary but still legitimate means of prediction is that in which we use the "natural character" of the aspects of movement of heavenly bodies to "investigate the changes which they bring about in that which they surround" ([20], p. 3). He argues that this method of prediction is possible because of the manifest effects of the sun, moon and planets on the earth, for example on weather and the tides.

The European Renaissance is heralded for the rise of modern science, but occult arts such as astrology and alchemy flourished as well. Arthur Koestler has described Kepler's interest in astrology: not only did astrology provide Kepler with a livelihood, he also pursued it as a serious interest, although he was skeptical of the particular analyses of previous astrologers ([13], pp. 244–248). Astrology was popular both among intellectuals and the general public through the seventeenth century. However, astrology lost most of this popularity in the eighteenth century, when it was attacked by such figures of the Enlightenment as Swift [24] and Voltaire [29]. Only since the 1930's has astrology again gained a huge audience: most people today know at least their sun signs, and a great many believe that the stars and planets exercise an important influence on their lives.

In an attempt to reverse this trend, Bart Bok, Lawrence Jerome and Paul Kurtz drafted in 1975 a statement attacking astrology; the statement was signed by 192 leading scientists, including 19 Nobel prize winners. The statement raises three main issues: astrology originated as part of a magical world view, the planets are too distant for there to be any physical foundation for astrology, and people believe it merely out of longing for comfort ([2], pp. 9f.). None of these objections is ground for condemning astrology as pseudoscience. To show this, I shall briefly discuss articles written by Bok [1] and Jerome [12] in support of the statement.

According to Bok, to work on statistical tests of astrological predictions is a waste of time unless it is demonstrated that astrology has some sort of physical foundation ([1], p. 31). He uses the smallness of gravitational and radiative effects of the stars and planets to suggest that there is no such foundation. He also discusses the psychology of belief in astrology, which is the result of individuals' desperation in seeking solutions to their serious personal problems. Jerome devotes most of his article to the origins of astrology in the magical principle of correspondences. He claims that astrology is a system of magic rather than science, and that it fails "not because of any inherent inaccuracies due to precession or lack of exact knowledge concerning time of birth or conception, but rather because its interpretations and predictions are grounded in the ancients' magical world view" ([12], p. 46). He does however discuss some statistical tests of astrology, which I shall return to below.

These objections do not show that astrology is a pseudoscience. First, origins are irrelevant to scientific status. The alchemical origins of chemistry ([11], pp. 10–18) and the occult beginnings of medicine [8] are as magical as those of astrology, and historians have detected mystical influences in the work of many great scientists, including Newton and Einstein. Hence astrology cannot be condemned simply for the magical origins of its principles. Similarly, the psychology of popular belief is also in itself irrelevant to the status of astrology: people often believe even good theories for illegitimate reasons, and even if most people believe astrology for personal, irrational reasons, good reasons may be available.[1] Finally the lack of a physical foundation hardly marks a theory as unscientific ([22], p. 2). Examples: when Wegener [31] proposed continental drift, no mechanism was known, and a link between smoking and cancer has been established statistically [28] though the details of carcinogenesis remain to be discovered. Hence the objections of Bok, Jerome and Kurtz fail to mark astrology as pseudoscience.

Now we must consider the application of the criteria of verifiability and falsifiability to astrology. Roughly, a theory is said to be verifiable if it is possible to deduce observation statements from it. Then in principle, observations can be used to confirm or disconfirm the theory. A theory is scientific only if it is verifiable. The vicissitudes of the verification principle are too well known to recount here ([9], ch. 4). Attempts by A. J. Ayer to articulate the principle failed either by ruling out most of science as unscientific, or by ruling out nothing. Moreover, the theory/observation distinction has increasingly come into question. All that remains is a vague sense that testability somehow is a mark of scientific theories ([9], ch. 4; [10], pp. 30–32).

Well, astrology *is* vaguely testable. Because of the multitude of influences resting on tendencies rather than laws, astrology is incapable of making precise predictions. Nevertheless, attempts have been made to test the reality of these alleged tendencies, using large scale surveys and statistical evaluation. The pioneer in this area was Michel Gauquelin, who examined the careers and times of birth of 25,000 Frenchmen. Astrology suggests that people born under certain signs or planets are likely to adopt certain occupations: for example, the influence of the warlike planet Mars tends to produce soldiers or athletes, while Venus has an artistic influence. Notably, Gauquelin found *no significant correlation* between careers and either sun sign, moon sign, or ascendant sign. However, he did find some statistically interesting correlations between certain occupations of people and the position of certain planets at the time of their birth ([5], ch. 11; [6]). For example, just as astrology would suggest, there is a greater than chance association of athletes and Mars, and a greater than chance association of scientists and Saturn, where the planet is rising or at its zenith at the moment of the individual's birth.

These findings and their interpretation are highly controversial, as are subsequent studies in a similar vein [7]. Even if correct, they hardly verify astrology, especially considering the negative results found for the most important astrological categories. I have mentioned Gauquelin in order to suggest that through the use of statistical techniques astrology is at least *verifiable*. Hence the verification principle does not mark astrology as pseudoscience.

Because the predictions of astrologers are generally vague, a Popperian would assert that the real problem with astrology is that it is not falsifiable: astrologers can not make predictions which if unfulfilled would lead them to give up their theory. Hence because it is unfalsifiable, astrology is unscientific.

But the doctrine of falsifiability faces serious problems as described by Duhem [4], Quine [21], and Lakatos [15]. Popper himself noticed early that no observation ever guarantees falsification: a theory can always be retained by introducing or modifying auxiliary hypotheses, and even observation statements are not incorrigible ([17], p. 50). Methodological decisions about what can be tampered with are required to block the escape from falsification. However, Lakatos has persuasively argued that making such decision in advance of tests is arbitrary and may often lead to overhasty rejection of a sound theory which *ought* to be saved by anti-falsificationist strategems ([15], pp. 112 ff.). Falsification only occurs when a better theory comes along. Then falsifiability is only a matter of replaceability by another theory, and since astrology is in principle replaceable by another theory, falsifiability provides no criterion for rejecting astrology as pseudoscientific. We saw in the discussion of Gauquelin that astrology can be used to make predictions about statistical regularities, but the nonexistence of these regularities does not falsify astrology; but here astrology does not appear worse than the best of scientific theories, which also resist falsification until alternative theories arise.[2]

Astrology cannot be condemned as pseudoscientific on the grounds proposed by verificationists, falsificationists, or Bok and Jerome. But undoubtedly astrology today faces a great many unsolved problems ([32], ch. 5). One is the negative result found by Gauquelin concerning careers and signs. Another is the problem of the precession of the equinoxes, which astrologers generally take into account when heralding the "Age of Aquarius" but totally neglect when figuring their charts. Astrologers do not always agree on the significance of the three planets, Neptune, Uranus and Pluto, that were discovered since Ptolemy. Studies of twins do not show similarities of personality and fate that astrology would suggest. Nor does astrology make sense of mass disasters, where numerous individuals with very different horoscopes come to similar ends.

But problems such as these do not in themselves show that astrology is either false or pseudoscientific. Even the best theories face unsolved problems throughout their history. To get a criterion demarcating astrology from science, we need to consider it in a wider historical and social context.

A demarcation criterion requires a matrix of three elements: [theory, community, historical context]. Under the first heading, "theory," fall familiar matters of structure, prediction, explanation and problem solving. We might also include the issue raised by Bok and Jerome about whether the theory has a physical foundation. Previous demarcationists have concentrated on this theoretical element, evident in the concern of the verification and falsification principles with prediction. But we have seen that this approach is not sufficient for characterizing astrology as pseudoscientific.

We must also consider the *community* of advocates of the theory, in this case the community of practitioners of astrology. Several questions are important here. First, are the practitioners in agreement on the principles of the theory and on how to go about solving problems which the theory faces? Second, do they care, that is, are they concerned about explaining anomalies and comparing the success of their theory to the record of other theories? Third, are the practitioners actively involved in attempts at confirming and disconfirming their theory?

The question about comparing the success of a theory with that of other theories introduces the third element of the matrix, historical context. The historical work of Kuhn and others has shown that in general a theory is rejected only when (1) it has faced anomalies over a long period of time and (2) it has been challenged by another theory. Hence under the heading of historical context we must consider two factors relevant to demarcation: the record of a theory over time in explaining new facts and dealing with anomalies, and the availability of alternative theories.

We can now propose the following principles of demarcation:

A theory or discipline which purports to be scientific is *pseudoscientific* if and only if:
(1) It has been less progressive than alternative theories over a long period of time, and faces many unsolved problems; but
(2) the community of practitioners makes little attempt to develop the theory towards solutions of the problems, shows no concern for attempts to evaluate the theory in relation to others, and is selective in considering confirmations and disconfirmations.

Progressiveness is a matter of the success of the theory in adding to its set of facts explained and problems solved ([15], p. 118; cf. [26], p. 83).

This principle captures, I believe, what is most importantly unscientific about astrology. First, astrology is dramatically unprogressive, in that it has changed little and has added nothing to its explanatory power since the time of Ptolemy. Second, problems such as the precession of equinoxes are outstanding. Third, there are alternative theories of personality and behavior available: one need not be an uncritical advocate of behaviorist, Freudian, *or* Gestalt theories to see that since the nineteenth century psychological theories have been expanding to deal with many of the phenomena which astrology explains in terms of heavenly influences. The important point is not that any of these psychological theories is established or true, only that they are growing alternatives to a long-static astrology. Fourth and finally, the community of astrologers is generally unconcerned with advancing astrology to deal with outstanding problems or with evaluating the theory in relation to others.[3] For these reasons, my criterion marks astrology as pseudoscientific.

This demarcation criterion differs from those implicit in Lakatos and Kuhn. Lakatos has said that what makes a series of theories constituting a research program scientific is that it is progressive: each theory in the series has greater corroborated content than its predecessor ([15], p. 118). While I agree with Lakatos that progressiveness is a central notion here, it is not sufficient to distinguish science from pseudoscience. We should not brand a nonprogressive discipline as pseudoscientific unless it is being maintained against more progressive alternatives. Kuhn's discussion of astrology focuses on a different aspect of my criterion. He says that what makes astrology unscientific is the absence of the paradigm-dominated puzzle solving activity characteristic of what he calls normal science ([14], p. 9). But as Watkins has suggested, astrologers are in some respects model normal scientists: they concern themselves with solving puzzles at the level of individual horoscopes, unconcerned with the foundations of their general theory or paradigm ([30], p. 32). Hence that feature of normal science does not distinguish science from pseudoscience. What makes astrology pseudoscientific is not that it lacks periods of Kuhnian normal science, but that its proponents adopt uncritical attitudes of "normal" scientists despite the existence of more progressive alternative theories. (Note that I am not agreeing with Popper [18] that Kuhn's normal scientists are unscientific; they can become unscientific only when an alternative paradigm has been developed.) However, if one looks not at the puzzle solving at the level of particular astrological predictions, but at the level of theoretical problems such as the precession of the equinoxes, there is some agreement between my criterion and Kuhn's; astrologers do not have a paradigm-induced confidence about solving theoretical problems.

Of course, the criterion is intended to have applications beyond astrology. I think that discussion would show that the criterion marks as pseudoscientific such practices as witchcraft and pyramidology, while leaving contemporary physics, chemistry and biology unthreatened. The current fad of biorhythms, implausibly based like astrology on date of birth, can not be branded as pseudoscientific because we lack alternative theories giving more detailed accounts of cyclical variations in human beings, although much research is in progress.[4]

One interesting consequence of the above criterion is that a theory can be scientific at one time but pseudoscientific at another. In the time of Ptolemy or even Kepler, astrology had few alternatives in the explanation of human personality and behavior. Existing alternatives were scarcely more sophisticated or corroborated than astrology. Hence astrology should be judged as not pseudoscientific in classical or Renaissance times, even though it is pseudoscientific today. Astrology was not simply a perverse sideline of Ptolemy and Kepler, but part of their scientific activity, even if a physicist involved with astrology today should be looked at askance. Only when the historical and social aspects of science are neglected does it become plausible that pseudoscience is an unchanging category. Rationality is not a property of ideas eternally: ideas, like actions, can be rational at one time but irrational at others. Hence relativizing the science/pseudoscience distinction to historical periods is a desirable result.

But there remains a challenging historical problem. According to my criterion, astrology only became pseudoscientific with the rise of modern psychology in the nineteenth century. But astrology was already virtually excised from scientific circles by the beginning of the eighteenth. How could this be? The simple answer is that a theory can take on the appearance of an unpromising project well before it deserves the label of pseudoscience. The Copernican revolution and the mechanism of Newton, Descartes and Hobbes undermined the plausibility of astrology.[5] Lynn Thorndike [27] has described how the Newtonian theory pushed aside what had been accepted as a universal natural law, that inferiors such as inhabitants of earth are ruled and governed by superiors such as the stars and the planets. William Stahlman [23] has described how the immense growth of science in the seventeenth century contrasted with stagnation of astrology. These developments provided good reason for discarding astrology as a promising pursuit, but they were not yet enough to brand it as pseudoscientific, or even to refute it.

Because of its social aspect, my criterion might suggest a kind of cultural relativism. Suppose there is an isolated group of astrologers in the jungles of South America, practicing their art with no awareness of alternatives. Are we to say that astrology is *for them* scientific? Or, going in the other direction, should we count as alternative theories ones which are available to extraterrestrial beings, or which someday will be conceived? This wide construal of "alternative" would have the result that our best current theories are probably pseudoscientific. These two questions employ, respectively, a too narrow and too broad view of alternatives. By an alternative theory I mean one generally available in the world. This assumes first that there is some kind of communication network to which a community has, or should have, access. Second, it assumes that the onus is on individuals and communities to find out about alternatives. I would argue (perhaps against Kuhn) that this second assumption is a general feature of rationality; it is at least sufficient to preclude ostrichism as a defense against being judged pseudoscientific.

In conclusion, I would like to say why I think the question of what constitutes a pseudoscience is important. Unlike the logical positivists, I am not grinding an anti-metaphysical ax, and unlike Popper, I am not grinding an anti-Freudian or anti-Marxian one.[6] My concern is social: society faces the twin problems of lack of public concern with the advancement of science, and lack of public concern with the important ethical issues now arising in science and technology, for example around the topic of genetic engineering. One reason for this dual lack of concern is the wide popularity of pseudoscience and the occult among the general public. Elucidation of how science differs from pseudoscience is the philosophical side of an attempt to overcome public neglect of genuine science.

NOTES

1. However, astrology would doubtlessly have many fewer supporters if horoscopes tended less toward compliments and pleasant predictions and more toward the kind of analysis included in the following satirical horoscope from the December, 1977, issue of *Mother Jones*: VIRGO (Aug. 23–Sept. 22). You are the logical type and hate disorder. This nit-picking is sickening to your friends. You are cold and unemotional and sometimes fall asleep while making love. Virgos make good bus drivers.

2. For an account of the comparative evaluation of theories, see [26].

3. There appear to be a few exceptions; see [32].

4. The fad of biorhythms, now assuming a place beside astrology in the popular press, must be distinguished from the very interesting work of Frank Brown and others on biological rhythms. For a survey, see [5].

5. Plausibility is in part a matter of a hypothesis being of an appropriate *kind,* and is relevant even to the acceptance of a theory. See [26], p. 90, and [25].

6. On psychoanalysis see [3]. I would argue that Cioffi neglects the question of alternatives to psychoanalysis and the question of its progressiveness.

The author is grateful to Dan Hausman and Elias Baumgarten for comments.

REFERENCES

[1] Bok, Bart J. "A Critical Look at Astrology." In [2]. Pages 21–33.

[2] _____, Jerome, Lawrence E., and Kurtz, Paul. *Objections to Astrology.* Buffalo: Prometheus Books, 1975.

[3] Cioffi, Frank. "Freud and the Idea of a Pseudoscience." In *Explanation in the Behavioral Sciences.* Edited by R. Borger and F. Cioffi. Cambridge: Cambridge University Press, 1970. Pages 471–499.

[4] Duhem, P. *The Aim and Structure of Physical Theory.* Translated by P. Wiener. New York: Atheneum, 1954. (Translated from 2nd edition of *La Théorie Physique: Son Object Sa Structure.* Paris: Marcel Rivière & Cie, 1914.)

[5] Gauquelin, Michel. *The Cosmic Clocks*. Chicago: Henry Regnery, 1967.

[6] _____. *The Scientific Basis of Astrology*. New York: Stein and Day, 1969.

[7] _____. "The Zelen Test of the Mars Effect." *The Humanist* 37 (1977): 30–35.

[8] Haggard, Howard W. *Mystery, Magic, and Medicine*. Garden City: Doubleday, Doran & Company, 1933.

[9] Hempel, Carl. *Aspects of Scientific Explanation*. New York: The Free Press, 1965.

[10] _____. *Philosophy of Natural Science*. Englewood Cliffs: Prentice-Hall, 1966.

[11] Ihde, Aaron, J. *The Development of Modern Chemistry*. New York: Harper and Row, 1964.

[12] Jerome, Lawrence E. "Astrology: Magic or Science?" In [2]. Pages 37–62.

[13] Koestler, Arthur. *The Sleepwalkers*. Harmondsworth: Penguin, 1964.

[14] Kuhn, T. S. "Logic of Discovery or Psychology of Research." In [16]. Pages 1–23.

[15] Lakatos, Imre. "Falsification and the Methodology of Scientific Research Programmes." In [16]. Pages 91–195.

[16] _____ and Musgrave, Alan, eds. *Criticism and the Growth of Knowledge*. Cambridge: Cambridge University Press, 1970.

[17] Popper, Karl. *The Logic of Scientific Discovery*. London: Hutchinson, 1959. (Originally published as *Logik der Forschung*. Vienna: J. Springer, 1935.)

[18] _____. "Normal Science and Its Dangers." In [16]. Pages 51–58.

[19] Ptolemy. *The Almagest (The Mathematical Composition)*. (As printed in Hutchins, Robert Maynard, ed. *Great Books of the Western World*, Volume 16. Chicago. Encyclopedia Britannica, Inc., 1952. Pages 1–478.)

[20] _____. *Tetrabiblos*. Edited and translated by F. E. Robbins. Cambridge: Harvard University Press, 1940.

[21] Quine, W. V. O. "Two Dogmas of Empiricism." In *From a Logical Point of View*. New York: Harper & Row, 1963. Pages 20–46. (Originally published in *The Philosophical Review* 60 (1951): 20–43.)

[22] Sagan, Carl. "Letter." *The Humanist* 36 (1976): 2.

[23] Stahlman, William D. "Astrology in Colonial America: An Extended Query." *William and Mary Quarterly* 13 (1956): 551–563.

[24] Swift, Jonathan. "The Partridge Papers." In *The Prose Works of Jonathan Swift*, Volume 2. Oxford: Basil Blackwell, 1940–1968. Pages 139–170.

[25] Thagard, Paul R. "The Autonomy of a Logic of Discovery." Forthcoming in the *Festschrift* for T. A. Goudge.

[26] _____. "The Best Explanation: Criteria for Theory Choice." *Journal of Philosophy* 75 (1978): 76–92.

[27] Thorndike, Lynn. "The True Place of Astrology in the History of Science." *Isis* 46 (1955): 273–278.

[28] U.S. Department of Health, Education and Welfare. *Smoking and Health: Report of the Advisory Committee to the Surgeon General of the Public Health Service.* Washington, D.C.: U.S. Government Printing Office, 1964.

[29] Voltaire. "Astrologie" and "Astronomie." *Dictionnaire Philosophique.* In *Oeuvres Complètes de Voltaire,* Volume XVII. Paris: Garnier Frères, 1878-1885. Pages 446-453.

[30] Watkins, J. W. N. "Against 'Normal Science.'" In [16]. Pages 25-37.

[31] Wegener, Alfred. "Die Entstehung der Kontinente." *Petermann's Geographische Mitteilung* 58 (1912): 185-195, 235-256, 305-309.

[32] West, J. A. and Toonder, J. G. *The Case for Astrology.* Harmondsworth: Penguin, 1973.

1. According to Popper, what is the problem of demarcation? State Popper's solution of that problem.

2. Show why, in Popper's views, psychoanalytic theories do not meet his criterion of demarcation whereas Einstein's theory of gravitation does.

3. In addition to the examples which he already gives, list several theories, statements, or enterprises which do *not* meet Popper's criterion of demarcation. Show *why* they do not do so.

4. Try to think of some theories or truths which (in your view) should be accorded scientific status but which do *not* meet Popper's criterion for being scientific. How would you revise—or replace—Popper's criterion in order to account for them?

5. State the definitions of science which Ziman rejects. Why does he reject them? State *his* definition. Why does he hold it to be superior to the others? Do you agree with him? Why or why not?

6. What is Ziman's criterion for distinguishing between science and non-science? Does it seem acceptable to you? Why or why not? How does Ziman's criterion differ from Popper's?

7. Defend or attack Ziman's thesis that the distinction between pure science and technological knowledge is difficult to make.

8. According to Feyerabend, *how* should one defend society against science? *Why* should one do so? Do you agree with him? Why or why not?

9. Why does Feyerabend reject Popper's criterion of falsifiability? Has he provided *good reasons* for rejecting it? Why or why not?

10. What thesis, if any, in Feyerabend's essay do you disagree with or find fault with? Why?

11. What is Thagard's criterion for distinguishing science from pseudoscience? Compare and contrast with the views of Popper, Ziman, and Feyerabend. Which of the criteria formulated by these four authors do you find most acceptable? Why?

12. On the basis of Thagard's criterion, why does he claim that astrology is a pseudoscience? Has he (in your view) successfully shown that it is a pseudoscience? Why or why not?

13. How do (explicitly or implicitly) Popper, Ziman, Feyerabend, and Thagard answer the question "What is science?"? Which answer seems preferable to you? Why?

14. Write a detailed and critical reply to the following letter (from J. Somerville, "Umbrellaology," *Philosophy of Science,* 1941).

Dear Sir:

I am taking the liberty of calling upon you to be the judge in a dispute between me and an acquaintance who is no longer a friend. The question at issue is this: Is my creation, umbrellaology, a science? Allow me to explain. . . . For the past eighteen years assisted by a few faithful disciples, I have been collecting materials on a subject hitherto almost wholly neglected by scientists, the umbrella. The results of my investigations to date are embodied in the nine volumes which I am sending to you under a separate cover. Pending their receipt, let me describe to you briefly the nature of their contents and the method I pursued in compiling them. I began on the Island of Manhattan. Proceeding block by block, house by house, family by family, and individual by individual, I ascertained (1) the number of umbrellas possessed, (2) their size, (3) their weight, (4) their color. Having covered Manhattan after many years, I eventually extended the survey to the other boroughs of the City of New York, and at length completed the entire city. Thus I was ready to carry forward the work to the rest of the state and indeed the rest of the United States and the whole known world.

It was at this point that I approached my erstwhile friend. I am a modest man, but I felt I had the right to be recognized as the creator of a new science. He, on the other hand, claimed that umbrellaology was not a science at all. First, he said, it was silly to investigate umbrellas. Now this argument is false, because science scorns not to deal with any object, however humble and lowly, even to the "hind leg of a flea." Then why not umbrellas? Next, he said that umbrellaology could not be recognized as a science because it was of no use or benefit to mankind. But is not the truth the most precious thing in life? Are not my nine volumes filled with the truth about my subject? Every word in them is true. Every sentence contains a hard, cold fact. When he asked me what was the object of umbrellaology I was proud to say, "To seek and discover the truth is object enough for me." I am a pure scientist; I have no ulterior motives. Hence it follows that I am satisfied with truth alone. Next, he said my truths were dated and that any one of my findings might cease to be true tomorrow. But this, I pointed out, is not an argument against umbrellaology, but rather an argument for keeping it up to date, which exactly is what I propose. Let us have surveys monthly, weekly, or even daily, to keep our knowledge abreast of the changing facts. His next contention was that umbrellaology had entertained no hypotheses and had developed no theories or laws. This is a great error. In the course of my investigations, I employed innumerable hypotheses. Before entering each new block and each new section of the city, I entertained an hypothesis as regards the number and characteristics of the umbrellas that would be found there, which hypotheses were either verified or nullified by my subsequent observations, in accordance with proper scientific procedure, as explained in authoritative texts. (In fact, it is of interest to note that I can substantiate and document every one of my replies to these objections by numerous quotations from standard works, leading journals, public speeches of eminent scientists, and the like.) As for theories and laws, my work represents an abundance of them. I will here mention only a few, by way of illustration. There is the Law of Color Variation Relative to Ownership by Sex. (Umbrellas owned by women tend to great variety of color, whereas those owned by men are almost all black.) To this law I have given exact statistical formulation. (See vol. 6, Appendix 1, Table 3, p. 582.) There are the curiously interrelated Laws of Individual Ownership of Plurality of

Umbrellas, and Plurality of Ownership of Individual Umbrellas. The inter-relationship assumes the form, in the first law, of almost direct ratio to annual income, and in the second, in almost inverse ratio to annual income. (For an exact statement of the modifying circumstances, see vol. 8, p. 350.) There is also the Law of Tendency Towards Acquisition of Umbrellas in Rainy Weather. To this law I have given experimental verification in chapter 3 of volume 3. In the same way I have performed numerous other experiments in connection with my generalizations.

Thus I feel that my creation is in all respects a genuine science, and I appeal to you for substantiation of my opinion. . . .

SELECTED BIBLIOGRAPHY

A. What is Science?

[1] Campbell, Norman. *What Is Science?* New York: Dover Publications, 1952.
[2] Cohen, Morris R., and Nagel, Ernest. *An Introduction to Logic and Scientific Method,* book 2. New York: Harcourt, Brace, 1934.
[3] Conant, James B. *On Understanding Science.* New Haven: Yale University Press, 1951.
[4] Feyerabend, Paul. *Against Method.* New York: Schocken Books, 1977.
[5] Kemeny, J. G. *A Philosopher Looks at Science.* Princeton: D. Van Nostrand, 1959.
[6] Kuhn, Thomas S. *The Structure of Scientific Revolutions.* 2nd ed. Chicago: University of Chicago Press, 1971.
[7] Nagel, E. *The Structure of Science.* New York: Harcourt, Brace & World, 1961.
[8] Popper, Karl. *Conjectures and Refutations.* New York: Harper & Row, 1968.
[9] _____. *The Logic of Scientific Discovery.* New York: Harper & Row, 1968.
[10] Toulmin, Stephen. *Foresight and Understanding.* London: Hutchinson, 1961.

B. Science and Nonscience

[1] Flew, Anthony. "ESP." *The New York Review of Books,* October 8, 1964, pp. 15–16.
[2] Gardner, Martin. *Fads and Fallacies in the Name of Science.* New York: Dover Publications, 1957. (See also the detailed appendix and notes to this volume.)
[3] _____. "Funny Coincidence." *The New York Review of Books,* May 26, 1966.
[4] _____. "What Hath Hoova Wrought?" *The New York Review of Books,* May 16, 1974, pp. 18–19.

[5] _____. "Paranonsense." *The New York Review of Books,* October 30, 1975, pp. 12–14.

[6] _____. "Supergull." *The New York Review of Books,* March 17, 1977, pp. 18–20.

[7] Hansel, C. E. M. *ESP: A Scientific Evaluation.* New York: Scribner's, 1966.

[8] Popper, Karl. *Conjectures and Refutation.* New York: Harper & Row, 1968.

[9] _____. *The Logic of Scientific Discovery.* New York: Harper & Row, 1968.

[10] Toulmin, Stephen. "Contemporary Scientific Mythology." In *Metaphysical Beliefs,* by S. Toulmin, et al. London: SCM Press, 1957.

Part Two

Explanation and Law

Part 2: Explanation and Law

Introduction

It has been often held that one of the aims of science is explanation. What constitutes a scientific explanation? How (if at all) does a scientific explanation differ from an ordinary explanation in everyday life or an explanation in (say) history? In a scientific explanation, what is it that is being explained and what "does" the explaining? What is the role of the laws of science in explanation? What constitutes a scientific law? These are some of the questions to be pursued in this part. Before turning to these questions let us consider some examples of purported explanations.

I. Some Examples

Aristotle explains what appears to be the circular motion of the planets by appealing to the "fact" that they are members of the celestial region—a perfect and permanent region. Circular motion was regarded as the only motion appropriate or natural in that region. Kepler, at one point in his curious career, explained the distances of the various planets from the sun by drawing an analogy between these distances and the way the five Platonic solids will fit inside one another. (Platonic solids are solid figures that have faces of identical equilateral plane figures, e.g., a square and a tetrahedron. There are only five such objects.)

Scientific creationists claim that the biblical account in Genesis offers an alternative and more plausible explanation than Darwinian evolution of the origins and diversity of life. In several states creationists are waging a serious political and legal effort to have creationism given a place alongside evolution in the schools' curricula.

B. F. Skinner thinks that to explain human behavior by appealing to what is inside the skin is pointless. In fact, it is claimed that inquiry along these lines will only lead to confusions and banalities, not real explanations.

Population geneticists have recently become quite self-conscious. What is it that population genetics explains? How must one understand fitness in order to have causal explanations of the fecundity of populations?

Over and above a desire to understand the quick of science, cases such as the above motivate an inquiry into the nature of explanation. The Aristotelian and Keplerian explanatory episodes strike us as bizarre. But there are important differences. The Aristotelian view achieved great prominence and endured for a considerable period of time. The Keplerian view was seen as quackery. The philosopher of science would like to know exactly why these explanations appear to us, not just false, but in a deeper sense misconceived. We also need an account of the perceived scientific respectability of Aristotle's effort and the lack of such respectability in Kepler's effort. This raises the further question whether the standards for explanation are in some sense historical. (Several essays in Part 1 bear on this issue.)

Regardless of one's religious views, it is clear that the creationist/ evolutionist dispute has at its heart claims about the nature of explanation. Legislators and others cannot hope to think clearly about this issue without first forming a general and competent view about explanation.

The schools of psychology have long been at war. The present battle is between a brand of cognitive psychology called information-processing psychology and behaviorism which Skinner prominently espouses. The literature on this clash is replete with claims about the "proper" way to explain human behavior.

A recent charge is that a central explanatory concept of population genetics, fitness, is tautological. In other words, fitness is used to explain changes in the gene frequency of a population but, as the charge goes, fitness is simply defined as some mathematical function of the change in gene frequency of a population. This issue has forced a rethinking of the explanatory adequacy of population genetics.

The above examples raise a number of specialized issues in explanation. The bibliography will help provide some access to the relevant literature. But one must begin somewhere. The appropriate place is with a general discussion of the fundamentals of explanation. The essays in this part are selected to initiate that general and fundamental discussion.

John Hospers's paper "What is Explanation?" is one of the early contemporary treatments of the problem of explanation. Since his paper is not parasitic on a body of previous literature, it is an excellent port of entry to the topic. The reader should begin by studying Hospers's paper and then return to this introduction.

II. Explanation

The view of explanation explored by Hospers has come to be called the Hempelian or deductive-nomological view of explanation. The view has justly been named after Carl Hempel, who in a series of papers has lucidly and powerfully argued for the view.

Hospers does not commit himself to the Hempelian view, but is content to present it, rather informally, and raise some of the important issues surrounding it.

It will be helpful to have a crisp statement of the Hempelian account before us. First a bit of terminology. Explanations are composed of a set of statements. The statement describing the event or law to be explained is called the explanandum. The statements which are put forward to account for the phenomena expressed in the explanandum are called the explanans. Now, for the account:

E, some purported explanation, is an explanation if and only if
(1) The explanandum of *E* is a deductive consequence of the explanans.
(2) The explanans of *E* contains at least one law of nature.
(3) The law(s) of nature in the explanans is required for the derivation of the explanandum.
(4) The sentences of the explanans are testable by observation.
(5) The sentences of the explanans are true.

Hospers's simple example of why this piece of copper conducts electricity illustrates these conditions.
(a) All copper conducts electricity.
(b) This wire is made of copper.
(c) Therefore, this wire conducts electricity.
(c), the explanandum, is a deductive consequence of the explanans, (a) and (b). (a) is a law of nature that is required for the derivation of (c), and it is testable. Supposing that "this" refers to a piece of copper, the explanans are true.

The definition and example above rely on technical notions from logic. An argument as understood by logicians is a set of statements. One of those statements is the conclusion, that which is to be proved or established. The remaining statements are the premises, those statements which provide the reasons or evidence for the conclusion.

A special class of arguments are those which are deductively valid. An argument is deductively valid if, and only if, it is impossible for the premises to be true and its conclusion false.

Explanations, on the view being discussed, are arguments. Furthermore, according to condition (1) they must be deductively valid arguments.

Condition (3) with its technical term, 'derivation,' requires that if the law(s) of nature in an explanation were deleted from the argument, the argument would no longer be deductively valid. Crudely put, this requirement ensures that the laws are not just fluff, but rather are essential to the explanation. In the example given, (a) is required since, if it were deleted, (b) and (c) would not form a deductively valid argument.

Much of the history of philosophical reflection on explanation has been various attempts to provide the rationale for (1)–(5) or to show that one or more of (1)–(5) are not necessary conditions for explanation or to show that (1)–(5) are not jointly a sufficient condition for explanation. Hospers's essay is teeming with suggestions that bear on all three of these projects.

In your examination of the above account, it will be helpful to keep an important distinction in mind. Hempelians use this distinction to forestall certain criticisms of their view.

Consider two examples: When asked, "Why does this conduct electricity?" one might respond, "It's copper." When explaining to a group of fourth graders why the pressure on the walls of a closed pot increased, one might give the kinetic theory and the fact that the temperature increased.

Our intuitions encourage us to believe that the former episode is an explanation although it does not meet Hempel's criteria. On the other hand, the latter episode, even if filled out in splendid Hempelian detail, fails to explain. After all, it is a rare fourth grader that has the slightest idea what the mean molecular speed squared is.

Hempelians have a response to these worries. We need to make a distinction between the logic of explanation and the pragmatics of explanation. An account of the logic of explanation explores the relation between two groups of sentences such that the former explains the latter. The pragmatics of explanation concerns the important considerations in explaining something to *someone*. Though the Hempelian account is not concerned with pragmatics, Hempelians do not deny the importance of such studies.

Using this distinction, we see that the last of the examples above is indeed an explanation, though not appropriate for young children. But appropriateness is an issue in the pragmatics, not logic, of explanation.

The conductor example can also be defused. Yes, we talk that way when we explain things to someone. But for it to really be an explanation, it must meet conditions (1)–(5). The reason one regards the episode as explanatory is that he is implicitly assuming certain rather obvious statements that enable the episode to meet Hempel's conditions.

The logic/pragmatics distinction has been drawn in a very rough way. It certainly deserves further clarification and perhaps criticism.

III. Laws of Nature

The Hempelian view is noticeably incomplete until supplemented with an account of the concept of a law of nature. As the Hempelian view has been presented, one could plug one of various views on the nature of laws into the account. But, at least as a matter of historical fact, the Hempelian view has been allied with a particular view on laws. Hospers's second essay spells out that view. (It will be helpful to the student to state briefly and carefully Hospers's necessary and sufficient conditions for a statement expressing a law of nature.)

Baruch Brody's paper should be understood as a critique of the sufficiency of the Hempelian view of explanation, by way of a critique of the account of law embedded therein. Furthermore, Brody has some exciting suggestions toward an alternative treatment of laws.

The paper by Brody raises in an indirect manner a classical problem in the study of explanation. A problem that bears on the aims of science which in turn is relevant to the issues discussed in Parts 1 and 6. Notice that the Hempelian model appears to provide one with predictions as well as explanations.

Consider an example. The general form of the law of conservation of momentum is as follows: The total momentum of an isolated system of bodies remains constant. Suppose we observe a car-truck crash and wish to explain why the combined velocity of the enmeshed car and truck is 6 meters/sec. We could do this by appealing to a specific form of the above law, $mv_c + mv_t = mv_{c+t}$, and the masses and velocities of the vehicles before the crash, car 600 kg. 20m/sec., truck 1400 kg. stationary. In Hempelian form the explanation would look like this:

(1) $mv_c + mv_t = mv_{c+t}$
(2) mass of t = 1400 kg.
(3) mass of c = 600 kg.
(4) velocity of t = 0
(5) velocity of c = 20 m/sec.
(6) therefore mv_{c+t} = 6 m/sec.

But it should be clear that we could use this same form of argument not to explain the momentum of the joined car and truck but to *predict* it. If we did not know mv_{c+t}, but knew that the crash would occur, we could use (1)–(5) to predict (6).

This example might lead one to suspect that the only important difference between explaining and predicting an event is when the event occurs relative to when the argument is constructed. If the event has already occurred, then one explains the event. If the event is yet to occur, one predicts the event. Perhaps from a structural or logical point of view there is no difference between explaining and predicting. Brody's examples will be useful in examining these suggestions.

IV. Types of Explanation

Both Hospers and Hempel appear to base their insights about explanation on examples from common sense and elementary physics. They have their critics even in this limited domain. But they also have critics of another sort. Some philosophers think that there are types of non-Hempelian explanation appropriate to certain subject matters. R. G. Collingwood was one such thinker. D. W. Dray in the essay "Historical Understanding as Re-thinking" tries to articulate Collingwood's view. The reader must evaluate whether the Collingwood-Dray position really is coherent and really is different from the Hempelian view.

A.D.K.

5

What Is Explanation?

John Hospers

I

We are sometimes presented with a statement describing some observed fact, and when we ask "Why?" we are presented with another statement which is said to constitute an explanation of the first. What is the relation between these two statements? What is it that makes the second statement an "explanation" of the first? By virtue of what does it explain? Though everyone is constantly uttering statements which are supposed in one way or another to explain, few persons are at all clear about what it is that makes such statements explanations. Nor is the situation clarified when it is declared on the one hand that science explains everything and on the other hand that science never explains at all but only describes.

The question "What is it to explain?" admits of no general answer, for the term "to explain" covers many activities: one may explain how, and why, and whither, and whence, and how much, and many other things. Very frequently when we ask someone to explain what he has just said we are merely asking him to restate his assertion in clearer or simpler words.

In this essay I shall treat only explaining *why*. Even within this area there are some cases with which we shall not be concerned: one may explain why the angles of a Euclidean triangle must equal 180°, and this is quite different from explaining why iron rusts. The latter is an event or a process, and I shall be concerned solely with explaining why in the special context of temporal events: roughly, why did event x happen, or why do events of class X happen? The illustration from geometry is, I should prefer to say, an example of giving *reasons* rather than explanations. Another example may further illustrate the point: If you ask me to explain why I hold a certain belief, I may reply by giving *reasons* for it—statements which I take to be evidence for the belief in question. Now, if I am rational, the fact that there is good evidence for *p* may explain why I believe *p*—that is, the reason for my believing *p* may also constitute an explanation of why I believe *p*. But this may not be so: the explanation of a person's believing in a benevolent Deity may be that he wants a father-substitute or that he needs a protector in a cold harsh world; but when asked to explain why he believes in a benevolent Deity he may cite reasons, e.g. the Argument from Design, which may have nothing to do with *why* he holds the belief. We shall be concerned here, then, with the explanation of events, not with reasons or evidences one might cite in favour of propositions.

II

What, then, is it to explain why an event occurs? (1) It has sometimes been said that we have explained it if we have stated its *purpose*. "Why did you walk through the snow for ten miles when you could have taken the bus?" "Because I wanted to win a wager." "Why does that dog scratch at the door?" "He's cold and he wants to get in." When such answers are given we are inclined to feel that our question has been answered and that the event has been satisfactorily explained; and it has been explained with reference to a purpose which some sentient being(s) had in attaining a certain end. This is the most primitive conception of explanation. People like to feel that there is a purposive explanation for everything: if not in terms of human or animal purposes, then of divine ones, or mysterious forces and powers. We tend to extend what holds true of some events to all events whatever; we know what conscious motivation is like from our own experience of it, and so we "feel at home" with this kind of explanation.

We shall examine the scope and legitimacy of purposive explanation later in this paper. It is enough to remark here that if explanation must always be in terms of purpose, then the physical sciences do not explain anything. The properties of uranium, the rise of aeroplanes, the phenomena of magnetism are not explained in terms of any purposes at all; biologists even avoid talking about animal events such as the hen sitting on eggs in terms of purpose. However animistically the nature of explanation may at one time have been conceived, purposiveness is certainly no essential part of its meaning now. The stone is no longer held to fall because it wants to get to the centre of the earth.

(2) Another account of the nature of explanation is that an event has been explained when it has been shown to be an instance of some class of events which is already familiar to us. For example, when a person's behaviour seems strange to us, we are satisfied when it is "explained" to us as being really impelled by the same sort of motives and desires as occur in us, and are therefore familiar to us. "Why is he introducing the man he hates to the woman he loves?" "Because he wants them to fall in love with each other" would not generally be accepted as an explanation, for this very reason. When

we observe that a balloon ascends rather than descends, unlike most objects, and it is made clear to us that air has weight and that the gas inside the balloon weighs less than would an equal volume of air, we are satisfied; the phenomenon has been "reduced" to something already familiar to us in everyday experience, such as a dense object sinking in water while a hollow one floats. The event is no longer unusual, strange, or unique; it has been shown to illustrate a principle with which we were already acquainted. When we want to know why gases diffuse when released into a chamber from which the air has been pumped out, the explanation offered by the kinetic theory of gases is satisfactory to us because it asserts that molecules behave *like* particles with which we are already acquainted in our everyday experience.

> Only those who have practiced experimental physics know anything by actual experience about the laws of gases; they are not things which force themselves on our attention in common life, and even those who are most familiar with them never think of them out of working hours. On the other hand, the behaviour of moving solid bodies is familiar to everyone; everyone knows roughly what will happen when such bodies collide with each other or with a solid wall, though they may not know the exact dynamical laws involved in such reactions. In all our common life we are continually encountering moving bodies, and noticing their reactions; indeed, if the reader thinks about it, he will realize that whenever we are passively affected by it, a moving body is somehow involved in the transaction. Movement is just the most familiar thing in the world; it is through motion that everything and anything happens. And so by tracing a relation between the unfamiliar changes which gases undergo when their temperature or volume is altered, and the extremely familiar changes which accompany the motions and mutual reactions of solid bodies, we are rendering the former more intelligible; we are explaining them. (Norman Campbell, *What Is Science?*, Dover, N.Y., p. 84).

Professor Bridgman holds that all explanation is of this kind: "I believe that examination will show that the essence of an explanation consists in reducing a situation to elements with which we are so familiar that we accept them as a matter of course, so that our curiosity rests" (P. W. Bridgman, *The Logic of Modern Physics,* p. 37).

And yet I am sure that such a view as this must be mistaken. In the *first* place, we may seek explanations for the most familiar events as well as of those unfamiliar to us. We may ask why stones fall as well as why aeroplanes rise, and be curious for an answer equally in both cases. True, our motivation for asking the latter question is probably greater because the kind of phenomenon in question is (or was) less familiar; most people would not think to ask it about stones because the falling of stones is familiar and usual—but the question can as legitimately be asked in the one case as in the other. In the *second* place, the explanation may not be familiar at all: it may be far less familiar than the event to be explained. The discoloration of a painted kitchen wall when gas heat is used may be a familiar phenomenon to the housewife—surely more familiar than its explanation in terms of the chemical combination of sulphur in the gas fumes with elements in the paint, producing a compound that is dark in colour. Yet this is the true explanation. If the explanation is not familiar, one is tempted to say, it ought to be, as long as it is true. Surely its familiarity is irrelevant to its validity as an explanation. Familiarity is, in any case, a subjective matter—what is familiar to you may not be familiar to me; and yet the explanation, if true, is as true for me as for you.

The only grain of truth in the view that explaining is rendering familiar seems to be this: the law that does the explaining may not be familiar, *but* the fact that the phenomenon in question, such as the flight of an aeroplane, *can* be subsumed under a law—the fact that the behaviour *is* lawlike and hence predictable—tends to make it less mysterious, less like a miracle, and thus in a sense more familiar. To show that the behaviour of something is lawlike is to show it to be a part of the order of nature, and in that sense familiar, although the particular law or laws stating the uniformity may be quite unfamiliar.

In what, then, *does* explanation consist? The answer, I think, is quite simple: (3) to explain an event is simply to bring it under a law; and to explain a law is to bring it under another law. It does not matter whether the law is one about purposes or not, or whether it is familiar or not; what matters is that if the explanation is to be *true* (and we surely seek true explanations, not false ones), the law invoked must be true: indeed, this is already implied in the use of the word "law," which refers to a true, i.e. a really existing, uniformity of nature; if the uniformity turned out to be only imaginary, or having exceptions, we would no longer call it a law.

In saying that explanation is in terms of laws, I use the word "law" in a wider sense than is sometimes employed: in the sense I mean, any uniformity of nature is a law. Thus, it is a law that iron rusts, and it is a law that iron is magnetic—although both of these are usually listed in textbooks as "properties of iron" rather than as laws. In this sense, it seems to me that explaining why something occurs always involves a law. If we ask, "Why don't the two liquids in the flask mix?" and someone answers, "Don't you see, the one is transparent and the other is red?" this does not strike us as an explanation (i.e. as a true explanation) of the phenomenon, because we know of no law according to which red liquids will not mix with transparent ones. But when we are told that the red liquid is coloured water and that the transparent liquid is gasoline, we consider the phenomenon to be explained, for we hold it to be a law of nature that water and gasoline do not mix. In the sense in which I am using the word "law," the non-mixture of water and gasoline is a law; and *only* if a law is brought in do we have an explanation of the phenomenon.

Sometimes, I should add, all we have available is a "statistical law"—a law not of the form "All A is B" or "Whenever A, then B," but, e.g., "75 per cent of A is B." Can such a "law" constitute an explanation? I should be inclined to say that it is, although we would still want an explanation of why 25 per cent of A's are *not* B's. If water did not always boil at 212° F. but did so only 75 per cent of the time, we might explain the boiling of this kettle of water by saying that its temperature had reached 212°, though we would still want an explanation of why the kettle of water next to it, which had also reached 212°, did not boil. In other words, our statistical law would still not answer the question, "Why this and not that?" and in order to answer *this* question, we would need a non-statistical law of the form, "Under such-and-such conditions, water always boils at 212° F., but under such-and-such other conditions, it does not." It would seem, then, that a statistical law has in turn to be explained by a non-statistical one, although of course we may not, at any given stage in the progress of science, know of any non-statistical law by which to explain the statistical one.

Another example: "Why does Johnny have a cold?" "Because Johnny has been playing with Roger, and Roger has a cold." It is not a law that everyone who plays with someone who has a cold also gets a cold; the best we can do here is to state a percentage of cases in which this happens. So far as it goes, this is satisfactory; some uniformity is better than none. And yet, surely, we do not rest satisfied with this; we want to go on and ask why it sometimes happens but sometimes not. And the answer to this question would be a non-statistical law: "People always get colds under such-and-such conditions." Whether a statistical law can *always* be explained in terms of a non-statistical one depends not only on our powers of discovery but upon the nature of the universe. It is certainly no *a priori* truth that nature's uniformities are all of the 100 per cent variety instead of 75 per cent.

One further qualification: We have said that we explain particular events in terms of laws, and laws in terms of wider laws. But sometimes we give at least tentative explanations of them in terms not of laws but of general *hypotheses*: if a law is a well-established statement of how nature works, a statement about nature's workings that is not well established, or perhaps not even probable but only possible, cannot be a law. And yet we can use it to explain a law. But to whatever degree the hypothesis is uncertain, to that degree the explanation is jeopardized. An explanation cannot be known to be true if it involves a hypothesis which (by the definition of "hypothesis") is *not* known to be true. Whether the explanation is a true explanation, then, depends on the fate of the hypothesis. (In the "higher reaches" of most sciences, where the most general laws are involved, the only explanations possible are usually those in terms of very general hypotheses.)

III

So much for a general statement of what explanation consists of. I should like now to append some comments and to answer some questions to which the above account may give rise.

1. Thus far we have been content to answer the question "Why does A do B?" by saying "Because all A's do B." But there are those who say that such an answer is no explanation at all. "To say that all gases expand when heated," says Norman Campbell (*What Is Science,* p. 80), "is not to explain

why hydrogen expands when heated; it merely leads us to ask immediately why all gases expand. An explanation which leads immediately to another question of the same kind is no explanation at all."

I want to insist that the answer given *is* an explanation of the occurrence in question; to say "Hydrogen is a gas, and all gases expand when heated" is a perfectly satisfactory answer to the question why hydrogen expands when heated. But it is *not,* of course, an answer to *another* question—Why do all gases expand when heated?—and this is probably the question which the person meant to ask in the first place. These questions must not be confused with each other; I believe Campbell's position is the result of this confusion. It is fatally easy to telescope (unconsciously) two questions into one, and then be dissatisfied with the answer. Distinguishing them, we get:

Question 1. Why does this gas expand when heated?
Explanation. It is hydrogen, and hydrogen expands when heated.
Question 2. Why does hydrogen expand when heated?
Explanation. Hydrogen is a gas, and all gases expand when heated.
Question 3. Why do all gases expand when heated?

Here we attempt to give an explanation in terms of the kinetic theory of gases. To criticize Answer 1 because it is not an answer to Question 2, or Answer 2 because it is not an answer to Question 3, is surely a confusion. I want to say that Answer 1 is a perfectly satisfactory explanation for the phenomenon referred to in Question 1, though of course not for those referred to in Questions 2 and 3. But there is a frequent tendency to telescope these questions and demand to Question 1 the answer to Question 3.

The situation may be illustrated in another way. If I ask, "Why did the water-pipes in my basement burst last night?" someone may answer that it is because the basement got too cold, and another may answer that it is because water expands when it freezes, while yet another may say that we do not know the "real explanation" unless we can state why water expands when it freezes. Here, again we must separate the questions:

Question 1. Why did the water-pipes break?
Explanation. They always do when the temperature falls to below 32°.
Question 2. Why do they break when the temperature falls . . . etc.?
Explanation. Because the water in them expands when it freezes, and the water on expanding breaks the pipes.
Question 3. Why does water expand when it freezes?
Explanation. Here we try to answer in terms of the structure of the water-molecule.

But to say that we have not explained (1) until we have explained (3) is grossly to underestimate the number of phenomena for which we do have perfectly satisfactory explanations. That is, we *do* have explanations for (1) and (2), and our having them is *not* contingent upon having an explanation for (3).

We could put our point in another way. *Logically* the answers given to each question in turn are satisfactory explanations; but *psychologically* they may not be equally satisfying, *depending on the previous knowledge of the questioner.* To the questioner who knew nothing about the relation of pipes bursting to temperature, the answer "Because they got cold" (to the first question) would be psychologically quite satisfactory, but not to the person who already knew that it had something to do with temperature, for the question *he* meant to ask was (2) or (3). Again: If I ask why this wire conducts electricity, it is a perfectly good explanation to answer "Because it is made of copper, and copper is a conductor of electricity." Psychologically, however, this answer would not be equally satisfying to everyone; it *would* be to the person who knew nothing of the properties of copper (or who did not know that this wire was copper), but it would *not* be to the person who already knew the properties of copper but was really enquiring as to why copper, unlike many other substances, is a conductor of electricity.

2. Can an event have *two* explanations? Why not? Let us suppose that we want to explain an event E, and that we have a law saying that every time conditions A are fulfilled, E happens, and another law saying that every time conditions B are fulfilled, E happens. A will then be a complete explanation for the occurrence of E, and B will also be a complete explanation. Whether any such state of affairs actually occurs in the world is, of course, another question. Most of the suggested double explanations of events are in fact parts of a single explanation. Thus, for example, if we are asked to explain why the burglar committed the robbery last night, the detective may explain it in terms of his expertness at picking locks, the butler may explain it in terms of the family being out of the room, the maid may say it was because the bedroom window was open, the policeman may say it was because the night was foggy and visibility at a minimum, the sociologist may explain it in terms of the criminal's background of slum conditions, and the psychologist may explain it in terms of pseudoaggressive impulses dating from a childhood period marked by intense family quarrels. All these explanations are probably correct enough as far as they go. It may well be that in the absence of any one of these factors the burglary would not have occurred. But these are, it would surely seem, parts and aspects of *one* complete explanation—and in explaining human actions the whole explanation may be inconceivably complex. Still, the possibility remains that in *some* cases there may be two separate and complete explanations for an occurrence; at least it cannot be ruled out *a priori.*

3. Must there be a *deductive* relation between the thing to be explained and the explanation, such that one can deduce the statement of the phenomenon to be explained from the explanation?

All copper conducts electricity.
This wire is made of copper.
Therefore, This wire conducts electricity.

Here the explanation yields the desired conclusion easily, and it is quite clear that what we have here is a genuine explanation. The question is, must all explanation conform to this model? Have we failed to give an explanation if we have failed to deduce the explanandum from the explanation?

Let us first note that in many cases, if this is required, the explanation would be bewilderingly complex, and the premises in the deduction extremely numerous. Consider the burglary example just cited. From the fact that the weather was foggy and that the man had tendencies to steal and that he had a poor background . . . etc., we cannot deduce the fact that he committed the theft. We cannot deduce it, indeed, from any set of premises known to be true. What we need for deducing it is a law, to the effect that if such-and-such conditions are fulfilled an act of this kind will always occur, and then a minor premise to the effect that these conditions were in fact fulfilled. The conditions would indeed have to be extremely numerous, and the statement of the law immensely complicated. Yet such a law is required if the desired conclusion is to be deduced.

We never in fact use a deductive model in cases like this one, and it is worthy of note that we do not deny ourselves the claim that we have explained the event because of this. What, therefore, are we to say of the deductive model as a *sine qua non* for all explanation? As I see it, we have two alternatives open to us:

(*a*) We can, in the light of such examples, scrap the deductive model entirely. We can say that often one can in fact deduce the explanandum from the explanation, but that this is not essential to explanation. We might add, as some do, that to perform the deduction is one way (the best way?) to *justify* an explanation we have put forward, but that the giving of a true explanation is not dependent on this.

(*b*) We can still insist that a complete explanation does involve the deduction, but that what we often give is in fact less than a complete explanation. We list, as in the burglary example, a few salient facts and either take the remainder as too obvious to mention or do not know what they are. But such measures are concessions of failure. The fact is that the only way to be sure of our explanation is to deduce the phenemenon in question from premises which we know to be true.

I merely wish here to state these alternatives, not to decide between them. It is, surely, a matter of how liberally or how strictly we wish to use the term "explanation"; and, though I incline toward the second alternative, I do not wish to champion without reserve a "puristic" account of explanation which is not in fact followed by anyone—at least anyone in the psychological and social sciences—and which, it is sometimes declared, is in practice almost useless and boringly academic.

Thus far in enquiring about the need for a deductive relationship, we have considered only the explanation of particular events: we have deduced them from two premises, one stating a law and the other stating a particular condition: "All copper conducts electricity; this is copper, therefore this conducts electricity." "All water freezes at 32° F., the water in the pond went below 32° last night; therefore the water in the pond froze." And so on. But, as we saw earlier, we not only explain particular events; we also explain *laws.* And the same question could be repeated here: is the deductive requirement necessary? There is no doubt that in the "neat, tidy" cases it is fulfilled: for example, Kepler's laws of planetary motion can be deduced from Newton's laws of motion together with the law of gravitation; and thus the latter clearly explain the former. But is this strictly a requirement for *all* explanation of laws? Again, some would say that it is—that anything short of this is not a full explanation. Others would say that it is not—that the deductive case is only the ideal one but that explanation does not require it. For example, a law can be explained in terms of a very general theory, from which the law cannot be strictly deduced, but which will nevertheless entitle the theory to be called an explanation. (The deductivist will reply that it is not *known* to be an explanation until the acid test, i.e. the deduction, is performed.)

4. In any case, whether deducibility is a necessary condition of explanation or not, it is not a sufficient condition. One can deduce that this watch will not work from the premises that watches will not work if gremlins get into them and that gremlins are in fact in this watch. Yet no one would accept this as an explanation for the misbehaviour of the watch. Similarly, one might deduce it from the premises that whatever God wills happens and that God has willed the misbehaviour of this watch. One can deduce anything if one selects one's premises carefully.

One might remark at this point that it is also necessary that the premises be *true,* and that this is the required addition. I would unhesitatingly agree that the premises must indeed be true—false statements cannot form parts of true explanations (indeed, if explanation is in terms of law, and a law is a true statement of a uniformity, i.e. one that actually occurs, then this proviso has already been implicit in our account of explanation). But suppose we make this proviso explicit—is it enough? I do not believe so. It might be true that God wills everything that happens, but as long as we have no means of knowing this, we cannot use it as a premise in our explanation. That is, we cannot use it as an explanation unless the proposition is not only true, but is *known to be so.*

Suppose, then, that we accept this last revision—will it do the trick? I hardly think so; it still misses the main point. Let us imagine a deeply religious scientist who holds that everything that happens is the result of divine will; he may yet reject the theological explanation as an account of *why* things happen as they do. The reason is surely fairly obvious: what the scientist wishes to discover is why this happened *rather than that,* and the theological explanation will not enable him to make this discrimination: *whatever* happens, one can deduce it from the premises that God willed it to happen and that whatever He wills happens.

What condition, then, remains to be supplied? The condition seems to be a rather simple one, yet one which it is difficult to state precisely. What we have in mind is this: we want to eliminate the indiscriminate "explanatory" power of the gremlin-hypothesis and the God-hypothesis, even though they slip through the deductive net, because they do not enable us to explain why this happens *rather than that*. "What explains everything explains nothing."

This *can* be put by saying that the explanation must have *predictive* value, but this is a bit misleading. For one thing, it places undue emphasis upon the future, whereas explanation of past is just as important as explanation of future; we would have, then, to use a tenseless sense of "predict." For another thing, there are many explanations which seem to be true but whose predictive power is minimal or at any rate difficult to see: many biological phenomena can be explained in terms of laws of mutations, for example, but it is not clear what these laws enable us to *predict*—certainly not where or when a mutation will occur or what kind it will be when it does arise.

Perhaps what we want to say can be best expressed by the simple proviso that the explanation must explain *other* phenomena than those it is invoked to explain, and yet, unlike the God-hypothesis, not just everything indiscriminately: in other words, it should explain other events (whether past, present, or future makes no difference), but it should all the same be *capable of disproof* by empirical observations, whether or not any actual empirical observations ever disprove it, it must be capable of testing. Without this condition it would not be considered an explanation in any science.

In fact all this is implicit in our requirement that an explanation be in terms of law or laws. A law is a universal proposition about all events or processes in a certain class, and if it holds for A, a member of the class (a present event), it also holds for members B, C, and D (future events); thus by the very nature of a law, laws explain more than a single event. The testability of explanations is also implicit in the concept of law, for a law is an empirical statement of a uniformity of nature, and, being contingent, it is always subject to disconfirmation by observation. Still, it is well to make the implicit explicit to show why the deductive requirement is not enough and what more is required of an explanation.

5. In evaluating the extent to which proffered explanations yield us genuine empirical knowledge (i.e. are real empirical laws), much care is required, for in this field the verbal booby-traps in our way are numerous and intricate.

If someone asked, "Why is this object spherical?" and the reply were given, "Because it's globular," everyone would recognize the answer to be trivial because it is analytic. Many so-called explanations do not give much more information than this, although even very bad ones are not usually quite as empty as this one. Even when one says that opium produces sleep because of its dormitive power, we are at least told that it is because of something within it that sleep is produced, not by some outside factor such as the atmosphere. When we ask why hydrogen combines with oxygen to form water, and are told that it is because hydrogen has an *affinity* for oxygen, again the reply is relatively empty: it tells us only that under certain conditions hydrogen does combine with oxygen but tells us nothing of why hydrogen rather than some other substances does this; but at least we know from the answer that there *is* a law relating the combination of elements to some set of conditions, though we do not yet know what this law is. And if we ask why the mother cat takes care of her kittens and fights to defend them, and are told that it is because she has a *maternal instinct,* at least we know that the activity is not a learned one—and this is indeed something—although again the answer may not give us the kind of thing we were asking for. Most explanations in terms of instinct, tendency, affinity, power, and faculty are of this next-to-worthless kind, conveying only a minimum of information, and leading us to ask a why-question of the explanation given.

Let us observe how easily the invention of a name may make us assume that an explanation has been given. If it is asked, "Why is iron magnetic?" and we answer, "Because iron, cobalt, and nickel are magnetic," no one would think much of this as an explanation; but the moment we give a name to the behaviour of these metals, and call them, say, "fero-affinitive," then when someone asks why iron is magnetic, we can say, "Why, because it's a fero-affinitive metal, that's why." And yet no more has been said in the second case than in the first. Similarly, if we had a name for the tendency of seeds to sprout upwards to reach the surface of the ground, people would be readier to say that their tendency to rise could be *explained* by the presence of this property. Yet a name for what it does is a different thing from an explanation of why it does what it does.

Not all examples are as simple as this. When external influences tend to reduce or raise the bodily temperature of an organism, various bodily mechanisms come into play to return the temperature to normal. This is known as "homeostasis." So far, we simply have a name for the phenomenon, and if someone volunteered it as an *explanation* he would surely be mistaken. But now suppose a bird finds its nest partially destroyed and it sets about rebuilding it to the way it was before; we ask why, and are told, "That's the bird's homeostatic tendency." Now the name "homeostasis" is no longer merely a label for the temperature-controlling mechanisms; it relates these mechanisms to a quite different thing, the bird's attempt to restore the *status quo.* In both examples there is an attempt to restore a state which has ceased to exist. Is "homeostasis" now an explanation, or is it simply a description-in-a-nutshell, a *generalized* description, of what the organism does, without attempting to explain why?

Observe, incidentally, how easily all these so-called explanations slip through the deductive net. We can deduce the required conclusion easily: "When organisms have homeostatic tendencies, they do so-and-so. This organism has homeostatic tendencies. Therefore, it does so-and-so." The deductive requirement will let good and bad explanations alike slip through like water through a sieve. This shows us again that, whether necessary to explanation or not, the deducibility requirement is not sufficient.

But let us return: Is homeostasis an explanation of the organism's behaviour or not? Before we say, "No, it isn't," let us reflect on this point: if appeal to homeostasis is simply a short way of saying that birds do this and people do that, is not the appeal to gravitation simply a short way of saying that apples do this and stars do that? And yet the Law of Gravitation is one of the most sacred of our explanatory principles. Perhaps, as Wisdom says, talking about gravitation is simply a way of saying that apples fall *and so on*; but then is not homeostasis simply a way of saying that birds rebuild their nests *and so on*?

It is, of course, incorrect to say that apples fall because of gravitation, if we mean by this that gravitation is some animistic force or pull, just as it would be wrong to say that birds behave so-and-so because of homeostasis, if we mean it to be a separate force or magnetism within birds. If we are so tempted, it is both useful and important to say that each of the explanations referred to is simply a way of saying "this happens *and so on.*" But it is, I should think, the *extent and range* of the "and so on" that matters here. What gives the Principle of Gravitation its remarkable explanatory power is not its appeal to an occult force but its bringing together under one formula an enormous range of diverse and complex phenomena. Because of this range, and the exactitude with which it can be applied to widely separate phenomena, the Law of Gravitation is the classical case of a law having predictive power—and it is extremely doubtful whether homeostasis possesses or ever will possess this. We rest, then, once again with this second and all-important necessary condition of explanation (the first being, at least in common opinion, the deducibility requirement): its power to explain a wide range of phenomena *other* than those it was invoked to explain.

6. No mention has thus far been made of explanation in terms of *purpose*. And yet this is the oldest concept of explanation and still the one most frequently employed by primitive peoples. And there are contexts in which we still employ the concept of purpose in giving explanations—for example, when we say that my purpose in going to the store was to do some Christmas shopping, and that this is *why* I went.

The word "purpose" is, of course, ambiguous. (*a*) Most frequently in ordinary usage a purpose is something of which I am conscious—a conscious intent to do something. The conscious intent is not the *whole* of the purpose: part of the criterion of whether it is my purpose to do X is whether I am disposed towards doing X, whether I take steps towards X and do X if I have the chance. (*b*) Some tendencies to act are not accompanied by any state of awareness; and here psychologists speak of *unconscious* purposes. We need not stop here over the exact interpretation of this way of speaking; let us simply say that one is said to have X as his unconscious purpose if he consistently acts, without intending it, so as to bring about X. (*c*) We speak of inanimate objects as having purposes—for example, the purpose of a hammer is to drive nails. This of course is not a purpose consciously envisaged by the hammer. All we mean here is that the mechanical object *reflects* the conscious purposes of its makers. *We* had a conscious purpose in making the hammer, and thus we speak elliptically of the hammer as having that purpose. Strictly speaking, of course, the purpose is ours and not the hammer's.

In all of these cases a purpose implies a purposer, or someone to have the purpose. We do sometimes use the word "purpose" in another sense which carries no such implication, (*d*) when we say, "What is the purpose of the heart?" "To pump blood through the body." Here purpose simply means function—i.e. what does it *do*? what part does it play in the bodily economy? If the word "purpose" is used here I would view it as a "degenerate" usage—a misleading locution in which another word, "function," would serve much better. It is true that someone, in asking the purpose of the heart, might have in mind a theological question, "What purpose did God have in endowing us with this organ?" but if this is meant, we are back again to purpose in sense 1, in which purpose implies a purposer and the word "purpose" refers to conscious intent—the only difference now being that it is God's intent and not ours that is in question. But this, of course, is not what medical men generally have in mind when they ask purpose-questions about parts of organisms; else every such medical question would be a disguised theological question.

Having disentangled these senses of "purpose," let us ask about the legitimacy of purposive explanations. Briefly I think it comes to this: explanations require laws, and if there are laws *about* purposes, there is no reason why they cannot figure in some explanations just as laws about falling bodies figure in other explanations. To the extent that laws about purposes have been established, they can be used as explanations like any other laws. Unfortunately the only laws (if any) that we are in a position to make about purposes are about human ones. Explanations in terms of divine purposes cannot be employed because no laws about divine purposes have ever been established. Even explanations of biological events in terms of animal purposes is frowned upon: we do not count it an explanation if it is said that the hen sits on her eggs *in order* to hatch chicks, because we have no indication that the hen does so with this purpose in mind; even if this is true, we do not know it, and therefore we cannot use it as a law in our explanation. In the human realm alone we know that purposes exist, and only there can we therefore employ them in explanations. We can even deduce conclusions from them, thus:

> People act so as to fulfil their purposes, unless prevented by external
> circumstances. My purpose was to go shopping, and I was not
> prevented . . . etc. Therefore, I went shopping.

This way of putting it may sound rather silly, as the deductive model often does, but at any rate a deduction can be achieved from premises which are in all probability true.

The chief mistake which people are in the habit of making with regard to purposive explanation is probably that of wanting an answer to a why-question in terms of purpose when the conditions under which a purpose-answer is legitimate are not fulfilled. People extend their questioning unthinkingly from areas in which purposive explanation is in order into areas in which it is not. Thus: "Why did he go to New York?" "Well, in response to impulses from certain centres in his brain, some muscles in his arms and legs started moving towards the airport and . . ." "No, that's not what I mean. I mean, why did he go? what did he go for? what purpose did he have in view?" "He went in order to see some operas." Contrast this with the following: "Why did he die?" "Well, a bullet entered his lung, puncturing some blood vessels, and the blood filled his lung so that he couldn't breathe any more, and . . ." "No, that's not what I mean. I mean, *why* did he die?" But here we can no longer give an answer in terms of purpose—unless, that is, our talk is rooted in a theological context and we are willing to say that, just as the first person went to New York because he wanted to see operas, so the second person died because God had some purpose (intent) in seeing to it that he was murdered. If this is what is meant, one could try to answer the question in the theistic context of divine purposes; but if this context is rejected, the why-question demanding an answer in terms of purpose is meaningless, because an answer is being demanded when the only conditions under which the question is meaningful are not fulfilled.

This point is worth emphasizing because it is so often ignored in practice. Having received answers to why-questions when these questions were meaningful and explanations could be given, people continue to use why-questions even when they no longer know what they are asking for. One need not be surprised that no answer is forthcoming to such questions. And in our discouragement with such questions we are all too prone to make a mistake ourselves and terminate an exasperating series of why-questions with a remark such as, "That's just something we don't know," as if it were like cases where something definite is being asked but we do not yet know the laws which explain the phenomena we are asking about. If something in the case is not known, there must be something in the case which we could fail to know. If we are to ask a meaningful question, we must know what it is that we are asking for; only then can we recognize an answer as being one when we do find it.

7. This leads us directly into an important question, How far can explanation go? We may explain an event in terms of a law, and this law in terms of other laws, and so on? but must we not finally come to a stop? The bursting of the pipes is explained by the expansion of water on freezing; let us assume that water expands on freezing because the water-molecule has such-and-such a structure; now why does the water-molecule have this structure? Perhaps this can some day be explained by reference to electron-proton arrangements within the atom, and this in turn by reference to the disposition of more minute particles (if they can be called such) yet to be discovered; but sooner or later must we not say, "That's just the way things are—this is just an ultimate law about the universe. We can explain other things in terms of it, but it we cannot explain?" Are there ultimate laws, laws which explain but cannot even in principle be explained?

In practice we come rather quickly to laws which cannot be explained further. Laws about atomic structure are typical of such laws. Laws of psycho-physical correlation are another example. *Why* do I have a certain colour-sensation which I call red, indescribable but qualitatively different from all others, when light within a certain range of wave-length impinges upon my retina, and another indescribably different sensation which I call yellow when rays of another wave-length strike the retina? That this wave-length is correlated with this visual experience seems to be sheer "brute fact"—a law[1] which cannot be explained in terms of anything more ultimate than itself.

At the same time, we should be careful in dismissing any uniformity we cannot explain as a "brute fact" or "basic law." Many things, such as why this element has this melting-point and these spectral lines, were once considered basic and unexplainable properties of the element, but have since been explained in terms of the intra-molecular structure of the element. No matter how much at a loss we may be for an explanation, we can always ask and speculate. If it had been accepted as a basic law that water starts to expand when it gets below 39° F., we would never have gone on to discover anything about the structure of the water-molecule. Fruitful scientific procedure depends on assuming that no given law is basic; if scientists did not continue always to ask the question "Why?" this process of scientific enquiry would stop dead in its tracks.

Thus, if there *are* basic laws, it seems that we cannot know of any given law that it is one. We can know that it is *not,* by explaining it in terms of other laws; but how could we know that it *is*? Discovering basic laws is epistemologically similar to discovering uncaused events: if there are uncaused events, we can never know that there are, for all we can safely say is that we have not yet found causes for them.

One further point about basic or ultimate laws: If a law is really a basic one, any request for an explanation of it is self-contradictory. To explain a law is to place it in a context or network of wider and more inclusive laws; a basic law is by definition one of which this cannot be done; therefore to ask of an admittedly basic law that it be explained is implicitly to deny that it is basic and thus to deny the very premise of the argument. It is a request for explanation in a situation where by one's own admission no more explaining can be done.

Like so many others, this point may seem logically compelling but psychologically unsatisfying. Having heard the above argument, one may still feel inclined to ask, "Why are the basic uniformities of the universe the way they are, and not some other way? Why should we have just *these* laws rather than other ones? I want an *explanation* of why they are as they are." I must confess here, as an autobiographical remark, that I cannot help sharing this feeling: I want to ask why the laws of nature, being contingent, are as they are, even though I cannot conceive of what an explanation of this would be like, and even though by my own argument above the request for such an explanation is self-contradictory.[2] The fact is, as we saw above, that why-questions have had answers so many times that we tend automatically to ask them here even when they can have no answers because we have ripped them out of the only context in which they have meaning—like the situation of the child who, being told what is above the table and above the ceiling of his room and above the house and above the earth, now asks what is above the universe. The question has now gone outside the context of meaningful discourse, and so has the request for the explanation of a basic law. We should remember: to explain is to explain *in terms of something,* and if *ex hypothesi* there is no longer any something for it to be explained in terms of, then the request for an explanation is self-contradictory: it demands on the one hand that you explain X in terms of a Y while insisting simultaneously that there is no Y.

8. One sometimes encounters the complaint that science does not really explain but only describes. "Science doesn't tell us *why* things happen," it is said, "it only tells us *how* things happen." Now it does often happen that the exact intention of the user of a why-question is not very clear—as we have already seen. But in the way in which the term "why" is most commonly used, science *does* explain why: for example, the bursting of the pipes, the formation of ice at the top of ponds rather than at the bottom, and many other phenomena, are explained by reference to the law that water expands when it freezes. (If someone says we have *not* explained why the pipes burst, then what does he mean by "why"? What sort of thing is he asking for? What *would* answer his question? Let him state in other terms what it is that he is asking for.)

"But is not explanation after all merely description?" It is all very well to say that when we explain something we actually describe—*e.g.* stating laws of nature is describing how nature works. But this does not preclude the fact that we *are* explaining. When the question is asked why pressing the button turns on the light, we explain by describing just what goes on—currents, open and closed circuits, conduction of electricity by wires, dynamos in the power plant, and so on. But have we not in so doing explained the phenomenon about which we were asking? We have explained *by* describing, if you will; but certainly we have explained. To say that because we are describing we cannot be explaining would be like saying that because an object is red it cannot also be coloured.

9. A similar complaint is sometimes voiced against scientific explanation, that it "explains things *away.*" Explaining something is interpreted as equivalent to explaining it away. Now the precise meaning of the phrase "explaining away" is one which I have never been able to discover. What is one supposed to be doing when he explains something away? Surely not to declare that it does not exist! Explanation deprives us of no facts we had before. To "explain colour" in terms of light-waves is not, of course (as should have been obvious), to take away the fact of colour-experiences. "Thinking is nothing but the occurrence of certain neural impulses" should be changed into "*When* thinking takes place (and that it does is just as incontrovertible a fact as the neurons are), there are neural impulses."

In the special context of beliefs, perhaps "explaining away" may mean impugning the truth of one's conclusions. If so, there are again no grounds for fear. To "explain away" someone's politically reactionary tendencies by saying, "He's old, and people always get conservative when they get old," does not for a moment take away whatever truth the person's opinions may have; at most, it only exposes part of the causal genesis of his having them. And if the views of this person were "explained away" by these biographical observations, the views of his opponent would be equally vulnerable: "You needn't pay any attention to that young upstart, they're all hot communists when they're young." Reference to biography may, together with laws of human nature (if any are known in this area), explain why a person held a certain belief at a certain time, but the truth or falsity of the belief is quite unaffected by this and, of course, is tested on different grounds entirely. The idea that reference to a person's mental or physical condition could "explain away" the truth of a belief is one of the most flagrant blunders of the materialistically minded laity of our day.

NOTES

1. A law which would, to be sure, have to be qualified to take care of abnormal cases, *e.g.* colour-blindness, jaundice, etc. The genesis of colour-sensations is complex and does not depend *merely* upon the kind of light-rays entering the eye.

2. Explanation in terms of divine purposes again will not help: if we are told that the laws of nature are as they are because God willed it so, we can ask why He should have willed it so; and if here again an answer is given, we can once again ask a why-question of this answer.

6

Law

John Hospers

By means of sense-experience we learn many things about the physical world—we perceive countless physical things, processes, and events, as well as the interaction of our own bodies with these things in nature. But if our knowledge ended there, we would have no means of dealing effectively with the world. The kind of knowledge we acquire through the sciences begins only when we notice *regularities* in the course of events. Many events and processes in nature occur the same way over and over again. Iron rusts, but gold does not. Chickens lay eggs, but dogs do not. Lightning is followed by thunder. Cats catch mice, but cows don't. (Even to speak of a cat or a cow is to have noted some regularity—that some characteristics regularly recur, or go together.) Amidst the constant diversity in our daily experience of nature, we try to find regularities: we trace "the thin red vein of order through the flux of experience."

If we were as interested in discovering *ir*regularities in our experience as we are in regularities, the task would be much easier. Some rocks are hard and some soft, some heavy and some light. Some rains are helpful, some ruinous. Some people are tall, some short. If *all* experiences were like this, we would not know what to expect next: each new situation would confront us as if no past situations had ever occurred, and years of experience would give us no hint about the way future events would occur. But nature is not like this; nature does contain regularities, difficult though they sometimes are to find.

Why are we interested in tracking down these regularities? Not, as a rule, because we enjoy contemplating them for their own sake but because we are interested in *prediction*. If we can rely on it that when we see a twister in the sky a tornado is approaching, then if we see one we may be able to take precautions by finding shelter before it strikes. If people who are in proximity to others who have colds get colds themselves, we may keep Johnny from getting a cold by keeping him temporarily away from Billy, who has a cold. We want some basis for prediction, so that we will not always be taken by surprise at the next series of events with which nature confronts us. And often when we can predict, we can also *control* the course of events; at least we are in a better position to control if we can first make a reliable prediction. We can reliably predict eclipses, but we cannot control their occurrence; but in many cases we can control as a result of our prediction: if we can predict reliably that after heavy rains the river will flood, we can get out of the way of the flood, or even (if it happens repeatedly) build a dam.

Most of the regularities that we find have many exceptions: they are not *invariants.* There is a certain regularity to children getting colds when they play with other children who already have colds, but it doesn't always happen that way. Chickens lay eggs and never cans of sardines, but how many eggs they do lay and at what intervals is extremely variable. Trees are more likely to fall during a severe storm, but they don't always: some do and some don't. The scientific enterprise could be described as the search for genuine invariants in nature, for regularities without exception, so that we are enabled to say, "Whenever such-and-such conditions are fulfilled, this kind of thing *always* happens." Many times we think we have found a genuine invariance, but we have not. We may have been sure that water always boils at 212° F., since we have tried it many times and it has always happened. But if we try it on a mountain top, we discover that the water there boils at a slightly lower temperature, so our hope that we had found a true invariance is upset. We try some more, however, and find that the temperature of water boiling depends not on the moisture in the air, not on the time of day, not on anything except the pressure of the surrounding air. We are thus able to say, "Water at the pressure found at sea level boils at 212°. Here at last we have a statement of genuine invariance; and behold, we have a *law of nature.*

Prescriptive v. descriptive laws. The word "law" is ambiguous, and the ambiguity can be extremely misleading if we are not aware of it.

(1) In daily life we most often use the word "law" in the context of "passing a law," the law prohibits you from . . . ," and so on. Law in this sense is *prescriptive*: it is a rule of behavior imposed by a monarch or passed by a legislative body, and enforced by the legal machinery of the state. Laws in this sense are not propositions, because they cannot be false (it is, however, true or false *that* certain laws have been passed); they are, rather, imperatives, in effect "Do this," "Don't do that." The law does not state that anything *is* the case; rather, it issues a command, a *pre*scription, usually with penalties attached for failure to obey. But this is not the sense of "law" that is involved in speaking of laws of nature.

(2) Laws of nature are *descriptive*: they describe the way nature works. They do not prescribe anything: Kepler's laws of planetary motion do not prescribe to the planets that they should move in such-and-such orbits, with penalties invoked if they fail; rather, Kepler's laws *de*scribe how planets actually *do*

move. Laws in this sense describe certain uniformities that exist in the universe. Sometimes, for the sake of simplicity, they describe only what would happen under certain ideal conditions: Galileo's Law of Falling Bodies describes only the velocities at which bodies fall in a vacuum. But such a law is still descriptive: it describes our universe (not any logically possible universe), and it prescribes nothing. Only conscious beings can prescribe, since only they are capable of giving orders. But the uniformities of nature would still occur even if there were no human beings to describe them.

Several confusions can be avoided if we keep this distinction in mind. (1) "Laws should be obeyed." Whether or not you should obey all the laws of the land is a problem in ethics. But a law of nature is not the sort of thing you can obey or disobey, since it is not an order or command anyone has given. What could you do if someone said to you, "Obey the Law of Gravitation?" Your motions, along with those of stones and every particle of matter in the universe, are *instances* of this law; but since the law only tells us how matter *does* behave, and cannot prescribe how things *should* behave, you cannot be said either to obey or disobey it. A prescriptive law, moreover, could still be said to exist even if it were universally disobeyed. (2) "Where there's a law, there's a lawmaker." Again this applies clearly to prescriptive law: if a course of action is prescribed, someone must have prescribed it. But the same consideration does not apply to laws of nature. Did someone make the planets move in a certain way? . . . It is sufficient to observe that "law implies lawmaker" is not a necessary proposition in regard to descriptive law as it is with prescriptive law. (3) "Laws are discovered, not made." This applies only to descriptive laws: we *discover* how nature works, we do not make it work that way. But statute laws are made, devised, passed by human beings in positions of authority. Such laws do not exist but for human beings, but laws of nature would—that is, the uniformities of nature would exist whether men were there to observe them or not, although the *formulation* of these uniformities is the work of men.

Laws of nature constitute a smaller class of propositions than empirical statements in general. Any statement whose truth can be tested by observation of the world is an empirical proposition. "Some chickens lay eggs," "World War I lasted from 1914 to 1918," "She fell ill with pneumonia yesterday," and "New York City contains approximately 8 million residents" are all empirical statements. Indeed, most of the statements we utter in daily life are empirical statements. But none of these is a law of nature: laws of nature are a special class of empirical statements. Since laws of nature are the very basis of the empirical sciences—physics, chemistry, astronomy, geology, biology, psychology, sociology, economics—it is important that we be clear about the principal defining characteristics of laws of nature.

The meaning of "law of nature." What, then, is a law of nature? What requirements must an empirical proposition fulfill in order to be a member of that select class of propositions which we call laws of nature?

1. It must be a true *universal* empirical proposition. To say that a proposition is universal is to say that it applies to *all* members of a given class without exception. That all iron rusts when exposed to oxygen is a universal proposition, but that *this* piece of iron rusts, or even that *some* iron rusts, is not a universal proposition.

a. A proposition about a single thing—"This piece of rock is metamorphic"—may be *material* for a law of nature, but it is not a law. Science does not consist of such singular propositions. Books on physics, the most developed of the empirical sciences, make no reference (except by way of example) to the motions of particular bodies, nor do chemistry books tell us about this piece of lead or that vessel of chlorine. But one does find many such references in psychology books (psychiatry division), for example in case-histories of patients. In this area few genuine laws have yet been discovered, so the psychologist must rely on individual case-histories as a means toward finding laws of human behavior. In this sense, psychology is still very much in a pre-scientific stage, a stage physics had already passed out of three centuries ago. But physics is in an advantageous position in that its laws are *simpler*—not in the sense of "easier to understand," for physics is more difficult for most students than any of the other empirical sciences, but in the sense that a law of physics can be stated in terms of the smallest number of conditions. In stating the velocity at which objects fall, one can ignore most of the universe: one can ignore the color of the object, its smell or taste, the temperature of the environment, the number of people watching the event, and so on for thousands of factors. By contrast, in dealing with human behavior it would be difficult to say what might *not* turn out to be relevant. A trivial event that occurred in your childhood, which neither you nor anyone else may remember, may still influence your behavior today and cause you to react differently to a given stimulus. The best we can do, usually, is to state certain general tendencies of human behavior, allowing for many exceptions. In psychology we hardly have laws at all, only tentative blueprints for laws; laws about human behavior that are both true and exceptionless have seldom been found.

The obvious examples of "laws of human nature" that come to mind all turn out on examination to be analytic. "People always act from the strongest motive" sounds like a plausible candidate for a law of human nature: people do a tremendous variety of things, but whatever they do, don't they always do what their strongest motive impels them to do? Waiving the fact that people don't always act from motives (they sometimes act from habit), the uncomfortable fact is that there seems to be no way of specifying what is meant by "strongest motive" except as the motive from which one acts. Thus: one acts from the motives from which one acts—true, but analytic. Similarly, "People always do what they most want to do" is either synthetic but false or true but analytic, depending on what one means by the sentence: in a familiar sense, we all do things (like coming to class) that we don't want to do; we often perform unpleasant chores even though we hate them. If one says even in these cases that we always do what we want to do, we must mean "want" in some unusual sense—and indeed this sense is not far to seek: for the only criterion for knowing what we "really want" to do turns out to be what we actually do. So once again we have "We do what we do," which is true but analytic.

The universal proposition constituting the law must, then, be an empirical truth: it must not be analytic. "All A is B" is true in the case of "All triangles have three sides," but this statement, being analytic, is not a law of nature. Nor is "All gold is yellow," if being yellow is considered a defining characteristic of gold; in that case it would have to be yellow in order to be called gold in the first place, and the statement would be analytic. But if gold is defined by other means (such as atomic number), then it is a law of nature that everything having this property is also yellow. The B in "All A is B" must be connected with the A as a matter of contingent fact, not a priori or of necessity, if "All A is B" is to count as a law of nature.

b. Even true propositions about *some* members of a class are not usually considered laws of nature, though sometimes they are given the honorary title of "statistical laws." If 90 per cent of the A's there are are B's, there is a considerable regularity between the two, and the statement is far from useless as a basis for prediction. But, we are led to ask, if only 90 per cent of A's are B's and the remaining 10 per cent are not, why are the 90 per cent B's and not the others? What we want to find is some uniformity of a universal character underlying the statistical one. In daily life, however, we are constantly confronted by such regularities that are not universal: People with a cold usually have the sniffles, but not always; if one person hits another in the nose, the second person often gets a nosebleed, but not always. We have not yet formulated any universal statement about the precise conditions under which people get nosebleeds when struck on the nose, though we have a fairly adequate idea on what factors it depends. There is some regularity here (the harder you hit him, the more likely he is to get a nosebleed, and so on), but no invariant relationship.

2. These universal propositions are hypothetical in form. Now, universal propositions, both in logic and in science, are usually interpreted *hypothetically*—that is, as propositions of the "if . . . then . . ." form. "All iron rusts when exposed to oxygen" would thus be translated as "*If* there is iron, it will rust when exposed to oxygen." Thus formulated, the proposition does not tell us that there *is* any iron (it makes no existential claim), but only what happens under certain circumstances *if* there is. "All bodies freely falling in a vacuum accelerate at the rate of 16 feet per second per second" does not imply that there actually were or are any bodies falling in a vacuum. "At 99.9 per cent of the velocity of light, organisms grow old far more slowly than do those traveling at slower velocities" is a universal proposition that scientists believe to be true, but no one would declare that any organism is now traveling at that velocity.

The hypothetical interpretation of laws, however, can get us into trouble. The hypothetical "If p is true, then q is true" in logic is equivalent to "It is not the case that p is true, but q is false." For example, "If there is friction, then there is heat" (a law of nature) would be translated "It is not the case that there is friction but no heat." But now let us take the proposition "All unicorns are white." This is translatable into "It is not the case that there is a unicorn that is not white." And since there are no unicorns the proposition is true: there are no nonwhite unicorns, for the excellent reason that there are no unicorns at all. Moreover, by the same reasoning, "All unicorns are green" would also be true, since there are no nongreen unicorns. When *p* is false, then anything whatever follows, so we can put what we like into *q*.

Of course, "All unicorns are white" would never be counted as a law of nature; yet it is a universal proposition, construed as a hypothetical. Why would the proposition about unicorns not be considered a law, whereas the propositions about friction and organisms at almost the velocity of light are so considered? The difference lies in the fact that there is evidence from *other* laws that these laws are true. Indeed, the proposition about aging more slowly as one approaches the velocity of light is a logical consequence of (deducible from) Einstein's relativistic laws of time, whereas the proposition about unicorns is connected with no laws at all.

But even this is not sufficient to characterize laws of nature.

3. There are many true, universal propositions of hypothetical form that do not pass as laws of nature. Suppose I were to say, "All the dogs in this kennel are black," and that my statement were true: it would still not qualify as a law of nature. It is limited to a definite area in space and time—this kennel today. Even if its scope were broader ("All the dogs I've ever had in my kennels are black"), it still would say nothing about *all* dogs, or even all dogs of a certain breed. But if I say that all crows are black, I mean that all crows, *wherever* they may be, and *whenever* they may exist or have existed or have yet to exist, are, were, and will be black. (Blackness is not here considered defining of crows, else the proposition would be analytic.) The law is "open-ended": it has an infinite range, both in time and in space. This does not mean that there is an infinite number of crows—nor indeed that there are any crows at all—but that it is an *open class,* with *no strictures of time and space* operating to limit the scope of the law. There is no time or place at which the law will not hold true: considerations of when and where are irrelevant to the application of the proposition. By contrast, the proposition about the dogs in my kennel will not pass as a law because (1) though universal in form, the universality is restricted to a specifically delimited time and space; (2) the number of things covered by the proposition is not only finite, but this finiteness may be inferred from the terms in the proposition itself; this is not so for Kepler's laws of planetary motion, for example: though there are a finite number of planets, this fact cannot be deduced from the law; and (3) the evidence for the proposition exhausts its domain of application—the proposition is simply a summative report of what *has been* observed to be the case.

Since laws of nature apply to all places and all times, their claim extends into the future. This is perhaps the most important single feature of laws, for it enables them to be made the basis for prediction. If the proposition merely read "All crows thus far have turned out to be black," one might say "So what?"—we could not deduce any predictions from it; but if we say that *all* A's, no matter when or where, are B's, we can deduce from it, plus the proposition that this is an A, that it will also be a B.

4. But even when all these conditions are met, a proposition may not be classified as a law of nature. "All crows are black" is unrestricted in time, place, and number of individuals in its domain of application. Yet it would generally not be counted as a law, because the only evidence for it is direct evidence, and a proposition is not usually accorded the status of a law unless there is some indirect evidence for it. This requires a word of explanation.

The laws of any science are not viewed in independence of one another. Together they form a vast body or system of laws, with each law fitting into a system including many other laws, each mutually reinforcing the others. The laws that scientists are most loath to abandon are those that form such an integral part of a system of laws that the abandonment of the one law would require the abandonment or alteration of a large number of other laws in the system. Thus an observation that directly confirms one law indirectly confirms a group of other laws, because of the interconnection of the laws in a system. (Physics is, again, the most systematized science. Biology was not very highly systematic until the present century: until then, it was for the most part a classificatory science, recording the properties of various species of creatures but discerning no interconnectedness among them. It was in much the state of chemistry before the rise of atomic theory.)

"All crows are mortal" is supported by much indirect evidence: the mortality of organisms in general, the biochemical deterioration of tissue, increase in auto-allergenic response, and so on. But "All crows are black" seems to relate to no other significant regularities, of either greater or lesser generality. A crow that was not black would change no other laws known to us; but a crow that was immortal (or even one that lived a thousand years) would excite considerable scientific surprise, because it might force us to reconsider the many other laws (about deterioration of tissue, etc.) with which it is interlocked. Whether or not something is called a law, then, depends to a large extent on how deeply embedded it is in a wider system of laws. A true universal proposition for which there was no indirect evidence would have little fundamentality in science: it could easily be abandoned without effect on the rest of the system. But "All metals are good conductors" is so fundamentally tied to other laws (of atomic structure) that a counterinstance would have far-reaching consequences.

5. Even when all this has been said, many propositions that satisfy all these criteria are often denied the status of laws. The difference seems to lie in the proposition's degree of *generality*. "All metals are good thermal conductors" and "Silver is a good thermal conductor" are both universal propositions, since both apply to all members of the given class; but the first statement is more general than the second, for it covers a wider scope. Universal propositions whose degree of generality is greater are more likely to pass as laws; thus

the statement about metals is considered a law, but the statement about silver is not. While "All rare-earth metals have higher melting points than the halogens" may pass as a law, one would not likely hear the fact that tungsten melts at 3,370° C. referred to as a law but only as a fact.

Sometimes one of these conditions works against another, and the outcome is not certain. Einstein refers to the constancy of the velocity of light in a vacuum as a law of nature, and this is referred to as a law, in spite of its limited generality, because of the fundamentality of this item in the system of physical laws. On the other hand, while the mass of the electron is an elemental fact of physical science, its precise value remains largely independent of the main body of scientific theory, and therefore it is not accorded the status of a law. . . .

7

Towards an Aristotelean Theory
of Scientific Explanation

B. A. Brody

Let us begin by considering the following explanation of why it is that sodium normally combines with chlorine in a ratio of one-to-one[1]:

(A) (1) sodium normally combines with bromine in a ratio of one-to-one
(2) everything that normally combines with bromine in a ratio of one-to-one normally combines with chlorine in a ratio of one-to-one

---

(3) therefore, sodium normally combines with chlorine in a ratio of one-to-one.

This purported explanation meets all of the requirements laid down by Hempel's covering law model for scientific explanation ([5], pp. 248–249). After all, the law to be explained is deduced from two other general laws which are true and have empirical content. Nevertheless, this purported explanation seems to have absolutely no explanatory power. And even if one were to say, as I think it would be wrong to say, that it does have at least a little explanatory power, why is it that it is not as good an explanation of the law in question (that sodium normally combines with chlorine in a ratio of one-to-one) as the explanation of that law in terms of the atomic structure of sodium and chlorine and the theory of chemical bonding? The covering law model, as it stands, seems to offer us no answer to that question.

A defender of the covering law model would, presumably, offer the following reply: both of these explanations, each of which meets the requirements of the model, are explanations of the law in question, but the explanation in terms of atomic structure is to be preferred to the explanation in terms of the way that sodium combines with bromine because the former contains in its

112

explanans more powerful laws than the latter. The laws about atomic structure and the theory of bonding are more powerful than the law about the ratio with which sodium combines with bromine because more phenomena can be explained by the former than by the latter.

I also find this answer highly unsatisfactory, partially because I don't see that (A) has any explanatory power at all. But that is not the real problem. The real problem is that this answer leaves something very mysterious. I can see why, on the grounds just mentioned, one would prefer to have laws like the ones about atomic structure rather than laws like the one about the ratio with which sodium and bromine combine. But why does that make explanations in terms of the latter type of laws less preferable? Or to put the question another way, why should laws that explain more explain better?

So much for my first problem for the covering-law model, a problem with its account of the way in which we explain scientific laws. I should now like to raise another problem for it, a problem with its account of the way in which we explain particular events. Consider the following three explanations:

(B) (1) If the temperature of a gas is constant, then its pressure is inversely proportional to its volume
    (2) at time t_1, the volume of the container c was v_1, and the pressure of the gas in it was p_1
    (3) the temperature of the gas in c did not change from t_1 to t_2
    (4) the pressure of the gas in container c at t_2 is $2p_1$

    (5) the volume of x at t_2 is $\frac{1}{2}v_1$.

(C) (1) if the temperature of a gas is constant, then its pressure is inversely proportional to its volume
    (2) at time t_1 the volume of the container c was v_1 and the pressure of the gas in it was p_1
    (3) the temperature of the gas in c did not change from t_1 to t_2
    (4) the volume of the gas in container c at t_2 is $\frac{1}{2}v_1$

    (5) the pressure of c at t_2 is $2p_1$.

(D) (1) if the temperature of a gas is constant, then its pressure is inversely proportional to its volume
    (2) at time t_1, the volume of the container c was v_1 and the pressure of the gas in it was p_1
    (3) the temperature of the gas in c did not change from t_1 to t_2
    (4) by t_2, I had so compressed the container by pushing on it from all sides that its volume was $\frac{1}{2}v_1$

    (5) the pressure of c at t_2 is $2p_1$.

All three of these purported explanations meet all of the requirements of Hempel's model. The explanandum, in each case, is deducible from the explanans which, in each case, contains at least one true general law with empirical content. And yet, there are important differences between the three. My intuitions are that (B) is no explanation at all (thereby providing us with a clear counter-example to Hempel's model), that (C) is a poor explanation, and that (D) is a much better one. But if your intuitions are that (B) is still an explanation, even if not a very good one, that makes no difference for now. The important point is that there is a clear difference between the explanatory power of these three explanations, and the covering law model provides us with no clue as to what it is.

These problems and counter-examples are not isolated cases. I shall give, later on in this paper after I offer my own analysis and solution of them, a recipe for constructing loads of additional problems and counter-examples. Now the existence of these troublesome cases led me to suspect that there is something fundamentally wrong with the whole covering law model and that a new approach to the understanding of scientific explanation is required. At the same time, however, I felt that this model, which fits so many cases and seems so reasonable, just couldn't be junked entirely. This left me in a serious dilemma, one that I only began to see my way out of after I realized that Aristotle, in the *Posterior Analytics,* had already seen these problems and had offered a solution to them, one that contained both elements of Hempel's model and some other elements entirely foreign to it. So let me begin my presentation of my solution to these problems by looking at some aspects of Aristotle's theory of scientific explanation.

Aristotle (*ibid.,* I, 13), wanted to draw a distinction between knowledge of the fact (knowledge that p is so) and knowledge of the reasoned fact (knowledge why p is so) and he did so by asking us to consider the following two arguments, the former of which only provides us with knowledge of the fact while the latter of which provides us with knowledge of the reasoned fact:

(E) (1) the planets do not twinkle
(2) all objects that do not twinkle are near the earth

   (3) therefore, the planets are near the earth.

(F) (1) the planets are near the earth
(2) all objects that are near the earth do not twinkle

   (3) therefore, the planets do not twinkle.

The interesting thing about this point, for our purposes, is that while both of these arguments fit Hempel's model,[2] only one of them, as Aristotle already saw, provides us with an explanation of its conclusion. Moreover, his account of why this is so seems just right:

> . . . of two reciprocally predictable terms the one which is not the cause may quite easily be the better known and so become the middle

term of the demonstration. . . . This syllogism, then, proves not the reasoned fact but only the fact; since they are not near because they do not twinkle. The major and middle of the proof, however, may be reversed, and then the demonstration will be of the reasoned fact . . . since its middle term is the proximate cause. (78ᵃ 28–78ᵇ 3)

In other words, nearness is the cause of not twinkling, and not vice versa, so the nearness of the planets to the earth explains why they do not twinkle, but their not twinkling does not explain why they are near the earth.

It is important to note that such an account is incompatible with the logical empiricists' theory of causality as constant conjunction. After all, given the truth of the premises of (E) and (F), nearness and nontwinkling are each necessary and sufficient for each other, so, on the constant conjunction account each is equally the cause of the other.[3] And even if the constant conjunction account is supplemented in any of the usual ways, nearness and nontwinkling would still equally be the cause of each other. After all, both "all near celestial objects twinkle" and "all twinkling celestial objects are near" contain purely qualitative predicates, have a potentially infinite scope, are deducible from higher-order scientific generalizations and support counterfactuals. In other words, both of these generalizations are lawlike generalizations, and not mere accidental ones, so each of the events in question is, on a sophisticated Humean account, the cause of the other. So Aristotle's account presupposes the falsity of the constant conjunction account of causality. But that is okay. After all, the very example that we are dealing with now, where nearness is clearly the cause of nontwinkling but not vice versa, shows us that the constant conjunction theory of causality (even in its normal more-sophisticated versions) is false.

Now if we apply Aristotle's account to our example with the gas, we get a satisfactory account of what is involved there. The decrease in volume (due, itself, to my pressing on the container from all sides) is the cause of the increase in pressure, but not vice versa, so the former explains the latter but not vice versa. And Aristotle's account also explains a phenomenon called to our attention by Bromberger [2], viz. that while we can deduce both the height of a flagpole from the length of the shadow it casts and the position of the sun in the sky and the length of the shadow it casts from the height of the flagpole and the position of the sun in the sky, only the latter deduction can be used in an explanation. It is easy to see why this is so; it is the sun striking at a given angle the flagpole of the given height that causes its shadow to have the length that it does, but the sun striking the flagpole when its shadow has the length of the flagpole is surely not the cause of the height of the flagpole.

Generalizing this point, we can add a new requirement for explanation: a deductive-nomological explanation of a particular event is a satisfactory explanation of the event when (besides meeting all of Hempel's requirements) its explanans contains essentially a description of the event which is the cause of the event described in the explanandum. If they do not then it may not be a satisfactory explanation. And similarly, a deductive-nomological explanation of a law is a satisfactory explanation of that law when (besides meeting all of Hempel's requirements) every event which is a case of the law to be explained is caused by an event which is a case of one (in each case, the same) of the laws contained essentially in the explanans.[4]

It might be thought that what we have said so far is sufficient to explain why it is that (A) is not an explanation and why it is that the explanation of sodium and chlorine's combining in a one-to-one ratio in terms of the atomic structure of sodium and chlorine is an explanation. After all, no event which is a case of sodium and chlorine combining in a one-to-one ratio is caused by any event which is a case of sodium and bromine combining in a one-to-one ratio. So, given our requirements, deduction (A), even though it meets all of Hempel's requirements, need not be (and indeed is not) an explanation. But every event which is a case of sodium and chlorine combining in a one-to-one ratio is caused by the sodium and chlorine in question having the atomic structure that they do (after all, if they had a different atomic structure, they would combine in a different ratio). So an explanation involving the atomic structure would meet our new requirement and would therefore be satisfactory.

The trouble with this account is that it incorrectly presupposes that it is the atomic structure of sodium and chlorine that causes them to combine in a one-to-one ratio. A whole essay would be required to show, in detail, what is wrong with this presupposition; I can, here, only briefly indicate the trouble and hope that this brief indication will be sufficient for now: a given case of sodium combining with chlorine is the same event as that sodium combining with that chlorine in a one-to-one ratio, and, like all other events, that event has only one cause.[5] It is, perhaps, that event which brings it about that the sodium and chlorine are in proximity to each other under the right conditions. That is the cause of the event in question, and not the atomic structure of the sodium and chlorine in question (which, after all, were present long before they combined). To be sure, these atomic structures help explain one aspect of the event in question, the ratio in which they combine, but that does not make them the cause of the event.[6]

To say that the atomic structure of the atoms in question is not the cause of their combining in a one-to-one ratio is *not* to say that a description of that structure is not an essential part of any causal explanation of their combining. It obviously is. But equally well, to say that a description of it is a necessary part of any causal explanation is *not* to say that it is (or is part of) the cause of their combining. There is a difference, after all, between causal explanations and causes and between parts of the former and parts of the latter. Similarly, to say that the atomic structure is not the cause of their combining is *not* to say that that event had no cause; indeed, we suggested one (the event which brought about the proximity of the atoms) and others can also be suggested (the event of the atoms acquiring certain specific electrical and quantum mechanical properties). It is only to say that the atomic structure is not the cause.

Keeping these two points in mind, we can see that all that we said before was that the perfectly satisfactory explanation, in terms of the atomic structure of the atoms, of their combining in a one-to-one ratio does not meet the condition just proposed because it contains no description of the event which caused the combining to take place. But since it obviously is a good explanation, some additional types of explanations must be allowed for.

So Aristotle's first suggestion, while quite helpful in solving some of our problems, does not solve all of them. There is, however, another important suggestion that he makes that will, I believe, solve the rest of them. Aristotle says:

> Demonstrative knowledge must rest on necessary basic truths; for the object of scientific knowledge cannot be other than it is. Now attributes attaching essentially to their subjects attach necessarily to them. . . . It follows from this that premises of the demonstrative syllogism must be connections essential in the sense explained: for all attributes must inhere essentially or else be accidental, and accidental attributes are not necessary to their subjects. (*Posterior Analytics* 74^b 5–12)

There are many aspects of this passage that I do not want to discuss now. But one part of it seems to me to suggest a solution to our problem. It is the suggestion that a demonstration can be used as an explanation (can provide us with "scientific knowledge") when at least one of the explanans essential to the derivation states, that a certain class of objects has a certain property, and (although the explanans need not state this) that property is possessed by those objects essentially.

Let us, following that suggestion, now look at our two proposed explanations as to why sodium combines with chlorine in a ratio of one-to-one. In one of them, we are supposed to explain this in terms of the fact that sodium combines with bromine in a one-to-one ratio. In the other explanation, we are supposed to explain this in terms of the atomic structure of sodium and chlorine. Now in both of these cases, we can demonstrate from the fact in question (and certain additional facts) that sodium does combine with chlorine in a one-to-one ratio. But there is an important difference between these two proposed explanations. The atomic structure of some chunk of sodium or mass of chlorine is an essential property of that object. Something with a different atomic number would be (numerically) a different object. But the fact that it combines with bromine in a one-to-one ratio is not an essential property of the sodium chunk, although it may be true of every chunk of sodium. One can, after all, imagine situations[7] in which it would not combine in that ratio but in which it would still be (numerically) the same object. Therefore, one of our explanans, the one describing the atomic structure of sodium and chlorine, contains a statement that attributes to the sodium and chlorine a property which is an essential property of that sodium and chlorine (even if the statement does not say that it is an essential property), while the other of our explanans, the one describing the way in which sodium combines with bromine, does not. And it is for just this reason that the former explanans, but not the latter, explains the phenomenon in question.

Generalizing this Aristotelean point, we can set down another requirement for explanations as follows: a deductive-nomological explanation of a particular event is a satisfactory explanation of that event when (besides meeting all of Hempel's requirements) its explanans contains essentially a statement attributing to a certain class of objects a property had essentially by that class of objects (even if the statement does not say that they have it essentially) and when at least one object involved in the event described in the explanandum is a member of that class of objects. If this requirement is unfulfilled, then it may not be a satisfactory explanation. And similarly, a deductive-nomological explanation of a law is a satisfactory explanation of that law when (besides meeting all of Hempel's requirements) each event which is a case of the law which is the explanandum, involves an entity which is a member of a class (in each case, the same class) such that the explanans contain a statement attributing to that class a property which each of its members have essentially (even if the statement does not say that they have it essentially).

It is important to note that such an account is incompatible with the logical empiricist conception of theoretical statements as instruments and not as statements describing the world. For after all, many of these essential attributions are going to be theoretical statements, and they can hardly be statements attributing to a class of objects an essential property if they aren't really statements at all. But that is okay, for it just gives us one more reason for rejecting an account, more notable for the audacity of its proponents in proposing it than for its plausability or for the illumination it casts.

There are two types of objections to essential explanations that we should deal with immediately. The first really has its origin in Duhem's critique of the idea that scientific theories explain the observable world ([3], Ch. 1). Duhem argued that if we view a theory as an explanation of an observable phenomenon, we would have to suppose that the theory gives us an account of the physical reality underlying what we observe. Such claims about the true nature of reality are, however, empirically unverifiable metaphysical hypotheses, which scientists should shun, and therefore we must not view a theory as an attempt to explain what we observe. Now contemporary theories of explanation, like the deductive-nomological model, avoid this problem, by not requiring of an explanation that its explanans describe the true reality underlying the observed explanandum. But if we now claim that a deductive-nomological explanation is a satisfactory explanation when (among other possibilities) its explanans describe essential properties of some objects involved in the explanandum event, aren't we introducing these disastrous, because empirically undecidable, issues about the true nature of the reality of these objects into science? After all, the scientist will now have to decide, presumably by nonempirical means, whether the explanans do describe the essence of the objects in question.

The trouble with this objection is that it just assumes, without any arguments, that claims about the essences of objects would have to be empirically undecidable claims, claims that could be decided only upon the basis of metaphysical assumptions. This presupposition, besides being unsupported, just seems false. After all, the claim that the essential property of sodium is its atomic number (and not its atomic weight, or its color, or its melting point) can be defended empirically, partially by showing that for this property, unlike the others just mentioned, there are no obvious cases of sodium which do not have it, and partially by showing how all objects that have this property behave alike in many important respects while objects which do not have this property in common do not behave alike in these important respects. Now a lot more has to be said about the way in which we determine empirically the essence of a given object (or a given type of object), and we will return to this issue below, but enough has been said, I think, to justify the claim that the idea that scientific explanation is essential explanation does not mean that scientific explanation involves empirically undecidable claims.

It should be noted, by the way, that this idea of the discovery of essences by empirical means is not new to us. It was already involved in Aristotle's theory of *epagoge* (*op. cit.,* II, 19). I do not now want to enter into the question as to exactly what Aristotle had in mind, if he did have anything exact in mind, when he was describing that process. It is sufficient to note that he, like all other true adherents to the theory of essential explanation, saw our knowledge of essences as the result of reflection upon what we have observed and not as the result of some strange sort of metaphysical knowledge.

The second objection to essential explanations has been raised by Popper. He writes:

> The essentialist doctrine that I am contesting is solely the doctrine that science aims at ultimate explanation; that is to say, an explanation which (essentially, or, by its very nature) cannot be further explained, and which is in no need of any further explanation. Thus my criticism of essentialism does not aim at establishing the nonexistence of essences; it merely aims at showing the obscurantist character of the role played by the idea of essences in the Galilean philosophy of science. ([6], p. 105)

Popper's point really is very simple. If our explanans contain a statement describing essential properties (e.g., sodium has the following atomic structure . . .), then there is nothing more to be said by way of explaining these explanans themselves. After all, what could we say by way of answering the question "why does sodium contain the atomic structure that it does?" so the use of essential explanations leads us to unexplainable explanans, and therefore to no new insights gained in the search for explanations of these explanans, and therefore to scientific sterility. Therefore, science should reject essential explanations.

There are, I believe, two things wrong with this objection. To begin with, Popper assumes that essential explanations will involve unexplainable explanans, and this is usually only partially true. Consider, after all, our explanation of sodium's combining with chlorine in a one-to-one ratio in terms of the atomic structure of sodium and chlorine. The explanans of that explanation, besides containing statements attributing to sodium and chlorine their essences (viz. their atomic number), also contain the general principles of chemical bonding, and these are not unexplainable explanans since they are not claims about essences. In general, even essential explanations leave us with some part (usually the most interesting part) of their explanans to explain, and they do not therefore lead to sterility in future enquiry. But, in addition, even if we did have an essential explanation all of whose explanans were essential statements and therefore unexplainable explanans, what are we to do according to Popper? Should we reject the explanation? Should we keep it but believe that it is not an essential explanation? Neither of these strategies seem very plausible in those cases where we have good reasons both for supposing that the explanation is correct and for supposing that the explanans do describe the essential properties of the objects in question. It cannot after all, be a good scientific strategy to reject what we have good reasons to accept. So even if Popper's claim about their sterility for future scientific enquiry is true for some essential explanations, I cannot see that it gives us any reasons for rejecting these explanations, or for rejecting their claim to be essential explanations, when these explanations and claims are empirically well supported.

There is, of course, a certain point to Popper's objection, a point that I gladly concede. As is shown by his example from the history of gravitational theory, people may rush to treat a property as essential, without adequate empirical evidence for that claim, and then it may turn out that they were wrong. They may even have good evidence for the claim that the property is essential and still be wrong. In either case, enquiry has been blocked where it should not have been blocked. We should certainly therefore be cautious in making claims about essential properties and should, even when we make them on the basis of good evidence, realize that they may still be wrong. But these words of caution are equally applicable to all scientific claims; the havoc wreaked by false theories that lead enquiry along mistaken paths can be as bad as the havoc wreaked by false essential claims that block enquiry. And since they are only words of caution, they should not lead us to give up either theoretical explanations in general or essential explanations in particular.

Let us see where we now stand. We have, so far, rejected Hempel's requirements for an explanation on the grounds that they are not sufficient and we have suggested two alternative Aristotelean conditions such that, for the set of explanations meeting Hempel's requirements, any explanation meeting either of these requirements is an adequate explanation.[8] . . .

It only remains, therefore, to consider the one serious objection to this whole Aristotelean theory, an objection that we have already touched upon when we dealt with Duhem. Given what we mean by 'causality' and 'essence,' can we ever know that e_1 is the cause of e_2 or that P is an essential property of e_1, and if so, how can we know this?

This problem can be sharpened considerably. There is no problem, in principle, about our coming to know that events of type E_1 are constantly conjoined with events of type E_2. All that we have to do is to observe that this is so in enough varied cases. And if 'e_1 causes e_2' only means that 'e_1 is an event of type E_2,' we can easily see how we could come to know that e_1 is the cause of e_2. But if, as our Aristotelean account demands, 'e_1 causes e_2' means something more than that, can we know that it is true, and if we can, how can we know that it is true? Similarly, there is no problem, in principle, about our coming to know that objects of type O_1 have a certain property P_1 in common. All that we have to do is to observe that this is so in enough varied cases. And if 'P_1 is an essential property of o_1' only means that 'o_1 is an object of type O_1 and all objects of type O_1 have P_1,' we can easily see how we could come to know that P_1 is an essential property of o_1. But if, as our Aristotelean theory demands, 'P_1 is an essential property of o_1' means something more than that, can we know that this is true, and if we can, how can we know that this is true?

There is an important difference between these questions. If we conclude that we cannot, or do not, know the truth of statements of the form 'e_1 causes e_2' or 'P_1 is an essential property of o_1' (where these statements are meant in the strong sense required by our theory), then our theory must be rejected. After all, knowledge of the truth of statements of that form is, according to our theory, a necessary condition for knowing that we have (although not for having) adequate explanations. And we obviously do know, in at least some cases, that a given explanation is adequate. So if we cannot, or do not, know statements of the above form, our theory is false. However, if we conclude that we can, and do, know the truth of statements of the above-mentioned form, but we don't know how we know their truth, then all that we have left is a research project, viz. to find out how we know their truth; what we don't have is an objection to our theory.

This is an extremely important point. I shall show, in a moment, that we do have, and a fortiori can have, knowledge of these statements. But, to be quite frank, I have no adequate account (only the vague indications mentioned above when talking about Duhem) of how we have this knowledge. So, on the basis of this last point, I conclude that the Aristotelean theory of explanation faces a research problem about knowledge (hence the title of this paper), but no objection about knowledge.

Now for the proof that we do, and a fortiori can, have knowledge of the above-mentioned type. Our examples will, I am afraid, be familiar ones. It seems obvious that we know that

(1) if the temperature of a gas is constant, then an increase in its pressure is invariably accompanied by an inversely proportional decrease in its volume

(2) if the temperature of a gas is constant, then a decrease in its volume is invariably accompanied by an inversely proportional increase in its pressure

(3) if the temperature of a gas is constant, then an increase in its pressure does not cause an inversely proportional decrease in its volume

(4) if the temperature of a gas is constant, then a decrease in its volume does cause an inversely proportional increase in its pressure.

Here we have causal knowledge of the type required, since the symmetry between (1) and (2) and the asymmetry between (3) and (4) show that the causal knowledge that we have when we know (3) and (4) is not mere knowledge about constant conjunctions. Similarly, it seems obvious that we know that

(1) all sodium has the property of normally combining with bromine in a one-to-one ratio

(2) all sodium has the property of having the atomic number 11

(3) the property of normally combining with bromine in a one-to-one ratio is not an essential property of sodium

(4) the property of having the atomic number 11 is an essential property of sodium.

Here we have essential knowledge of the type required, since the symmetry between (1) and (2) and the asymmetry between (3) and (4) show that the essential knowledge that we have when we know (3) and (4) is not mere knowledge about all members of a certain class having a certain property.

I conclude, therefore, that we have every good reason to accept, but none to reject, the Aristotelean theory of explanation sketched in this paper. And I also conclude that it therefore behooves us to find out how we have the type of knowledge mentioned above, the type of knowledge that lies behind our knowledge that certain explanations that we offer really are adequate explanations.

NOTES

1. I first called attention to the problems raised by this type of explanation in my [1].

2. Leaving aside the question, irrelevant for us now, about the truth of these premises, we shall throughout this discussion just assume that they are true.

3. Unless one adds the requirement that the cause must be before the effect, the normal way of drawing an asymmetry between causes and effects when each are necessary and sufficient for the other, in which case neither is the cause of the other and Aristotle's account still won't do.

4. We have made this condition sufficient, but not necessary, for reasons that will emerge below. It will also be seen there that Aristotle, who had a broader notion of cause, could have made it necessary as well.

5. To be sure, e_1 can have as causes both e_2 and e_3 (where $e_2 \neq e_3$) but only when either e_2 is the cause of e_3 or e_3 the cause of e_2. That exception is of no relevance here.

6. It might, at least, be maintained that they are still the cause of that aspect of the event. But that is just a confusion—it is events, and not their aspects, that have causes.

7. Even ones in which all currently believed scientific laws hold, but in which the initial conditions are quite different from the ones that now normally hold.

8. We have not, however, required as a necessary condition that any explanation must meet one of these two conditions. This is so, partially because of the problem of statistical explanations, but partially because of the possibility, raised by Aristotle, that there are additional types of explanations. After all, our two conditions let in explanations in terms of Aristotle's efficient and formal causes. We still have to consider, but will not in this paper, possible explanations in terms of what he would call material and final causes.

REFERENCES

[1] Brody, B. A. "Natural Kinds and Real Essences." *Journal of Philosophy* (1967).

[2] Bromberger, S. "Why Questions." *Introductory Readings in the Philosophy of Science.* Edited by B. A. Brody. Englewood Cliffs: Prentice Hall, 1969.

[3] Duhem, P. *La Théorie Physique, Son Objet et sa Structure.* Paris: Chevalier et Rivière, 1914.

[4] Eberle, R. A., Kaplan, D., and Montague, R. "Hempel and Oppenheim on Explanation." *Philosophy of Science* (1961).

[5] Hempel, C. G. *Aspects of Scientific Explanation.* New York: Free Press, 1955.

[6] Popper, K. R. *Conjectures and Refutations.* London: 1963.

8

Historical Understanding as Re-thinking

W. H. Dray

The theory of historical understanding set forth in R. G. Collingwood's *Idea of History*[1] has produced some sharp divisions of opinion among philosophers of history. While the "nays" have usually been written off by their opponents as insensitive positivists, obsessed by a model of inquiry derived from the natural sciences, the "ayes" have been considered, in their turn, as woolly-minded idealists, unable to distinguish between the essential logical structure of an enquiry and its psychological and methodological frills. Even those who have accepted Collingwood's doctrine, however, have often found it difficult to say exactly what they think it commits them to. For *The Idea of History* is an irritating, as well as an exciting book. It is full of paradoxes and apparent contradictions, which in some cases appear to be not entirely unconnected with a certain contempt which Collingwood from time to time displayed towards his philosophical opponents (indeed, his readers do not always escape): a contempt well exemplified by his remark, at one point in response to the objections of an imaginary opponent: "I am not arguing; I am telling him" (p. 263). In many cases, no doubt, a kinder explanation might be sought; for the book was left unfinished at its author's death, and some parts of it apparently record the results of fairly raw, if vigorous and stimulating, reflection.

Since some of the papers incorporated into *The Idea of History* were published by Collingwood during his lifetime, however, it seems reasonable to regard them as especially authoritative on points in his general thesis which have been disputed. The best short summary of his theory of understanding is, in fact, to be found in one of these: a lecture entitled "Human Nature and Human History," which Collingwood delivered before the British Academy in 1936. The theory which he there elaborates, in the space of a few pages (pp. 213-15), could be reduced to the following three propositions: first, that human action, which is the proper concern of history, cannot be described *as* "action" at all, without mentioning the thought which it expresses—it has, in Collingwood's terms, a "thought-side"; second, that once the thought in question has been grasped by the historian, the action is understood in the sense appropriate to actions, so that it is unnecessary to go on to ask for the cause which produced it, or the law which it instantiates; third, that the understanding of action in terms of thought requires the re-thinking of the thought in question by the historian, so that, in essence, all history is "the re-enactment of past thought in the historian's own mind." No doubt there is much more to Collingwood's account than can be stated in such a condensed, schematic way. But the propositions stated do seem to contain the core of the theory.

In what follows, I propose to discuss each of Collingwood's three propositions in turn, clarifying it where that is possible, amending it where that seems necessary. The discussion cannot, of course, hope to deal with more than a few of the many objections which have been advanced against his theory in recent years. But it may perhaps serve to clear the ground a little for a sympathetic consideration of what he had to say.

I

That history is concerned with human actions, perhaps few would want to dispute; but that actions necessarily have what Collingwood calls a "thought-side" may not pass quite so easily without challenge. For exactly what is meant here by "thought"? And how is the relation between such thought and the action itself to be conceived?

It must be admitted that Collingwood's treatment of these questions is often puzzling. In the paper entitled "The Subject Matter of History," for example, it seems to be his view that the thought-side of an action is an activity of reflection[2]—as if, in order to act, an agent must first consider what to do, and then act in accordance with his reflection. If history is said to be concerned exclusively with human actions, its field of study, on such a view of the nature of action, must appear ridiculously narrow. Even a sympathetic commentator like W. H. Walsh has compared Collingwood unfavourably with the German philosopher, Willhelm Dilthey, in his apparent attempt to confine the historian's attention to "intellectual operations."[3] And Arnold Toynbee, in one of his latest volumes, visibly smarting under Collingwood's opinion of his own vast enterprise as a kind of "pigeon-holing," has expressed great amusement at Collingwood's supposed ignorance of the fact that politicians, whose activities he had represented as especially suitable to historical investigation, behave in precisely the "intellectually horrifying way" which seems, for him, to place a subject matter out of the reach of the historian's techniques.[4] Such naïvete, Toynbee maliciously suggests, could be found only in a philosopher's "theorizing."

If the paper on subject matter contained Collingwood's only discussion of the thought-side of actions, it would be difficult to escape the conclusion that, as a philosopher, he sets up a definition of history so restrictive and arbitrary that it would rule out much of what he himself wrote as an historian. But the paper in question, which is one of the least finished ones, must be weighed carefully against what is said in other parts of *The Idea of History*. If this is done, I believe that a view of the relation between thought and action will emerge which is not open to the charge of intellectualism in any damaging sense, although it safeguards, nevertheless, the distinction which Collingwood was so anxious to maintain, between historical and scientific modes of understanding.

A hint of the rather innocuous character of Collingwood's supposed intellectualism might be gleaned from a remark he makes in discussing the weaknesses of Graeco-Roman historiography. If we look back "over our actions, or over any stretch of past history," he tells us, it becomes obvious that "to a very great extent people do not know what they are doing until they have done it, if then" (p. 42). That we may nevertheless *discover,* by careful inquiry, what they were doing—including its thought-side—is clearly implied in the British Academy Lecture already referred to, in which Collingwood argues that the historical understanding which we achieve of the actions of people remote in time from ourselves is identical in kind with that which we obtain every day of the actions of our friends and neighbours. We discover their thoughts by considering their actions. The difference between an action and a mere physical movement, according to Collingwood, is that the former is the "outward expression" of a thought.[5] We discover the thought by interpreting the action *as* an expression.

It might perhaps be argued that such an account of the relation of thought to action is quite compatible with the claim that when we recognize an action as expressing a certain thought, this is equivalent to treating it as evidence that some prior activity called "reflection" must have gone on, although, since the agent's "stream of consciousness" is not open to us, we are unable to verify this directly. But when Collingwood is found to extend his analysis to knowledge of our *own* thoughts, such an interpretation of his meaning becomes very unplausible. In "Human Nature and Human History" Collingwood begins his discussion with the claim that historical inquiry can provide us with a kind of self-knowledge which a natural science of human nature fails to afford. The conclusion which his argument eventually reaches is that *all* knowledge of the human mind is historical. "It is only by historical thinking," he declares, "that I can discover what I thought . . . five minute ago, by reflecting on an action that I then did, which surprised me when I realized what I had done" (p. 219). Elsewhere he makes the same point by distinguishing between memory knowledge of myself and autobiography—the latter being the discovery, by the techniques of the historian, of the thoughts actually expressed in my past actions, by contrast with what, at the time of acting, I (perhaps mistakenly) assumed my thoughts to be (pp. 295–96). As his equally celebrated successor in the chair of metaphysical philosophy at Oxford University, Professor Gilbert Ryle, was to put it: Self-knowledge is obtained, not by *introspection,* but by *retrospection.*[6]

Why does Collingwood insist that knowledge of even my own thoughts must come by a process of interpreting "expressions," after the fact? It is surely because he would agree with Ryle and Wittgenstein that having a certain thought is not, in essence, a matter of reciting certain propositions to oneself, or focusing certain images on one's internal cinema screen, or "going over" in any such way what one is about to do.[7] The being of a thought is in its expression; and thoughts are expressed in actions as well as in those internal monologues and private screenings which are associated with reflecting and planning. Covert activities of this sort cannot, on Collingwood's account, be regarded as thinking *by contrast with* overt ones. The fact that thought has *inward* expressions as well as outward ones does not give the former any privileged status as "the thought." If this is so, there is no reason to assume that reflection—however brief—must precede, or even accompany, an action for the latter to be properly regarded as an expression of thought, and hence as having a thought-side. For if this were Collingwood's view, he could scarcely claim that our knowledge of the thought-side of our own actions is by historical analysis, after the fact, since we have privileged access to the stream of consciousness in which the activity of reflection presumably goes on.

In claiming that *all* knowledge of mind is historical, however, Collingwood appears to have overstated his case. For if we allowed this claim, the following difficulty would arise. The doctrine, as he states it, is that if, in fact, I think a certain thought, x, at a certain time, t, I cannot know that I think x at t, but only at, say, t plus five minutes, when I can retrospect what I did and experienced at t. But if I say at t *plus 5* I can know, by historical inquiry, that at t I thought x, then I am claiming to know what I think, not only at t, but also at t *plus 5*; and on Collingwood's theory this is something I cannot justifiably do. To know what I think at t *plus 5*, I must make still another historical investigation at a later time, and so on ad infinitum. The regress which opens up would, of course, make it impossible to claim that I could *ever* know what I think at any time whatever. To stop the regress, it is necessary to claim that at some time or other I can know *all the time* what I think. Collingwood appears to have recognized this in the paper entitled "History as Re-enactment of Past Experience," when, having asked himself whether "a person who performs an act of knowing" knows "that he 'is performing or has performed' that act," concludes that he can know both (p. 292).

But although Collingwood apparently abandons, in the end, his claim that all knowledge of mind is historical, what he says about self-knowledge in the passage quoted from the British Academy Lecture makes it clear that he believes some to be. And if it is, it will be unplausible to contend that what we discover by interpreting "expressions" is an activity of reflection which we engaged in at some earlier time without being aware of it. The claim that Collingwood did not confine the thought-side of actions to "intellectual operations" gains additional support from what he says at many other points. In discussing, for example, the historian's investigation of the mind "of a community or an age"—a case where deliberate planning or reflecting would be out of the question—we are told once again that the problem is essentially one of the historical interpretation of "expressions" (p. 219). And elsewhere, in attacking what he calls "scissors and paste" conceptions of the use of documents, Collingwood claims that it is possible for the historian to discover not only what has been completely forgotten—that is, is nowhere recorded in any extant document—but also "what, until he discovered it, no one ever knew to have happened at all" (p. 238). Since, for Collingwood, historical facts always concern actions, and actions always have a thought-side, the conclusion would appear to follow that historians can discover *thoughts* which were unknown, not only to any contemporary eyewitness of the actions concerned, but even to the agents themselves. And it seems to me that this claim is perfectly justified.

It may be worth adding that, on the interpretation of Collingwood's meaning which I have argued for here, even thought*less* actions may express a thought, and hence be properly regarded as having a thought-side. For to act thoughtlessly is not necessarily to act to no purpose; it is to act, rather, without taking into account certain considerations which ought to have been taken into account. Similarly, impulsive actions are in no way extruded from the historian's concerns on the ground that they lack a thought-side; for although to do a thing on impulse is incompatible with doing it as a matter of policy, or in order to implement a plan, it is not necessarily to do it for no reason. W. H. Walsh has distinguished usefully, in this connection, between saying that a person acted with something *before his mind,* and saying that he had something *in mind*; and he points out that the historian is as interested in cases falling into the second category as into the first.[8] In arguing, however, that actions in the second category may also be said to express a thought, Walsh offers what he regards as an amendment required to bring Collingwood's theory to terms with historical inquiry as it now exists. What I have tried to show is that such cases are already provided for by a theory of mind and thought which there are grounds for believing that Collingwood actually held. In arguing thus, I am, of course, appealing without much scruple from Collingwood drunk to Collingwood sober; from "The Subject Matter of History" to "Human Nature and Human History," on the assumption that the latter can be considered as the more authoritative. That the criticisms of Walsh and Toynbee find justification in the contents of the former paper, I should not attempt to deny.

II

If we grant the foregoing analysis of what it is that the historian, in order to know another's thought, must set himself to discover, what are we to say of Collingwood's second proposition: that once the thought has been discovered, the action can be understood without our having also to know the law under which it falls, or the cause which brought it about? Collingwood expresses his claim in characteristically provocative fashion when he asserts that, in history, once we know *what* happened, we already know *why* it happened (p. 214). He says this, presumably, because what happened is an action, and this, on his view, includes an explanatory thought-side. But does the thought of the agent really have this explanatory force? Can it provide us by itself with a *complete* explanation of what was done?

In asking ourselves whether an action can be completely or satisfactorily explained in terms of the agent's thought, it will be helpful, at the outset, to remind ourselves of a peculiar feature of the question "Why?"—namely, that a person can go on asking it as long as he pleases. Any theory of the nature of explanation, therefore, which does not rule out in advance objections arising from this possibility, can never hope to gain acceptance by a really determined opponent. Thus, if a historian explains Caesar's crossing of the Rubicon by referring to his determination to oust Pompey from the capital, it cannot be taken as a defect of Collingwood's theory that this explanation is incomplete—if by this we mean only that we can still ask why Caesar wanted to get rid of Pompey. It is not a defect of Collingwood's theory because, if a defect at all, it would be a defect of *any* theory of what counts as an answer to the question "Why?" Explanations can be regarded as given at successive levels of inquiry; and the farther we carry the questioning process, the deeper the explanation may be said to be. What Collingwood has to say, however, can be fairly assessed only as a theory of what counts as an answer to the question "Why?" at a single level.

What we must ask, therefore, is whether a thought has full explanatory force—by contrast with great explanatory depth—even at one level of inquiry. And it will be seen at once that if we merely cited Caesar's goal or purpose in crossing the Rubicon, the "thought" in question would not achieve such explanatory completeness by itself. To achieve this we should have to fill in many other things which, in the course of his narrative, a historian would ordinarily be content to leave implicit or understood. We should have to take account of such other thoughts as, for example, Caesar's belief that Pompey was in Rome and would remain there; and before we were through, we should have collected a sizable and complex group of what might be called the considerations which moved Caesar to action. But although Collingwood would certainly admit that the original explanation, if criticized on the ground of its incompleteness, would have to be added to in some such way, he would insist that such addition need not refer beyond the action itself to laws or causes. The explanation is completed by rounding out the thought-side, not by adding something of an altogether different kind.

But what is the criterion of completeness, in this second, and relevant sense of the term? Collingwood himself speaks of the explanatory thought as what "moves" or "determines" the agent to act.[9] And I think it would do no violence to his meaning if we said that until the thought-side is rounded out to the point where we can see that it *necessitated* the action, no complete explanation will have been given. Perhaps this will appear, at first sight, to be too rigorous a demand. But upon reflection, I think we should have to agree that a set of considerations which shows only that a certain course of action was one of a number which *might* have been pursued by the agent, yet falls short of showing that he *had* to choose the one he did, would not explain his doing what he did. No doubt such considerations would explain something. They might explain, perhaps, how it was possible for the agent to have done what he did. But they could scarcely be regarded as answering the question "Why?"

Unfortunately, Collingwood was not as careful as he should have been to draw such distinctions between kinds of explanatory problems. In the passage of the British Academy Lecture already referred to, he gives as an example of an explanatory thought, Caesar's "defiance of Republican law"; and although this does indeed appear to be a thought which the crossing of the Rubicon could be said to express, it is difficult to see how it could be an answer, or even part of an answer, to the question: "Why did Caesar cross the Rubicon?" Caesar *could,* of course, have crossed it in order to defy Republican law; but if his purpose had been only to carry out a demonstration of this sort, there would have been no need for him to march on Rome. The thought elicited in this case seems to me to be much more plausibly represented as an answer to an entirely different sort of question: a question about the *significance* of Caesar's action.[10] Our concern here, however, is with "explanation why." The fact that, in Collingwood's example, "defiance of Republican law" does not appear to have been one of the considerations which "moved" Caesar to action, is therefore irrelevant to our assessment of the claim that a complete set of explanatory "thoughts" would be one which represented what was done as necessarily done.

If it is still not clear that this is what Collingwood is saying, the fault may lie in a certain lack of precision which should be noticed in the claim that once the thought-side of an action has been discovered, the conditions of understanding have been satisfied. For, strictly speaking, an action cannot be explained in terms of its own thought-side. As Collingwood originally states his theory, the thought-side of an action is something that must be included in the very *specification* of the action (p. 213). The action is not a "doing" at all, but simply a "movement," unless it is an expression of thought. The thought which is required to make it the action it is, therefore, cannot be considered as something logically distinct from it, by reference to which the action itself can be explained. The thought-side of an action might, of course, be called upon to explain the mere movement which is its own other "side": We might say, for instance, that Caesar's wanting to get to the other shore explains his physical progress across the river. But it is clear enough that in history, the problem will almost always be the explanation of actions, not mere movements; and in order to explain an action in terms of thought it will be necessary to refer beyond the action to a thought which is not itself part of the action as specified.

It would always be possible, of course, to incorporate this *further* thought into the specification of the action, thus expanding its thought-side. If we explained Caesar's crossing of the Rubicon, for example, in terms of his wanting to oust Pompey from the capital, we might say that his crossing expressed the latter thought as well as his merely wanting to get to the other shore. But in assessing the applicability of a theory of explanation or understanding, it is important to keep the questions, as well as the answers, straight; and it should be clear that if we allow such incorporation, we can no longer claim that Caesar's wanting to oust Pompey explains *the action*; for the action to be explained is now, not the crossing of the Rubicon, but the crossing of the Rubicon to oust Pompey. To insist on such logical niceties may appear unduly pedantic. Yet unless they are recognized, Collingwood's theory of explanation in terms of thought lies open to the charge—commonly made—that he believes human actions, by contrast with natural events, to be *self*-explanatory. And in spite of the misleading way in which he sometimes states the "inside-outside" theory, it is difficult to believe, in the light of his general view of historical thinking, that he intended to make any such claim.

If a complete explanation of an action is said to be achieved when a set of considerations, or "thoughts," not themselves included in the specification of the action, can be seen to have necessitated it, the question arises whether Collingwood can maintain the distinction between his own account and that of his positivist opponents. For it is a central doctrine of the positivists whom he attacks—a doctrine derived ultimately from Hume's classical discussion of causation—that necessities asserted of the real world are read into it on the basis of a belief in general laws. On such a view, the only way to vindicate my claim that, given x, y necessarily happened, is to argue: "Whenever x then y"; so that to say that the historian's explanation represents the action as necessitated by the thought which explains it, would commit the historian to the view that it does so by virtue of some law. Now Collingwood, at this point, is indeed questioning this positivist assumption quite directly: But it is important to see exactly how he is doing it. For his frequent remarks about the limited applicability of the generalizations which might be derived from a study of human activities in any historical period, and his denial that man's reason is itself subject to laws at all, may make it appear that his point is simply that human actions cannot be explained on the model of the natural sciences because they are not necessitated.[11] His point would, in fact, appear to be quite otherwise. For, as a study of "Human Nature and Human History" will confirm, his claim is rather that, whether "positive" laws of human nature are discoverable or not, they are not *required* for historical understanding (pp. 214, 223). They are not required because human actions can be understood when they are seen to have been necessitated in an entirely different way.

The issue between Collingwood and his opponents thus turns on the question: "How can an action be necessitated except in terms of natural law?" Collingwood's answer would be that it can be necessitated in the sense of its being rationally required. A set of antecedent conditions which explains a consequent event by virtue of a law of nature, shows the event to have been necessary in the sense of being "the thing to have expected, the laws of nature being what they are." The thoughts or considerations which explain an action in Collingwood's context of discussion show the action to have been necessary in the sense of being "the thing to have done, the principles of reason being what they are."[12] We could put the point by saying that the necessity which is required for the explanation of action in history, according to Collingwood, is a *rational* rather than a *natural* necessity. If something happens in spite of natural necessity, we call it a miracle. If an action is done in spite of rational necessity, we call it a stupidity, a mistake, an irrationality. It is Collingwood's claim that if, and only if, rational necessity can be shown, then we understand what the agent did. And he adds that such understanding does not require the further demonstration, by the methods of natural science, that what happened was a natural necessity as well.

The exact role which the concept of rational necessity plays in Collingwood's theory will appear more clearly when we go on to investigate the notion of "re-thinking." But a word should perhaps be added first about the concept of causation; for it may seem, once again, that Collingwood is simply flying in the face of the facts when he says that historians are not, or should not be, interested in the causes of actions. But his claim turns out, on closer examination, to be only that historians—their concept of understanding being what it is—do not need to discover *natural* causes for the actions they claim to understand. That historians commonly and legitimately use causal language, he does not attempt to deny. But he regards the historian's use of the term as a special one; for when it is said that it was his wanting to get rid of Pompey that caused Caesar to march on Rome, the word *cause* here implies a rational connection of the sort already examined, between what Caesar did and the considerations said to explain it (pp. 214–15). Collingwood himself speaks at various points of a thought not only causing an action but also determining or inducing it. But I think it is clear that he does not regard these locutions as reintroducing the concept of natural necessitation in terms of general laws of human mind or behaviour, which his general theory of historical understanding has ruled out as irrelevant.

III

We have still to examine Collingwood's third proposition: that discovering the explanatory thought requires its re-thinking in the historian's mind. From some critics, this feature of the theory has encountered the objection that such re-thinking is impossible; from others the complaint that it is unnecessary. It seems to me that neither objection is valid if Collingwood's point is taken in its proper sense. Let me therefore attempt, once again, to separate what I think is the basic doctrine from the misconceptions which may perhaps arise out of unfortunate forms of expression.

At least one of the reasons for questioning the possibility of re-thinking disappears in the light of our discussion of Collingwood's general conception of thought. For if thinking is logically distinguished from private imaginings and recitations, and if it is allowed that a person may sometimes not know what he thinks, there is no need to boggle at the notion of the historian *re*-thinking thoughts which may not have been fully articulated or explicitly recognized by the agent himself—even when those thoughts can be expressed by the historian as an argument. Since there are at least as many ways of thinking the same thought as there are ways of expressing it, there is nothing in Collingwood's theory which requires the historian to go through the same overt actions, or undergo the same private experiences, as the subject of his inquiry.[13] The fact that the original thought may not have been thought propositionally would therefore be no barrier to its being re-thought propositionally—the way we should naturally expect the historian to re-think it. It does not seem to be absolutely necessary, however, that the historian's re-thinking should take this form. He might, for example, re-think Nelson's thoughts with models of his ships on a table of naval operations, for a start—although he would, of course, have to express the thoughts propositionally when writing his monograph on "Nelson at Trafalgar."

The demand that the agent's thoughts be re-thought has also been seen to raise a different sort of problem. For, as Collingwood himself declares, "the mere re-enactment of another's thought does not make historical knowledge; we must also know that we are re-enacting it" (p. 289). How can the historian know that the thought he thinks is identical with the agent's? Collingwood unfortunately does not give this question the kind of consideration we should have liked it to get from a philosopher with his experience in historical research. Apart from a denial that historical arguments are either deductive or inductive in the ordinary sense,[14] and a lively dissection of the reasoning implicit in crime detection—illustrated by a detective yarn of his own invention (pp. 266 ff.)—we are left with little more than the rather dogmatic assurance that, in so far as there is knowledge at all, there are occasions on which the historian can be sure of his conclusions. Perhaps this contention would have seemed more acceptable if Collingwood had not, in an apparently unguarded moment, claimed that historical conclusions can, in some cases, be known as certainly as a proof in mathematics (p. 262). Since history proceeds, on his own showing, by the interpretation of evidence, not by formal deduction, the precise meaning of this curious claim is not obvious. But that a "weight of evidence" may, from time to time, justify a non-mathematical kind of certainty in history, would appear to be deniable only by someone whose skepticism embraced a good deal more than historical arguments.

If we grant that the historian *can* re-think his subject's thoughts, must we go on to agree that historical understanding *requires* this? Collingwood is most insistent that it does. "To know 'what someone is thinking' (or 'has thought')," he writes, "involves thinking it for oneself" (p. 288). And it is clear that he means this quite literally. In order to grasp the agent's thought, and see that it really does explain his action, the historian must do more than merely reproduce the agent's argument, whether implicit or explicit; he must also *draw his conclusion*. It is not enough merely to examine a report of the agent's "thought-process"; the historian must, on inspecting the thoughts, and treating them as premises of practical deliberation, actually *think that* the conclusion follows. The historian's "seeing" the connection between the agent's "considerations" and his action entails his *certification* of the connection between them—this entailment being a logical one. If the attempt to re-think, and thus to certify, the agent's thought-action complex breaks down—as Collingwood admits it has done in the case of certain early Roman emperors—then we have a dark spot, an unintelligibility, a failure to understand.[15]

Collingwood sometimes puts the point in terms of a distinction between knowing the agent's thoughts as "object" and as "act." In order to understand an action, he declares, "it is necessary to know 'what someone else is thinking,' not only in the sense of knowing the same object that he knows, but in the further sense of knowing the act by which he knows it" (p. 288). And again: "The act of thinking can be studied only as an act" (p. 293). To put the point in these terms is perhaps to be unnecessarily obscure. But to represent re-thinking as a sharing in an activity does serve to bring out the requirement that the connection between thought and action be *tested* by the historian, if he wishes to understand what was done. The doctrine may appear less formidable when Collingwood goes on to consider examples. In order to understand the Theodosian Code, we are told, the historian "must envisage the situation with which the emperor was trying to deal, and he must envisage it just as the emperor envisaged it. Then he must see for himself, just as if the emperor's situation were his own, how such a situation might be dealt with; he must see the possible alternatives, and the reasons for choosing one rather than another; and thus he must go through the process which the emperor went through in deciding on this particular course" (p. 283). To see that the emperor had reasons for doing what he did is thus to understand his doing it. And to see that he had reasons is, at the same time, to certify the reasons *as* reasons.

The explanation must thus be said to succeed to the extent to which it reveals the rationality of the agent. An action is said to be understood, on Collingwood's view, when it is seen to have been rationally necessary. Does this way of putting it impute to Collingwood a rationalism as divorced from the realities of historical inquiry as the intellectualism we have already denied? Some of Collingwood's dicta might seem to support such a conclusion; others point the other way. The strategy of Admiral Villeneuve at Trafalgar, he tells us, cannot possibly be understood; for the fact that he lost the battle shows that Villeneuve's strategy was wrong—as if an action must be rationally appropriate to the *actual* situation or be unintelligible.[16] On the other hand, in discussing Hegel's philosophy of history, Collingwood welcomes the "fertile and valuable principle" asserted there that "every historical character in every historical situation thinks and acts rationally as that person in that situation *can* think and act."[17] What was said above about the understanding of the Theodosian Code should be sufficient to show that in Collingwood's hands, such a principle attributes no absolute rationality to the agents whose actions historians claim to understand. For what renders an action "rational," in the sense of "understandable," is its being required by the situation as it was envisaged, not by the situation as it actually was. The historian's certification might thus be said to be a "hypothetical" one. Its implication is simply that *if* the situation had been as the agent envisaged it, then what he did would have been the rational thing to do.

But even if we interpret Hegel's "fertile principle" in terms of such a "hypothetical" rationality, the principle may possibly lead us astray. For it apparently suggests to Collingwood that when understanding fails, we can only attribute this failure to a breakdown of historical analysis—"It is the historian himself who stands at the bar of judgment . . ." (p. 219). Historical understanding being the sort of thing it is, it seems to me that a number of other possibilities would have to be taken into account.

We must allow, for example, for cases where the agent's reasoning about his situation is itself mistaken. For an agent may be in error, not only in the way he envisages his situation, but also in concluding that what he did was required by the situation as he envisaged it. Such an agent would have to be regarded by the historian as "irrational" in a stronger sense than the agent who, although he misconceived his situation, nevertheless acted in the way required by the situation as he conceived it. The thought of the latter agent can be re-thought by the historian, albeit only hypothetically; but the thought of the former cannot be re-thought at all. In so far as a person makes a mistake in the reasoning which represents what he did as rationally required, his action can only be explained *as* mistaken—a word whose function here is to rule out the possibility of explaining what was done in terms of reasons. Such an action could therefore not be "understood," in Collingwood's sense of the word.

We must allow also for cases where what is done was not intended by the agent: cases of accidental or inadvertent action, illustrated, for example, by a cabinet minister's subversion of his own government's position by an unwise speech made in its defence. It would appear to be cases of this kind which Collingwood has in mind when he denies that "every agent is wholly and directly responsible for everything that he does" (p. 41). It might be argued, I suppose, that in such cases, although there is doubtless an action (giving the speech) which is explicable in the ordinary way, what is specified as requiring explanation (bringing down the government) is not really the agent's *action* at all. Yet what was done *was,* in this case, something which the agent could be held responsible for; and his colleagues would not be slow to say that he *did* it. If for these reasons we called it an action, then it must be admitted that this, too, is a kind of action which could not be understood in Collingwood's sense.

Still another sort of case which Hegel's principle must allow for is that of arbitrary or capricious action: cases where an agent may have known certain things about his situation, and may have realized that a certain response was, in reason, demanded, yet deliberately acted otherwise. The possibility of such action has sometimes been denied. It has been alleged that what we do, we always do for some reason—even if it takes a psychoanalyst to discover it. I cannot undertake to argue the contrary here in any detail. But I think we might perhaps be a trifle wary of a theory of action which, on some occasion, might condemn a historical agent to the fate of Buridan's ass, who, on finding himself stationed equidistant between two equally succulent bundles of hay, followed reason and starved to death. The ass, admittedly, had every reason to eat one bundle before the other; and if he had eaten them successively, *something* could doubtless have been explained—for example, that he ate any hay at all, or even that he ate one bundle before the other. But it would clearly have been impossible, *ex hypothesi,* to explain in Collingwood's sense why he chose to eat the one he did before eating the other. I do not, of course, suggest that the historian will very often be confronted with an agent placed in the unhappy position of such a philosophical ass. But it is hard to see, if the ass's philosophy be not refuted, how Hegel's principle, even in its attenuated Collingwoodian form, can be accepted as universally valid.

NOTES

1. Oxford, 1946.

2. *Idea of History,* pp. 305–15.

3. *Introduction to Philosophy of History* (London, 1951), p. 50.

4. *A Study of History* (London, 1954), IX, 722.

5. *Idea of History,* p. 217. See also pp. 212, 214, 220.

6. *The Concept of Mind* (London, 1949), Chap. vi. Rylian language is employed in the two paragraphs following.

7. *Idea of History,* pp. 285, 301, 303.

8. *Philosophy of History,* p. 54.

9. E.g., *Idea of History,* pp. 216–17.

10. For a different solution, see Toynbee, *Study of History,* pp. 720–21. According to him, Caesar's action expresses "will," not "thought."

11. See *Idea of History,* pp. 220, 224.

12. I am discussing here the general *concept* of rational necessity. The qualifications requiring to be made when it is employed in historical understanding are introduced in Sec. III.

13. Collingwood's term "re-enactment" may appear more troublesome in this connection than "re-thinking"; but for Collingwood, thinking is an activity.

14. *Idea of History,* pp. 261–62. Mr. Alan Donagan has discussed this aspect of Collingwood's theory in an illuminating way in "The Verification of Historical Theses," *Philosophical Quarterly* (July 1956).

15. *Idea of History,* p. 310.

16. R. G. Collingwood, *An Autobiography* (Oxford, 1941), p. 70. T. M. Knox, in the "Editor's Preface" to *The Idea of History,* assumes that this is Collingwood's meaning. I think it possible that what he had in mind was rather the extent to which the evidence, upon which conclusions about Villeneuve's actual strategy during the battle would have been based, was destroyed by Nelson's victory.

17. *Idea of History,* p. 116.

STUDY QUESTIONS FOR PART TWO
"What Is Explanation" — J. Hospers

1. Which of these sentences is a request for an explanation of the sort Hospers is interested in? How can you distinguish that type of request from the others?

What does 'banal' mean?

Why did that piece of copper turn green?

How do you get to Dayton's department store?

Why do you still believe that Nixon did not know about the Watergate caper?

2. What is wrong with the claim that explaining is taking an event that is unfamiliar to us and showing that it is really a member of a class of events that is familiar to us?

3. According to Hospers, what does explanation consist of?

4. Give a counterexample to the claim that a *sufficient* condition for something being an explanation is that the explanandum be a deductive consequence of the explanans.

5. Discuss the issue of whether the condition above ought to be a *necessary* condition.

6. What might be wrong with the explanation given in the following dialogue?

Announcer: Why are the Bullets playing so much better this quarter?
Color man: Well, it's obvious that the momentum is on their side now.

7. In the sentences that follow, what are the important differences in the meaning of 'purpose'?

a. His purpose in wearing that hat was to attract her attention.

b. The purpose of that tube is to return unburned gases to the carburetor.

8. Give the best argument you can for (a) there being basic laws but (b) our being unable to know what they are.

9. How would you respond to a request for the explanation of a basic law?

10. Often it is claimed that all science does is tell us *how* things happen or, alternatively, that science just *describes*. Part of this position seems to be that explaining is incompatible with telling us or describing. Evaluate this view.

"Law" — J. Hospers

1. What are the essential differences between descriptive and prescriptive laws?

2. What confusions are embedded in talk of nature obeying laws?

3. What are the necessary features of a law of nature? Give a brief argument for why each is necessary.

4. Hospers assumes that laws are statements. How can Hospers or you argue against the claim that Hospers confuses *laws* with *descriptions of laws*?

"Towards an Aristotelean Theory of Scientific Explanation"—B. Brody

1. Brody's "sodium-bromine" and "temperature-pressure" examples are supposed to meet all of Hempel's requirements and yet not be explanations. Do they meet the requirements? What is counterintuitive about them?

2. Explain the difference between Aristotle's "knowledge of the fact" and "knowledge of the reasoned fact."

3. Be sure you see and can show that Aristotle's view of causation is different from the constant conjunction view. According to that latter view to say that *a* causes *b* is to say that *a* and *b* occurred and that from a description of *a* and some law, *b,* a description of *b* can be deduced. "Law" on this view is spelled out as Hospers did.

4. What is the first new requirement Brody adds to the Hempelian conditions? Show how it accounts for our intuitions about the temperature-pressure cases.

5. Why won't the new requirement handle the "sodium-bromine" case?

6. What is Brody's second requirement? Show how it explains what is wrong with the "sodium-bromine" example.

7. Give some examples of accidental and essential properties. Does Brody give any hints at a criterion for individuating them?

8. Evaluate Brody's "proof" that we have knowledge of Aristotelean type causation and essential properties.

"Historical Understanding as Re-thinking"—W. Dray

1. According to Collingwood, action, understood as necessarily having a thought-side, is the subject of historical inquiry. What does he mean by this? With what are actions in this sense contrasted?

2. When a historical agent acts, must he be aware of his thought-side? To answer this question you need to formulate a view on what it is to have a thought.

3. What, according to Collingwood, are the conditions for explaining an action? How do they differ from the Hempelian conditions? Give a nice example of a Collingwoodian explanation.

4. What does it mean to say one must fill out the thought-side until it *necessitates* the action? In Hempelian explanations, in what, if any, sense is the explanandum necessitated? Is the event described by the explanandum necessitated in any sense?

5. What does Collingwood seem to have in mind by his rethinking method for discovering the thought-side of another's action?

6. Is there anything about Collingwood's account of *historical* explanation that makes it *historical,* i.e., if it explains actions at all, why doesn't it work on the actions of one's friends as well as those of historical figures?

SELECTED BIBLIOGRAPHY

[1] Achinstein, P. *Law and Explanation.* Oxford: Clarendon Press, 1971. Does not sever the logical from the pragmatic aspects of explanation.
[2] Brodbeck, M. *Readings in the Philosophy of the Social Sciences.* New York: Macmillan, 1968, sec. 3–5.
[3] Brody, B., ed. *Readings in the Philosophy of Science.* Englewood Cliffs, N.J.: Prentice-Hall, 1970, pt. 1. Contains the classic papers by Hempel and critical responses.
[4] Dray, W., ed. *Philosophical Analysis and History.* New York: Harper & Row, 1966. Contains essays on historical explanation.
[5] Hull, D. *Philosophy of Biological Science.* Englewood Cliffs, N.J.: Prentice-Hall, 1974, chap. 3.
[6] Scheffler, I. *The Anatomy of Inquiry.* New York: Knopf, 1969, pt. 1. Sympathetic and careful look at the Hempelian view.

Part Three

Theory and Observation

Part 3: Theory and Observation

Introduction

I. The Problem

Our student newspaper recently reported two episodes. In one, a biology student was reported to have said, in defense of some religious view of his, that the theory of evolution was just that—a theory. His professor supposedly shouted back, "It's not a *theory*!" In another story, a mathematician used, as an example of one of man's greatest cognitive achievements, Einstein's *theory* of special relativity.

Though the student and biologist have some disagreements, they do not appear to be over what a theory is. But the mathematician seems to have a different concept of what it is to call something a theory. For the student and biologist to call something a theory is to reduce its cognitive authority. The mathematician thinks that at least one theory is without cognitive superiors. Someone (perhaps all three) is confused.

In this part our goal is to begin to form a coherent and informed view on the nature of theories. Before attempting to characterize the structure of a theory, let's consider a pair of examples. The explanans of two of the examples in Part 2, when considered along with the wider set of laws and assumptions in which they are embedded, provide splendid examples of theories.

Aristotle's explanation of the circular movements of the planets is just a part of his wider astronomical theory. His theory "tells" us much more about the universe than why planets move in circular orbits; e.g., it tells us that the stationary earth is at the center of the universe; it gives the spatial order of the planets and their relation to the stars; and it provides for predictions of the movements of the planets.

There is an evolutionary *explanation* of certain increases in the "complexity" of life-forms as evidenced in the fossil record. But the wider *theory* of evolution provides explanations and predictions of a plethora of phenomena. For example, the theory allows us to explain why only certain species of moths in Birmingham, England, survived the pollution of that city brought on by industrialization. The theory provides us with insights into how we should go about creating more rust-resistant grains. The theory provides the causal mechanism for understanding why the descendants of certain mutants in a population come to dominate the population. And we could go on and on.

Our philosophical problem can be simply stated: What is a scientific theory? We shall see that the answer is considerably more complex.

II. The Structure of Theories

The topic of theories is perhaps the most difficult one in the philosophy of science. It is certainly the most difficult of those found in this anthology. Like the mind-body problem and the problem of perception, the topic of theories is really a constellation of intertwined problems. Furthermore, as in the area of explanation, there is a historically important view. Though this view is widely recognized as problematic, the reader of current literature often must have a substantial acquaintance with it. But unlike the Hempelian view on explanation, which can be presented to a student with minimal background, the standard view on theories cannot. Even elementary presentations require training in symbolic logic and the philosophy of language.

Fortunately, much about theories can be discussed without taking on all the details of the standard account. A place to begin is with a rough and tentative characterization of theories.

Structurally, theories are sets of statements, some of which state laws, others of which are singular factual or existential claims; e.g., that electrons exist and that electrons have a charge of minus one. Furthermore, theories contain some terms that refer to unobservable entities or properties. The statements of a theory are interrelated in such a way as to embody certain virtues: generality or comprehensiveness of explanatory and predictive power, ability to unify diverse phenomena and laws, depth of explanatory power. Theories explain not a particular law or phenomenon but whole ranges of each—this range is typically called the domain of the theory. Theories show how apparently diverse or unrelated phenomena and laws really are related, at least for explanatory purposes. Theories aim at a deep understanding of phenomena; i.e., theories often describe the causal mechanism behind regularities or appeal to microstructure in accounting for the macroscopic properties of objects.

We can consider in a little detail an example—the kinetic theory of matter and heat. In the mid-nineteenth century, J. P. Joule synthesized the work of

Rumford, Bernoulli, and others to form the kinetic theory. This theory exercised the talents of many of the great nineteenth-century physicists.

Structurally, the theory contains: (a) a set of laws—in particular, what we now call the laws of classical mechanics, the most central of which are Newton's laws of motion; (b) a set of singular existential and factual statements. We now refer to this set as the model of a gas. It includes the following claims: gases consist of molecules; the size of molecules is negligible; the number of molecules is very large; molecules are in random motion; molecular collisions are perfectly elastic. Notice that some of the terms in these statements refer to unobservable entities and properties.

It was claimed earlier that theories have certain virtues. The generality of the explanatory power of the kinetic theory is witnessed by the fact that it explains the gas laws (laws that relate to the pressure, temperature and volume of a gas), laws about the specific heats of gases, laws about the rates of diffusion of one gas in another, and so on. The theory unifies apparently diverse phenomena. It shows, for example, that the fact that an odiferous substance released, say, in the corner of a room spreads at the rate it does throughout the room and the fact that the earth has not lost its atmosphere are, from the point of view of the theory, rather similar phenomena. To use the gas-diffusion example again, the theory does, of course, explain this well-known phenomenon, but it also illuminates the underlying mechanism by which it occurs. The theory also takes common notions such as temperature and pressure and shows one what they "really" are. It gives the microstructure that "underlies" these properties.

This is a good point at which to read the selection by Hospers entitled "Theories," though a final point or two is in order. One should look rather carefully at Hospers's characterization of hypotheses. We suspect that his criterion for distinguishing hypotheses from theories actually contradicts his earlier explication of "theory." Furthermore, we suggest to you, that the difference between theories and hypotheses is not structural but epistemic. That is, it has to do with our confidence in the *truth* of the statements not with whether they have a general or singular form.

III. Observation

Most philosophers agree that an appeal to unobservables is an important mark of scientific theories. But here the agreement ends. When one considers the details of this rather vague criterion, a plethora of issues are forced upon us. Exactly how are observables and unobservables to be distinguished? Is what is observable subject to changes in historical contingencies—the invention of a new instrument, for example? Is what is observable subject to theoretical contingencies; that is, given two theories, T and T', at time t, can an entity or property be observable relative to T and unobservable relative to T'?

Consider the following view that has three parts: (i) At any time, t, all the nonlogical terms of science (terms other than 'all,' 'and,' 'or,' 'a,' and so on) can be uniquely parsed into one of two classes: (a) those that refer to observable entities or properties, e.g., 'red', 'circular', 'cup', 'cat'. (These are called observation terms.) (b) Those that refer to unobservable entities or properties, e.g., 'valence', 'charge', 'electron'. (These are called theoretical terms.) (ii) If a term is correctly classified as an observation or theoretical term it is permanently so classified. (iii) An entity or property is an observable if and only if normal observers in standard conditions can ascertain the presence of the entity or property by direct observation. Otherwise the entity or property is unobservable.

Something like the view above is embedded in the standard account of theories. Though, in fairness, it must be admitted to be an extreme and perhaps unsympathetic reconstruction of the standard view. Nevertheless, it will serve for anchoring our discussion.

If the outlined three-part account is to have a chance of success one must fend off the threatening circularity in the third condition. This, at least, requires spelling out what is meant by direct. Norwood Hanson's essay ("Observation") will be helpful along these lines. He tries to characterize the standard view on observation before developing a line of criticism which bears on all of the three conditions of our stated view.

Bluntly put, Carl Kordig in "The Theory-Ladenness of Observation" thinks that Hanson has gone too far. In fact, so far that from Hanson's alternative view on observation absurdities can be deduced.

In so far as you find Kordig's rebuttal compelling, you might be motivated to look again at the standard view. Perhaps one can modify the central claims of the standard view in such a way as to take account of Hanson's important insights or reconcile his insights with the standard view. It is hoped that one of these tacks can be accomplished while escaping the absurdities Kordig points out.

IV. Status of Unobservables

Supposing that a satisfactory way has been found to make the observable/unobservable distinction, another issue beckons us. Are unobservables real? Do they actually exist? This may strike one as an odd question. It is the case from a grammatical point of view that theoretical terms appear to refer and describe just as observation terms do. But as we shall see, philosophers have thought that the actual function of theoretical talk is quite different.

A question closely related to whether unobservables exist is whether theoretical statements are true. Theoretical statements are statements that contain only theoretical and logical terms, e.g., 'electrons have a charge of minus one.' Analogously, observation statements are statements that contain only observation and logical terms, e.g., 'that table is red.'

There are several distinct views on the questions we have been asking. It will be helpful to chart a taxonomy of the more important views. The chart can be generated by asking three questions: (Q1) Do unobservables exist? (Q2) Are theoretical statements true? (Q3) Are observation statements true? Do not be misled by (Q2) and (Q3). The question is not: Are all the theoretical or observational sentences of theories true or are some of them true; or do we know that some of them are true? The question is: Are they candidates for being true? They will not be candidates if there is some general philosophical reason for suspecting the truth of them as a class. The chart of views follows:

	Super-realism	Realism	Instrumentalism	Descriptivism
(Q1) Do unobservables exist?	Yes	Yes	No	No
(Q2) Are theoretical statements true?	Yes	Yes	No	Yes
(Q3) Are observation statements true?	No	Yes	Yes	Yes

There are other possibilities, of course. The trinity, No, No, No, is an obvious one. This possibility has been taken seriously by some philosophers. But the arguments for it are not usually produced from a close look at science. They are often generated from a global attack on the possibility of knowledge. We could construct a more elaborate and finer-grained taxonomy. But since those above are the most discussed possibilities and since our purposes are merely to initiate the discussion, the above major possibilities and grain will do.

Realists are fond of regarding their view as common sense. Whether they are right or not it is true that the dialectic of the issue is such that it is tacitly assumed that the burden of proof rests with the anti-realist. We can say from personal experience that scientists often talk the instrumentalist's line and that a little philosophy often jars students from a realist stance. Whether a little philosophy is a dangerous thing or not you will have to decide.

We do not wish to review in detail the reasons for the four views—the readings by Stace and Maxwell do that. We do want to say enough about the views so that they are at least on the surface distinguishable.

Sir Arthur Eddington (1881–1944), the famous English astronomer, held the view I have dubbed super-realism. Eddington begins the presentation of his view in a now famous passage that both excites and teases the reader.

> I have settled down to the task of writing these lectures and have drawn up my chairs to my two tables. Two tables! Yes; there are duplicates of every object about me—two tables, two chairs, two pens.[1]

The first table, the table of common sense, is extended, permanent, colored, and substantial. The second table, the table of physics, is mostly emptiness. It is a vast number of electric charges speeding about.

For our purposes, what is interesting about Eddington is that he denies that the first table exists, much less that there is a colored table there. In his words, "Modern physics has by delicate test and remorseless logic assured me that my second table is the only one which is really there—whatever 'there' may be."[2]

Exactly how physics teaches us, if it does, that observation sentences are false is a very complicated matter. If you are interested in pursuing super-realism, [2] and [3] of the bibliography may help.

In our selection by W. T. Stace, "Science and the Physical World," he argues for descriptivism. Stace claims that we have absolutely no reason to believe that the theoretical "entities" and "properties" of a theory exist. This leads Stace to propose a new function for theoretical talk. Theoretical statements supposedly provide us with what Stace calls "shorthand formulae." These formulae prove convenient for organizing observables and making predictions about observables, but they should not be taken as refer-ring to entities or properties over and above observables.

A simple example will help clarify Stace's point. The caterer of a wedding might find it convenient to plan in terms of the "average wedding-goer." Such a concept will be helpful in summarizing certain data from past weddings and predicting features of the wedding to be. But the caterer does not think average wedding-goers are real or exist. After all, not only does the average wedding-goer drink 3.2 whiskey sours and eat 6 cocktail knishes, but he has 2.1 children.

In the same way that the average wedding-goers do not exist over and above John, Mildred, and so on, and their actual and dispositional behavior, elec-trons do not exist over and above the complex of actual and possible data they stand in for.

Notice that, for the descriptivist, theoretical statements are true. This, at the outset, might appear strange since theoretical entities do not exist. But given Stace's view, we see that theoretical statements are just very complex, though briefly stated, observation statements. Since observation statements are true, so are theoretical statements.

Though descriptivism and instrumentalism are often confused they are distinct views. The "no" answer for the instrumentalist to (Q2) may be misleading—misleading in the same way as your answering "no" if a child asked you if you ever sprained your kidney. A better answer would be to point out that kidneys are not the kinds of things that can be sprained or for that matter not sprained. Similarly, for the instrumentalist, theoretical statements are not the kinds of things that are true or false. Instrumentalists often refer to theoretical statements as rules, or inference tickets, or predictive instruments. Just as the rules of chess, e.g., the rule that pawns can only move forward, are neither true nor false but useful for certain purposes, playing an interesting game, so the theoretical sentences of science are neither true nor false but useful for certain purposes, making predictions. (Of course, "'pawns can only move forward' is a rule of chess" is true or false.)

The essay by Grover Maxwell, "The Ontological Status of Theoretical Enti-ties," states and critiques both descriptivism and instrumentalism on the way to defending realism. But there is another important theme in Maxwell's paper. We began this section suggesting that the status and definitional questions on unobservables are independent. They are not. Maxwell's work makes explicit the interplay between the two issues.

A.D.K.

NOTES

1. A. Eddington. *The Nature of the Physical World*. New York: Cambridge University Press, 1929. p. ix.
2. Ibid. p. xii.

9

Theory

John Hospers

Not long after we have observed certain invariant relations in nature, we are led to construct theories to explain them. The distinction between a theory and a law is somewhat vague, but it is very important: in general, we *construct* or devise theories, but we *discover* laws of nature. A scientific theory always contains some term that does not denote anything that we can directly observe. If we can observe something only through a telescope or microscope, we are still said to observe it. But if there are no conditions under which we can observe it, it is a theoretical entity; and when the theory-word is a part of a statement, that statement is said to be a theory. Thus the proposition that there are protons and electrons is a theory. These entities cannot be observed, though we do observe many things that are presumed to be effects of them. Statements about protons and electrons (together with their progeny, such as neutrons and neutrinos) are theory, not law. This belief in the "ultimate constituents of matter" is perhaps the most comprehensive and thoroughly worked-out theory to be found in the empirical sciences.

How did such a theory arise? It began in ancient times with some rather obvious observations. From time immemorial, certain empirical truths had been observed that seemed to call for an explanation in terms of what could not be observed. The stone steps wear away bit by bit, year after year. Put a few drops of berry-juice into a glass of water, and in a moment the entire liquid has become red. Or put some sugar into it, and immediately the sweet taste pervades the entire liquid. How can these and countless other things be explained unless by the existence of very small particles, invisible to the unaided eye? The stone steps are composed of these particles, which wear away gradually one by one until after years of wear we can finally notice the difference. The berry juice is composed of very small particles that spread

148

throughout the entire liquid and color it red. The same with the sugar we dissolve in water, which makes the entire liquid taste sweet. Besides, the things we observe must be composed of something. I can cut this piece of chalk in half and rub against it with my fingers, with the result that pieces of it color my fingers white. But these small bits (it was reasoned) must in turn be composed of smaller ones, and these in turn of still smaller ones. At the end of this process, however, there must be particles that cannot be split any further, the ultimate constituents of matter (atoms, from the Greek *"atomos,"* meaning "unsplittable"). All the things we see and touch are composed of these very small particles that can no longer be subdivided. We cannot observe them with our eyes; but if we assume that they exist, we can account for an enormous number of different things that we do observe.

So went the reasoning of Democritus (born about 460 B.C.) and Lucretius (ca. 96–55 B.C.). Their atomic theory was primitive, but the principle involved was no different from modern theories: the unobserved was invoked to explain the observed. More refined atomic theories today have explained countless phenomena undreamed of by the ancients: why element A combines with elements B and C but not with D and E (and some with none at all), why certain elements and compounds have the properties that they do, why they evaporate or ignite at the temperatures they do, freeze at other temperatures, and so on. Virtually all the facts of modern chemistry have been explained in terms of atomic theory. But it is theory, not observed fact. (Certain complex molecules have now been seen through electron microscopes, and thus they no longer belong to theory. But atoms and electrons, together with the other and more minute "particles" that physics now deals in, remain unobservable.)

Do these tiny "particles" exist? Are they particles at all, like tiny marbles, or should they be called something else? There is scarcely a physicist who would deny that these entities exist, and that it is only the hypothesis that they exist that explains why certain observed facts are as they are. It has sometimes been suggested that they do not really exist but are simply "convenient fictions" by means of which we explain a diverse range of phenomena. But if they are only convenient fictions, how is it that the tremendous variety of occurrences they explain regularly happen as they do? Would it not be one vast coincidence if there are not really atoms and electrons that have the explanatory properties we attribute to them? Why should things behave just *as if* they are composed of very tiny particles if no such particles actually exist?

Some scientific theories, however, do involve concepts that are merely convenient fictions. Freudian psychology has as its first premise that there is a vast reservoir of *unconscious* mental events consisting of three departments: id, superego, and ego. These are theory, since no such inhabitants of the human psyche can be observed. Yet in postulating these entities, Freudian psychology endeavors to explain a vast number of psychological phenomena (mental conflicts, neuroses and psychoses, dreams, slips of the tongue, moods, depression) on the basis of a comprehensive theory involving these concepts. The id is the vast reservoir of human desires, most of them prohibited, so that they are repressed from the conscious into the unconscious; the superego is the prohibitor or nay-sayer, which refuses the granting of many of these desires; the ego is the adjudicator of the conflicting claims of these two parties, providing a defense for one or the other party in response to its claims. When one becomes acquainted with the literature of psychoanalysis, one is struck by the enormous variety of explanations of human behavior that are provided by this conceptual framework. Yet no one believes that there are really three people inside one's head; it is just as if there were, but of course there are not. Here the theory is an elaborate "as if," yet the theory has very great explanatory power (though it is not the only theory that attempts to explain human behavior, as the existence of numerous and conflicting schools of psychiatry indicates). Similarly, when one speaks of valence in chemistry, it is just "as if" the atoms had little hooks on them, one on hydrogen (for example) and two on oxygen, so that the two hooks on an atom of oxygen latch into the one hook on each of two atoms of hydrogen to form H_2O, or water. Yet no one believes that the atoms really have little hooks (but perhaps they have something like them?).

In either case, however, it is important to remember that the theory contains more (has more content) than the observed facts that are explained by means of it. A theory is not just a summary of facts already observed; it is not merely a shorthand way of referring to a diverse collection of facts: it involves concepts from which new and hitherto unknown facts can be inferred. This is as true of the "as-if" theory in psychoanalysis as it is of the "really-exists" theory of atomic structure. A theory that was merely a *summary* of the observed facts already known would have no explanatory power whatever.

Some statements we make *are* merely "summary-statements." When we say that there is a current in the wire, it is plausible to analyze this statement (though some have contested this) as a statement about a diverse group of observable phenomena: the wire affects voltmeters, gives us a shock when we touch it, throws off sparks, runs batteries, and so on. Saying that there is a current in the wire *is* just saying that the wire does these various things. But theories in science are not of this kind: a theory must always explain more facts than it was invoked to explain; the scientific potency of a theory is in direct proportion to the quantity and (more important) the *range* of facts it explains, particularly those that had not been known when the theory was devised. In this respect, both the atomic theory and the theory of the unconscious have remarkable explanatory power.

Hypotheses. It is important in this connection to distinguish theories from hypotheses. In daily life these two words are used interchangeably, but in science they are distinguished. Theories are continuous with laws: they both involve general statements about some aspect of the world, differing only in whether their key terms refer to what is observable in the world. But a hypothesis is not a universal proposition at all; it is a particular statement that *in conjunction with* laws or theories can be used to explain certain occurrences. Thus we speak of atomic *theory,* but we speak of various *hypotheses* to explain the origin of the solar system. We devise hypotheses every day of our lives:

we arise in the morning and notice that the street is wet, so we devise the hypothesis that rain has fallen during the night; we press the starter of the car and hear that ominous growling sound, followed by silence, and we devise the hypothesis that the battery is dead; we see the dog hovering around his dish and showing signs of restlessness, and immediately we devise the hypothesis that he is hungry and hasn't been fed; we see that Johnny is doubled up with stomach pains, and that the entire mincemeat pie is gone, and we devise the hypothesis that he has the stomach-ache from eating the entire pie. A hypothesis is a particular fact (or presumed fact) that, if true, explains, *together with certain laws or theories,* why something is as it is. A hypothesis may be improbable or even outlandish (such as the astrological hypothesis that you are due for bad luck today because the planets are in a certain conjunction), or it may be so probable that you take it as virtually certain (such as the hypothesis that the streets are wet because it has rained during the night). The degree of probability has nothing to do with whether it is a hypothesis, only with whether it is an acceptable or satisfactory hypothesis.

Moreover, a hypothesis is not concerned with something you have observed: if you have seen the rain falling, you would not call "Rain has fallen" a hypothesis but an observed fact. Normally, however, a hypothesis is about something observ*able*: rainfall, the dog not having nourishment for 24 hours, Johnny eating the pie—all these could have been observed, even if they were not. But some hypotheses do involve the unobservable, such as "The ominous feeling I have is a warning from God that I shouldn't do this."

Both law-theories and hypotheses have their part to play in the main function of the scientific enterprise, *explanation.*

10

Observation

N. R. Hanson

Were the eye not attuned to the Sun,
The sun could never be seen by it.

GOETHE[1]

A

Consider two microbiologists. They look at a prepared slide; when asked what they see, they may give different answers. One sees in the cell before him a cluster of foreign matter: it is an artefact, a coagulum resulting from inadequate staining techniques. This clot has not more to do with the cell, *in vivo,* than the scars left on it by the archaeologist's spade have to do with the original shape of some Grecian urn. The other biologist identifies the clot as a cell organ, a "Golgi body." As for techniques, he argues: "The standard way of detecting a cell organ is by fixing and staining. Why single out this one technique as producing artefacts, while others disclose genuine organs?"

The controversy continues.[2] It involves the whole theory of microscopical technique; nor is it an obviously experimental issue. Yet it affects what scientists say they see. Perhaps there is a sense in which two such observers do not see the same thing, do not begin from the same data, though their eyesight is normal and they are visually aware of the same object.

Imagine these two observing a Protozoon—*Amoeba.* One sees a one-celled animal, the other a non-celled animal. The first sees *Amoeba* in all its analogies with different types of single cells: liver cells, nerve cells, epithelium cells. These have a wall, nucleus, cytoplasm, etc. Within this class *Amoeba* is distinguished only by its independence. The other, however, sees *Amoeba*'s homology not with single cells, but with whole animals. Like all animals *Amoeba* ingests its food, digests and assimilates it. It excretes, reproduces and is mobile—more like a complete animal than an individual tissue cell.

This is not an experimental issue, yet it can affect experiment. What either man regards as significant questions or relevant data can be determined by whether he stresses the first or the last term in "unicellular animal."[3]

Some philosophers have a formula ready for such situations: "Of course they see the same thing. They make the same observation since they begin from the same visual data. But they interpret what they see differently. They construe the evidence in different ways."[4] The task is then to show how these data are moulded by different theories or interpretations or intellectual constructions.

Considerable philosophers have wrestled with this task. But in fact the formula they start from is too simple to allow a grasp of the nature of observation within physics. Perhaps the scientists cited above do not begin their inquiries from the same data, do not make the same observations, do not even see the same thing? Here many concepts run together. We must proceed carefully, for wherever it makes sense to say that two scientists looking at *x* do not see the same thing, there must always be a prior sense in which they do see the same thing. The issue is, then, "Which of these senses is most illuminating for the understanding of observational physics?"

These biological examples are too complex. Let us consider Johannes Kepler: imagine him on a hill watching the dawn. With him is Tycho Brahe. Kepler regarded the sun as fixed: it was the earth that moved. But Tycho followed Ptolemy and Aristotle in this much at least: the earth was fixed and all other celestial bodies moved around it. *Do Kepler and Tycho see the same thing in the east at dawn?*

We might think this an experimental or observational question, unlike the questions "Are there Golgi bodies?" and "Are Protozoa one-celled or non-celled?" Not so in the sixteenth and seventeenth centuries. Thus Galileo said to the Ptolemaist ". . . neither Aristotle nor you can prove that the earth is *de facto* the centre of the universe. . . ."[5] "Do Kepler and Tycho see the same thing in the east at dawn?" is perhaps not a *de facto* question either, but rather the beginning of an examination of the concepts of seeing and observation.

The resultant discussion might run:

"Yes, they do."

"No, they don't."

"Yes, they do!"

"No, they don't!" . . .

That this is possible suggests that there may be reasons for both contentions.[6] Let us consider some points in support of the affirmative answer.

The physical processes involved when Kepler and Tycho watch the dawn are worth noting. Identical photons are emitted from the sun; these traverse solar space, and our atmosphere. The two astronomers have normal vision; hence these photons pass through the cornea, aqueous humour, iris, lens and vitreous body of their eyes in the same way. Finally their retinas are affected.

Similar electro-chemical changes occur in their selenium cells. The same configuration is etched on Kepler's retina as on Tycho's. So they see the same thing.

Locke sometimes spoke of seeing in this way: a man sees the sun if his is a normally-formed retinal picture of the sun. Dr. Sir W. Russell Brain speaks of our retinal sensations as indicators and signals. Everything taking place behind the retina is, as he says, "an intellectual operation based largely on non-visual experience. . . ."[7] What we *see* are the changes in the *tunica retina*. Dr. Ida Mann regards the macula of the eye as itself "seeing details in bright light," and the rods as "seeing approaching motor-cars." Dr. Agnes Arber speaks of the eye as itself s\~eing.[8] Often, talk of seeing can direct attention to the retina. Normal people are distinguished from those for whom no retinal pictures can form: we may say of the former that they can see whilst the latter cannot see. Reporting when a certain red dot can be seen may supply the occulist with direct information about the condition of one's retina.[9]

This need not be pursued, however. These writers speak carelessly: seeing the sun is not seeing retinal pictures of the sun. The retinal images which Kepler and Tycho have are four in number, inverted and quite tiny.[10] Astronomers cannot be referring to these when they say they see the sun. If they are hypnotized, drugged, drunk or distracted they may not see the sun, even though their retinas register its image in exactly the same way as usual.

Seeing is an experience. A retinal reaction is only a physical state—a photochemical excitation. Physiologists have not always appreciated the differences between experiences and physical states.[11] People, not their eyes, see. Cameras, and eye-balls, are blind. Attempts to locate within the organs of sight (or within the neurological reticulum behind the eyes) some nameable called "seeing" may be dismissed. That Kepler and Tycho do, or do not, see the same thing cannot be supported by reference to the physical states of their retinas, optic nerves or visual cortices: there is more to seeing than meets the eyeball.

Naturally, Tycho and Kepler sees the same physical object. They are both visually aware of the sun. If they are put into a dark room and asked to report when they see something—anything at all—they may both report the same object at the same time. Suppose that the only object to be seen is a certain lead cylinder. Both men see the same thing: namely this object—whatever it is. It is just here, however, that the difficulty arises, for while Tycho sees a mere pipe, Kepler will see a telescope, the instrument about which Galileo has written to him.

Unless both are visually aware of the same object there can be nothing of philosophical interest in the question whether or not they see the same thing. Unless they both see the sun in this prior sense our question cannot even strike a spark.

Nonetheless, both Tycho and Kepler have a common visual experience of some sort. This experience perhaps constitutes their seeing the same thing. Indeed, this may be a seeing logically more basic than anything expressed in the pronouncement "I see the sun" (where each means something different by "sun"). If what they meant by the word "sun" were the only clue, then Tycho and Kepler could not be seeing the same thing, even though they were gazing at the same object.

If, however, we ask, not "Do they see the same thing?" but rather "What is it that they both see?," an unambiguous answer may be forthcoming. Tycho and Kepler are both aware of a brilliant yellow-white disc in a blue expanse over a green one. Such a "sense-datum" picture is single and uninverted. To be unaware of it is not to have it. Either it dominates one's visual attention completely or it does not exist.

If Tycho and Kepler are aware of anything visual, it must be of some pattern of colours. What else could it be? We do not touch or hear with our eyes, we only take in light.[12] This private pattern is the same for both observers. Surely if asked to sketch the contents of their visual fields they would both draw a kind of semicircle on a horizon-line.[13] They say they see the sun. But they do not see every side of the sun at once; so what they really see is discoid to begin with. It is but a visual aspect of the sun. In any single observation the sun is a brilliantly luminescent disc, a penny painted with radium.

So something about their visual experiences at dawn is the same for both: a brilliant yellow-white disc centred between green and blue colour patches. Sketches of what they both see could be identical—congruent. In this sense Tycho and Kepler see the same thing at dawn. The sun appears to them in the same way. The same view, or scene, is presented to them both.

In fact, we often speak in this way. Thus the account of a recent solar eclipse:[14] "Only a thin crescent remains; white light is now completely obscured; the sky appears a deep blue, almost purple, and the landscape is a monochromatic green . . . there are the flashes of light on the disc's circumference and now the brilliant crescent to the left. . . ." Newton writes in a similar way in the *Opticks*: "These Arcs at their first appearance were of a violet and blue Colour, and between them were white Arcs of Circles, which . . . became a little tinged in their inward Limbs with red and yellow. . . ."[15] Every physicist employs the language of lines, colour patches, appearances, shadows. In so far as two normal observers use this language of the same event, they begin from the same data: they are making the same observation. Differences between them must arise in the interpretations they put on these data.

Thus, to summarize, saying that Kepler and Tycho see the same thing at dawn just because their eyes are similarly affected is an elementary mistake. There is a difference between a physical state and a visual experience. Suppose, however, that it is argued as above—that they see the same thing because they have the same sense-datum experience. Disparities in their accounts arise in *ex post facto* interpretations of what is seen, not in the fundamental visual data. If this is argued, further difficulties soon obtrude.

B

Normal retinas and cameras are impressed similarly by fig. 1.[16] Our visual sense-data will be the same too. If asked to draw what we see, most of us will set out a configuration like fig. 1.

Do we all see the same thing?[17] Some will see a perspex cube viewed from below. Others will see it from above. Still others will see it as a kind of polygonally-cut gem. Some people see only criss-crossed lines in a plane. It may be seen as a block of ice, an aquarium, a wire frame for a kite—or any of a number of other things.

Do we, then, all see the same thing? If we do, how can these differences be accounted for?

Fig. 1

Here the "formula" re-enters: "These are different *interpretations* of what all observers see in common. Retinal reactions to fig. 1 are virtually identical; so too are our visual sense-data, since our drawings of what we see will have the same content. There is no place in the seeing for these differences, so they must lie in the interpretations put on what we see."

This sounds as if I do two things, not one, when I see boxes and bicycles. Do I put different interpretations on fig. 1 when I see it now as a box from below, and now as a cube from above? I am aware of no such thing. I mean no such thing when I report that the box's perspective has snapped back into the page.[18] If I do not mean this, then the concept of seeing which is natural in this connexion does not designate two diaphanous components, one optical the other interpretative. Fig. 1 is simply seen now as a box from below, now as a cube from above; one does not first soak up an optical pattern and then clamp an interpretation on it. Kepler and Tycho just see the sun. That is all. That is the way the concept of seeing works in this connexion.

"But," you say, "seeing fig. 1 first as a box from below, then as a cube from above, involves interpreting the lines differently in each case." Then for you and me to have a different interpretation of fig. 1 just *is* for us to see something different. This does not mean we see the same thing and then interpret it differently. When I suddenly exclaim "Eureka—a box from above," I do not refer simply to a different interpretation. (Again, there is a logically prior sense in which seeing fig. 1 as from above and then as from below is seeing the same thing differently, i.e. being aware of the same diagram in different ways. We can refer just to this, but we need not. In this case we do not.)

Besides, the word "interpretation" is occasionally useful. We know where it applies and where it does not. Thucydides presented the facts objectively; Herodotus put an interpretation on them. The word does not apply to everything—it has a meaning. Can interpreting always be going on when we see? Sometimes, perhaps, as when the hazy outline of an agricultural machine looms up on a foggy morning and, with effort, we finally identify it. Is this the "interpretation" which is active when bicycles and boxes are clearly seen? Is it active when the perspective of fig. 1 snaps into reverse? There was a time when

Herodotus was half-through with his interpretation of the Graeco-Persian wars. Could there be a time when one is half-through interpreting fig. 1 as a box from above, or as anything else?

"But the interpretation takes very little time—it is instantaneous." Instantaneous interpretation hails from the Limbo that produced unsensed sensibilia, unconscious inference, incorrigible statements, negative facts and *Objektive*. These are ideas which philosophers force on the world to preserve some pet epistemological or metaphysical theory.

Only in contrast to "Eureka" situations (like perspective reversals, where one cannot interpret the data) is it clear what is meant by saying that though Thucydides could have put an interpretation on history, he did not. Moreover, whether or not an historian is advancing an interpretation is an empirical question: we know what would count as evidence one way or the other. But whether we are employing an interpretation when we see fig. 1 in a certain way is not empirical. What could count as evidence? In no ordinary sense of "interpret" do I interpret fig. 1 differently when its perspective reverses for me. If there is some extraordinary sense of that word it is not clear, either in ordinary language, or in extraordinary (philosophical) language. To insist that different reactions to fig. 1 *must* lie in the interpretations put on a common visual experience is just to reiterate (without reasons) that the seeing of *x must* be the same for all observers looking at *x*.

"But 'I see the figure as a box' means: I am having a particular visual experience which I always have when I interpret the figure as a box, or when I look at a box. . . ." ". . . if I meant this, I ought to know it. I ought to be able to refer to the experience directly and not only indirectly. . . ."[19]

Ordinary accounts of the experiences appropriate to fig. 1 do not require visual grist going into an intellectual mill: theories and interpretations are "there" in the seeing from the outset. How can interpretations "be there" in the seeing? How is it possible to see an object according to an interpretation? "The question represents it as a queer fact; as if something were being forced into a form it did not really fit. But no squeezing, no forcing took place here."[20]

Consider now the reversible perspective figures which appear in textbooks on Gestalt psychology: the tea-tray, the shifting (Schröder) staircase, the tunnel. Each of these can be seen as concave, as convex, or as a flat drawing.[21] Do I really see something different each time, or do I only interpret what I see in a different way? To interpret is to think, to do something; seeing is an experiential state.[22] The different ways in which these figures are seen are not due to different thoughts lying behind the visual reactions. What could "spontaneous" mean if these reactions are not spontaneous? When the staircase "goes into reverse" it does so spontaneously. One does not think of anything special; one does not think at all. Nor does one interpret. One just sees, now a staircase as from above, now a staircase as from below.

The sun, however, is not an entity with such variable perspective. What has all this to do with suggesting that Tycho and Kepler may see different things in the east at dawn? Certainly the cases are different. But these reversible perspective figures are examples of different things being seen in the same configuration, where this difference is due neither to differing visual pictures, nor to any "interpretation" superimposed on the sensation. . . .

Fig. 2

A trained physicist could see one thing in fig 2: an X-ray tube viewed from the cathode. Would Sir Lawrence Bragg and an Eskimo baby see the same thing when looking at an X-ray tube? Yes, and no. Yes—they are visually aware of the same object. No—the *ways* in which they are visually aware are profoundly different. Seeing is not only the having of a visual experience; it is also the way in which the visual experience is had.

At school the physicist had gazed at this glass-and-metal instrument. Returning now, after years in University and research, his eye lights upon the same object once again. Does he see the same thing now as he did then? Now he sees the instrument in terms of electrical circuit theory, thermodynamic theory, the theories of metal and glass structure, thermionic emission, optical transmission, refraction, diffraction, atomic theory, quantum theory and special relativity.

Contrast the freshman's view of college with that of his ancient tutor. Compare a man's first glance at the motor of his car with a similar glance ten exasperating years later.

"Granted, one learns all these things," it may be countered, "but it all figures in the interpretation the physicist puts on what he sees. Though the layman sees exactly what the physicist sees, he cannot interpret it in the same way because he has not learned so much."

Is the physicist doing more than just seeing? No, he does nothing over and above what the layman does when he sees an X-ray tube. What are you doing over and above reading these words? Are you interpreting marks on a page? When would this ever be a natural way of speaking? Would an infant see what you see here, when you see words and sentences and he sees but marks and lines? One does nothing beyond looking and seeing when one dodges bicycles, glances at a friend, or notices a cat in the garden.

"The physicist and the layman see the same thing," it is objected, "but they do not make the same thing of it." The layman can make nothing of it. Nor is that just a figure of speech. I can make nothing of the Arab word for *cat,* though my purely visual impressions may be distinguishable from those of the Arab who can. I must learn Arabic before I can see what he sees. The layman must learn physics before he can see what the physicist sees.

If one must find a paradigm case of seeing it would be better to regard as such not the visual apprehension of colour patches but things like seeing what time it is, seeing what key a piece of music is written in, and seeing whether a wound is septic.[23]

Pierre Duhem writes:

> Enter a laboratory; approach the table crowded with an assortment of apparatus, an electric cell, silk-covered copper wire, small cups of mercury, spools, a mirror mounted on an iron bar; the experimenter is inserting into small openings the metal ends of ebony-headed pins; the iron oscillates, and the mirror attached to it throws a luminous band upon a celluloid scale; the forward-backward motion of this spot enables the physicist to observe the minute oscillations of the iron bar. But ask him what he is doing. Will he answer "I am studying the oscillations of an iron bar which carries a mirror?" No, he will say that he is measuring the electric resistance of the spools. If you are astonished, if you ask him what his words mean, what relation they have with the phenomena he has been observing and which you have noted at the same time as he, he will answer that your question requires a long explanation and that you should take a course in electricity.[24]

The visitor must learn some physics before he can see what the physicist sees. Only then will the context throw into relief those features of the objects before him which the physicist sees as indicating resistance.

This obtains in all seeing. Attention is rarely directed to the space between the leaves of a tree, save when a Keats brings it to our notice.[25] (Consider also what was involved in Crusoe's seeing a vacant space in the sand as a footprint.) Our attention most naturally rests on objects and events which dominate the visual field. What a blooming, buzzing, undifferentiated confusion visual life would be if we all arose tomorrow without attention capable of dwelling only on what had heretofore been overlooked.[26]

The infant and the layman can see: they are not blind. But they cannot see what the physicist sees; they are blind to what he sees.[27] We may not hear that the oboe is out of tune, though this will be painfully obvious to the trained musician. (Who, incidentally, will not hear the tones and *interpret* them as being out of tune, but will simply hear the oboe to be out of tune.[28] We simply see what time it is; the surgeon simply sees a wound to be septic; the physicist sees the X-ray tube's anode overheating.) The elements of the visitor's visual field, though identical with those of the physicist, are not organized for him as for the physicist; the same lines, colours, shapes are apprehended by both, but not in the same way. There are indefinitely many ways in which a constellation of lines, shapes, patches, may be seen. *Why* a visual pattern is seen differently is a question for psychology, but *that* it may be seen differently is important in any examination of the concepts of seeing and observation. Here, as Wittgenstein might have said, the psychological is a symbol of the logical.

You see a bird, I see an antelope; the physicist sees an X-ray tube, the child a complicated lamp bulb; the microscopist sees coelenterate mesoglea, his new student sees only a gooey, formless stuff. Tycho and Simplicius see a mobile sun, Kepler and Galileo see a static sun.[29]

It may be objected, "Everyone, whatever his state of knowledge, will see fig. 1 as a box or cube, viewed as from above or as from below." True; almost everyone, child, layman, physicist, will see the figure as box-like one way or another. But could such observations be made by people ignorant of the construction of box-like objects? No. This objection only shows that most of us — the blind, babies, and dimwits excluded — have learned enough to be able to see this figure as a three-dimensional box. This reveals something about the sense in which Simplicius and Galileo do see the same thing (which I have never denied): they both see a brilliant heavenly body. The schoolboy and the physicist both see that the X-ray tube will smash if dropped. Examining how observers see different things in *x* marks something important about their seeing the same thing when looking at *x*. If seeing different things involves having different knowledge and theories about *x,* then perhaps the sense in which they see the same thing involves their sharing knowledge and theories about *x*. Bragg and the baby share no knowledge of X-ray tubes. They see the same thing only in that if they are looking at *x* they are both having some visual experience of it. Kepler and Tycho agree on more: they see the same thing in a stronger sense. Their visual fields are organized in much the same way. Neither sees the sun about to break out in a grin, or about to crack into ice cubes. (The baby is not "set" even against these eventualities.) Most people today see the same thing at dawn in an even stronger sense: we share much knowledge of the sun. Hence Tycho and Kepler see different things, and yet they see the same thing. That these things can be said depends on their knowledge, experience, and theories.

. . . The elements of their experiences are identical; but their conceptual organization is vastly different. Can their visual fields have a different organization? Then they can see different things in the east at dawn.

It is the sense in which Tycho and Kepler do not observe the same thing which must be grasped if one is to understand disagreements within microphysics. Fundamental physics is primarily a search for intelligibility — it is philosophy of matter. Only secondarily is it a search for objects and facts (though the two endeavours are as hand and glove). Microphysicists seek new modes of conceptual organization. If that can be done the finding of new entities will follow. Gold is rarely discovered by one who has not got the lay of the land.

To say that Tycho and Kepler, Simplicius and Galileo, Hooke and Newton, Priestley and Lavoisier, Soddy and Einstein, De Broglie and Born, Heisenberg and Bohm all make the same observations but use them differently is too easy.[30] It does not explain controversy in research science. Were there no sense in which they were different observations they could not be used differently. This may perplex some: that researchers sometimes do not appreciate data in the same way is a serious matter. It is important to realize, however, that sorting out differences about data, evidence, observation, may require more than simply gesturing at observable objects. It may require a comprehensive reappraisal of one's subject matter. This may be difficult, but it should not obscure the fact that nothing less than this may do. . . .

NOTES

1.   Wär' nicht das Auge sonnenhaft,
                     Die Sonne könnt' es nie erblicken;
        Goethe, *Zahme Xenien* (Werke, Weimar, 1887–1918), Bk. 3, 1805.

2. Cf. the papers by Baker and Gatonby in *Nature,* 1949–present.

3. This is not a *merely* conceptual matter, of course. Cf. Wittgenstein, *Philosophical Investigations* (Blackwell, Oxford, 1953), p. 196.

4. (1) G. Berkeley, *Essay Towards a New Theory of Vision* (in *Works,* vol. I (London, T. Nelson, 1948-56), pp. 51 ff.

(2) James Mill, *Analysis of the Phenomena of the Human Mind* (Longmans, London, 1869), vol. I, p. 97.

(3) J. Sully, *Outlines of Psychology* (Appleton, New York, 1885).

(4) William James, *The Principles of Psychology* (Holt, New York, 1890-1905), vol. II, pp. 4, 78, 80 and 81; vol. I, p. 221.

(5) A Schopenhauer, *Satz vom Grunde* (in *Sämmtliche Werke,* Leipzig, 1888), ch. IV.

(6) H. Spencer, *The Principles of Psychology* (Appleton, New York, 1897), vol. IV, chs. IX, X.

(7) E. von Hartmann, *Philosophy of the Unconscious* (K. Paul, London, 1931), B, chs. VII, VIII.

(8) W. M. Wundt, *Vorleşungen über die Menschen und Thierseele* (Voss, Hamburg, 1892), IV, XIII.

(9) H. L. F. von Helmholtz, *Handbuch der Physiologischen Optik* (Leipzig, 1867), pp. 430, 447.

(10) A. Binet, *La psychologie du raisonnement, recherches expérimentales par l'hypnotisme* (Alcan, Paris, 1886), chs. III, V.

(11) J. Grote, *Exploratio Philosophica* (Cambridge, 1900), vol. II, pp. 201 ff.

(12) B. Russell, in *Mind* (1913), p. 76. *Mysticism and Logic* (Longmans, New York, 1918), p. 209. *The Problems of Philosophy* (Holt, New York, 1912), pp. 73, 92, 179, 203.

(13) Dawes Hicks, *Arist. Soc. Sup.* vol. II (1919), pp. 176-8.

(14) G. F. Stout, *A Manual of Psychology* (Clive, London, 1907, 2nd ed.), vol. II, 1 and 2, pp. 324, 561-4.

(15) A. C. Ewing, *Fundamental Questions of Philosophy* (New York, 1951), pp. 45 ff.

(16) G. W. Cunningham, *Problems of Philosophy* (Holt, New York, 1924), pp. 96-7.

5. Galileo, *Dialogue Concerning the Two Chief World Systems* (California, 1953), "The First Day," p. 33.

6. "'Das ist doch kein Sehen!'—'Das ist doch ein Sehen!' Beide müssen sich begrifflich recht-fertigen lassen" (Wittgenstein, *Phil. Inv.* p. 203).

7. Brain, *Recent Advances in Neurology* (with Strauss) (London, 1929), p. 88. Compare Helmholtz: "The sensations are signs to our consciousness, and it is the task of our intelligence to learn to understand their meaning" (*Handbuch der Physiologischen Optik* (Leipzig, 1867), vol. III, p. 433).

See also Husserl, "Ideen zu einer Reinen Phaenomenologie," in *Jahrbuch für Philosophie,* vol. I (1913), pp. 75, 79, and Wagner's *Handwörterbuch der Physiologie,* vol. III, section I (1846), p. 183.

8. Mann, *The Science of Seeing* (London, 1949), pp. 48–9. Arber, *The Mind and the Eye* (Cambridge, 1954). Compare Müller: "In any field of vision, the retina sees only itself in its spatial extension during a state of affection. It perceives itself as . . . etc." (*Zur vergleichenden Physiologie des Gesichtesinnes des Menschen und der Thiere* (Leipzig, 1826), p. 54).

9. Kolin: "An astigmatic eye when looking at millimeter paper can accommodate to see sharply either the vertical lines or the horizontal lines" (*Physics* (New York, 1950), pp. 570 ff.).

10. Cf. Whewell, *Philosophy of Discovery* (London, 1860), "The Paradoxes of Vision."

11. Cf. e.g. J. Z. Young, *Doubt and Certainty in Science* (Oxford, 1951, The Reith Lectures), and Gray Walter's article in *Aspects of Form,* ed. by L. L. Whyte (London, 1953). Compare Newton: "Do not the Rays of Light in falling upon the bottom of the Eye excite Vibrations in the Tunica Retina? Which Vibrations, being propagated along the solid Fibres of the Nerves into the Brain, cause the Sense of seeing" (*Opticks* (London, 1769), Bk. III, part I).

12. "Rot und grün kann ich nur sehen, aber nicht hören" (Wittgenstein, *Phil. Inv.* p. 209).

13. Cf. "An appearance is the same whenever the same eye is affected in the same way" (Lambert, *Photometria* (Berlin, 1760)); "We are justified, when different perceptions offer themselves to us, to infer that the underlying real conditions are different" (Helmholtz, *Wissenschaftliche Abhandlungen* (Leipzig, 1882), vol. II, p. 656), and Hertz: "We form for ourselves images or symbols of the external objects; the manner in which we form them is such that the logically necessary *(denknotwendigen)* consequences of the images in thought are invariably the images of materially necessary (*naturnotwendigen*) consequences of the corresponding objects" (*Principles of Mechanics* (London, 1889), p. 1).

Broad and Price make depth a feature of the private visual pattern. However, Weyl (*Philosophy of Mathematics and Natural Science* (Princeton, 1949), p. 125) notes that a single eye perceives qualities spread out in a *two*-dimensional field, since the latter is dissected by any one-dimensional line running through it. But our conceptual difficulties remain even when Kepler and Tycho keep one eye closed.

Whether or not two observers are having the same visual sense-data reduces directly to the question of whether accurate pictures of the contents of their visual fields are identical, or differ in some detail. We can then discuss the publicly observable pictures which Tycho and Kepler draw of what they see, instead of those private, mysterious entities locked in their visual consciousness. The accurate picture and the sense-datum must be identical; how could they differ?

14. From the B.B.C. report, 30 June 1954.

15. Newton, *Opticks,* Bk. II, part I. The writings of Claudius Ptolemy sometimes read like a phenomenalist's textbook. Cf. e.g. *The Almagest* (Venice, 1515), VI, section II, "On the Directions in the Eclipses," "When it touches the shadow's circle from within," "When the circles touch each other from without." Cf. also VII and VIII, IX (section 4). Ptolemy continually seeks to chart and predict "the appearances"—the points of light on the celestial globe. *The Almagest* abandons any attempt to explain the machinery behind these appearances.

Cf. Pappus: "The (circle) dividing the milk-white portion which owes its colour to the sun, and the portion which has the ashen colour natural to the moon itself is indistinguishable from a great circle" (*Mathematical Collection* (Hultsch, Berlin and Leipzig, 1864), pp. 554–60).

16. This famous illusion dates from 1832, when L. A. Necker, the Swiss naturalist, wrote a letter to Sir David Brewster describing how when certain rhomboidal crystals were viewed on end the perspective could shift in the way now familiar to us. Cf. *Phil. Mag.* III, no. I (1832), 329–37, especially p. 336. It is important to the present argument to note that this observational phenomenon began life not as a psychologist's trick, but at the very frontiers of observational science.

17. Wittgenstein answers: "Denn wir sehen eben wirklich zwei verschiedene Tatsachen" (*Tractatus*, 5. 5423).

18. "Auf welche Vorgänge spiele ich an?" (Wittgenstein, *Phil. Inv.* p. 214).

19. *Ibid.* p. 194 (top).

20. *Ibid.* p. 200.

21. This is *not* due to eye movements, or to local retinal fatigue. Cf. Flugel, *Brit. J. Psychol.* VI (1913), 60; *Brit. Psychol.* V (1913), 357. Cf. Donahue and Griffiths, *Amer. J. Psychol.* (1931), and Luckiesch, *Visual Illusions and their Applications* (London, 1922). Cf. also Peirce, *Collected Papers* (Harvard, 1931), 5, 183. References to psychology should not be misunderstood; but as one's acquaintance with the psychology of perception deepens, the character of the conceptual problems one regards as significant will deepen accordingly. Cf. Wittgenstein, *Phil. Inv.* p. 206 (top). Again, p. 193: "Its causes are of interest to psychologists. We are interested in the concept and its place among the concepts of experience."

22. Wittgenstein, *Phil. Inv.* p. 212.

23. Often "What do you see?" only poses the question "Can you identify the object before you?" This is calculated more to test one's knowledge than one's eyesight.

24. Duhem, *La théorie physique* (Paris, 1914), p. 218.

25. Chinese poets felt the significance of "negative features" like the hollow of a clay vessel or the central vacancy of the hub of a wheel (cf. Waley, *Three Ways of Thought in Ancient China* (London, 1939), p. 155).

26. Infants are indiscriminate; they take in spaces, relations, objects and events as being of equal value. They still must learn to organize their visual attention. The camera-clarity of their visual reactions is not by itself sufficient to differentiate elements in their visual fields. Contrast Mr. W. H. Auden who recently said of the poet that he is "bombarded by a stream of varied sensations which would drive him mad if he took them all in. It is impossible to guess how much energy we have to spend every day in not-seeing, not-hearing, not-smelling, not-reacting."

27. "He was blind to the *expression* of a face. Would his eyesight on that account be defective?" (Wittgenstein, *Phil. Inv.* p. 210) and "Because they seeing see not; and hearing they hear not, neither do they understand" (Matt. xiii. 10–13).

28. "Es hört doch jeder nur, was er versteht" (Goethe, *Maxims* (*Werke*, Weimar, 1887–1918)).

29. Against this Professor H. H. Price has argued: "Surely it appears to both of them to be rising, to be moving upwards, across the horizon . . . they both see a moving sun: they both see a round bright body which appears to be rising." Philip Frank retorts: "Our sense observation shows only that in the morning the distance between horizon and sun is increasing, but it does not tell us whether the sun is ascending or the horizon is descending . . ." (*Modern Science and its Philosophy* (Harvard, 1949), p. 231). Precisely. For Galileo and Kepler the horizon drops; for Simplicius and Tycho the sun rises. This is the difference Price misses, and which is central to this essay.

30. This parallels the too-easy epistemological doctrine that all normal observers see the same things in *x*, but interpret them differently.

11

The Theory-Ladenness of Observation
Carl R. Kordig

. . . .
A.

Hanson's remarks about observation in science are indeed quite provocative. Unlike Kant he maintains that even *particular* theories determine what is seen. He admits, however, that scientists in different traditions see the same thing in one sense of 'see' ([1], pp. 5, 7, 8, 18, 20). For example, *something* about the visual experiences of Johannes Kepler and Tycho Brahe, when on a hill watching the dawn, is the same for both. Namely, both have a visual experience of a brilliant yellow-white disc centered between green and blue color patches ([1], pp. 8, 18). And for both, the distance between this disc and the horizon is increasing ([1], p. 182, note 6).

Hanson, however, claims that this type of seeing does not exhaust the concept. He feels, with Kuhn, that there is also a sense in which two observers do not see the same thing, do not begin from the same data, even though they are visually aware of the same object ([1], p. 18). He presents us with several Gestalt figures to illustrate this second sense of seeing—"seeing-as" ([1], pp. 8–18).

Well, just what have Gestalt examples to do with science? One might agree that Gestalt examples provide genuine examples of *seeing*; yet one might still doubt that observations relevant· and important to science and scientific disputes are of this kind. Hanson wants to remove this doubt. He wants to claim that examples of *seeing-as,* analogous to his Gestalt examples, have occurred within the history of science; moreover, he wants to claim that they were more important than neutral observations for understanding scientific change and controversy.

There are then for Hanson these two senses of 'see.' In one sense (the neutral one) scientists, but presumably not *qua* scientists, see the same thing. In the other they do not. And Hanson, with Kuhn, urges it is in the latter sense, that we get *scientific* "data" ([1], pp. 4–5, 17; [2], p. 134). Thus, he concludes that Tycho and Kepler do not begin their inquiries from the same data, do not make the same observations, do not even see the same thing ([1], p. 5). Together on a hill Tycho and Kepler, according to Hanson, see (in the non-neutral sense of 'see') different things in the east at dawn. How could it be otherwise? "Practicing in different worlds, the two . . . scientists see different things when they look from the same point in the same direction" (Kuhn, [2], p. 149). "Kepler and Tycho are to the sun as we are to Fig. 1, when I see the bird and you see only the antelope" ([1], p. 18).

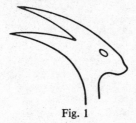

Fig. 1

A difficulty arises here. Hanson says you and I *both* see his Fig. 1, although we don't both see a bird. If then the sun is to be equated with Fig. 1, as Hanson has just suggested, *both* Tycho and Kepler should see it. Hanson's analogy, if accurate, seems to undercut his own position; he maintains that it is the sense in which Tycho and Kepler do *not* observe the same thing which must be grasped if one is to understand fundamental disagreements within science.

Let us, however, proceed to examine Hanson's argument that Tycho and Kepler see different things in a sense important and relevant to science.[1] The argument starts from two virtually incontestable premises,

(1) m_0 (Tycho) sees X_0 (the sun Tycho sees).

(2) m_1 (Kepler) sees X_1 (the sun Kepler sees).

Now for Hanson to *see* an object X is to *see that* if $A_1, \ldots A_n$, were done to X then $B_1, \ldots B_n$ respectively would result ([1], pp. 20–21, 22–23, 29–30, 58–59, 97).[2] In short, it is to see that x behaves in certain essential ways, i.e., has certain deep dispositional properties. For one to see Tycho's sun (X_0) is for him to at least see that from some celestial vantage point, the sun is such that it could be watched circling our fixed earth ([1], p. 23). That is, he would see a sun which is essentially mobile ([1], p. 17, 23–24, 182). "Watching the sun at dawn through Tychonic spectacles would be to see it in something like this way" ([1], p. 23). Hanson, therefore, adds (3) as a further premiss to his argument.

(3) If anyone sees X_0 then he sees that $P_0(X_0)$.

where 'P_0' stands for 'is mobile,' a predicate in Tycho's system which holds essentially of the sun.

The situation is, however, different for Kepler's sun. For one to see Kepler's sun (X_1) is for him at least to see that the sun does not behave in the above "Tychonic" way (cf. [1], p. 23). Why? Well, Kepler sees a static sun, whereas Tycho sees a mobile sun ([1], pp. 17, 23–24, 182). When Kepler sees the sun, he sees the horizon dipping or turning away from it. When Tycho sees the sun, he sees it ascending or rising ([1], pp. 23, 182). "The shift from sunrise to horizon-turn is analogous to the shift-of-aspect phenomena [viz., the Gestalt examples] already considered . . ." ([1], pp. 23–24). That is, Hanson adds another premise, (4), to his argument.

(4) If anyone sees X_1 then he sees that $P_1(X_1)$.

where 'P_1' stands for 'is static,' a predicate in Kepler's system which holds essentially of the sun.

From (1) and (3), Hanson validly infers (5),

(5) m_0 sees that $P_0(X_0)$.

And from (2) and (4), he similarly infers (6),

(6) m_1 sees that $P_1(X_1)$.

Now m_0 and m_1 presumably know that anything that is static is not mobile. Therefore, Hanson obtains (7) from (6),

(7) m_1 sees that $\sim P_0(X_1)$.

From (5) and (7) he then concludes (8),

(8) $X_0 \neq X_1$.

(8) says Tycho and Kepler do not see the same sun. But the validity of the inference from (5) and (7) to (8) is questionable. It depends on the meaning of 'sees that.' We will examine this shortly. Let us for the moment assume it to be valid. Then one link remains in Hanson's argument. Why is this sense of 'see,' and not the neutral sense, important and relevant to Kepler's and Tycho's observations? Simply because the differences in what they see amounts to the difference between an essentially geocentric and essentially heliocentric universe. And this *is* the essential and deep difference between Tycho's and Kepler's physical theories.

The preceding is Hanson's argument. I think it is fallacious. Hanson implicitly uses the phrase 'sees that' to mean either 'knows that' (cf., e.g., [1], pp. 18, 20–22) or 'believes that' (cf., e.g., [1], pp. 23–24). In either case his argument is unacceptable.

B.

Consider the case in which 'sees that' means 'knows that.' In this case Hanson's argument is valid, but leads to absurd consequences; and we shall discover that this is because its premises are false. From (5) and the fact that we are taking 'sees that' to mean 'knows that' we can infer (9),

(9) m_0 knows that $P_0(X_0)$.

Now, whatever else it may be, what is known is at least true. That is, if m_0 knows that $P_0(X_0)$, then we can conclude that $P_0(X_0)$ is the case. Therefore (10) follows,

(10) $P_0(X_0)$.

From (7) we can similarly infer (11),

(11) $\sim P_0(X_1)$

To show Hanson's argument is valid we will show that (8), his conclusion, follows from (10) and (11). And this is readily done: for if $X_0 = X_1$, by (10), we would have $P_0(X_1)$ which contradicts (11). Thus, Hanson's argument is valid.

It is, nevertheless, fallacious. The careful reader may have already noted the oddity of (10). Yet (10) was needed to make Hanson's argument valid; and it was indeed obtainable on the supposition that 'sees that' means 'knows that.' (10) is strange. Tycho's theory is false. He was mistaken about the status of the sun in the solar system. He mistakenly thought it was mobile. Either we hold (10) or not. If we hold (10), we will have to give up the widely accepted view that Tycho was wrong in thinking his sun revolved around the earth. On the other hand, if we give up (10), we will be forced to give up either (1) or (3). They were the premises which entailed (10). One could not, therefore, rationally hold them and deny (10). Now, the truth of (1) is, I think, beyond question. To deny it amounts to the claim that Tycho didn't see the sun he saw, which is absurd. If 'sees that' means 'knows that' Hanson's premiss (3) should, therefore, be abandoned if we deny (10).

Hanson would, given his philosophical position, have to maintain that a geocentric world picture really isn't wrong after all. However, Hanson's own point of view is also that of modern science. Thus, he should say that a geocentric world picture is wrong, and indeed was wrong before Kepler. He should therefore maintain that a geocentric point of view is both wrong and not wrong if his own point of view includes both modern science and his own philosophical position. At the expense of the viewpoint of modern science he might maintain that his philosophical position is the single, preferred, or absolute point of view. But then he would be open to the charge that, in spite of his explicit disclaimers, he is not engaging in a *descriptive* philosophy of science; he would be, rather, philosophically legislating to science—telling scientists that what they regard as correct scientific beliefs (e.g., that the solar system has never been geocentric) are, in fact, wrong. And this charge *Hanson* would have to take seriously. He thinks that a principal merit of his own approach is its descriptivism and its avoidance of the normative excesses of positivism. . . .

NOTES

1. Not only do Tycho and Kepler see different things, but laymen are compared to infants and idiots in that they are held to all be literally blind to what the physicist sees ([1], pp. 17, 20, 22).

2. Hanson implicitly claims "seeing X as Y" is the same as "seeing an object X." Namely, seeing X as Y *is to* see that if $A_1, \ldots A_n$ were done to X then $B_1, \ldots B_n$ respectively would result (cf. [1], p. 21). This is, unfortunately, incompatible with Hanson's explicit claim that he is not identifying *seeing* with *seeing-as*: "I do not mean to identify seeing with seeing as" ([1], p. 19). Since 'sees that' has the same sense in both claims, so would Hanson's two types of 'seeing.'

REFERENCES

[1] Hanson, N. R., *Patterns of Discovery* (Cambridge: The University Press, 1958).
[2] Kuhn, T. S., *The Structure of Scientific Revolutions* (Chicago: University of Chicago Press, 1962).

12

Science and the Physical World

W. T. Stace

So far as I know scientists still talk about electrons, protons, neutrons, and so on. We never directly perceive these, hence if we ask how we know of their existence the only possible answer seems to be that they are an inference from what we do directly perceive. What sort of an inference? Apparently a causal inference. The atomic entities in some way impinge upon the sense of the animal organism and cause that organism to perceive the familiar world of tables, chairs, and the rest.

But is it not clear that such a concept of causation, however interpreted, is invalid? The only reason we have for believing in the law of causation is that we *observe* certain regularities or sequences. We observe that, in certain conditions, *A* is always followed by *B*. We call *A* the cause, *B* the effect. And the sequence *A-B* becomes a causal law. It follows that all *observed* causal sequences are between sensed objects in the familiar world of perception, and that all known causal laws apply solely to the world of sense and not to anything beyond or behind it. And this in turn means that we have not got, and never could have, one jot of evidence for believing that the law of causation can be applied *outside* the realm of perception, or that that realm can have any causes (such as the supposed physical objects) which are not themselves perceived.

Put the same thing in another way. Suppose there is an observed sequence *A-B-C*, represented by the vertical lines in the diagram below.

The observer X sees, and can see, nothing except things in the familiar world of perception. What *right* has he, and what *reason* has he, to assert causes of A, B, and C, such as a′, b′, c′, which he can never observe, behind the perceived world? He has no *right*, because the law of causation on which he is relying has never been observed to operate outside the series of perceptions, and he can have, therefore, no evidence that it does so. And he has no *reason*

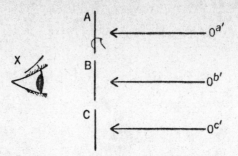

because the phenomenon C is *sufficiently* accounted for by the cause B, B by A, and so on. It is unnecessary and superfluous to introduce a *second* cause b′ for B, c′ for C, and so forth. To give two causes for each phenomenon, one in one world and one in another, is unnecessary, and perhaps even self-contradictory.

Is it denied, then, it will be asked, that the star causes light waves, that the waves cause retinal changes, that these cause changes in the optic nerve, which in turn causes movement in the brain cells, and so on? No, it is not denied. But the observed causes and effects are all in the world of perception. And no sequences of sense-data can possibly justify going outside that world. If you admit that we never observe anything except sensed objects and their relations, regularities, and sequences, then it is obvious that we are completely shut in by our sensations and can never get outside them. Not only causal relations, but all other observed relations, upon which *any* kind of inferences might be founded, will lead only to further sensible objects and their relations. No inference, therefore, can pass from what is sensible to what is not sensible.

The fact is that atoms are *not* inferences from sensations. No one denies, of course, that a vast amount of perfectly valid inferential reasoning takes place in the physical theory of the atom. But it will not be found to be in any strict logical sense inference from *sense-data to atoms.* An *hypothesis* is set up, and the inferential processes are concerned with the application of the hypothesis, that is, with the prediction by its aid of further possible sensations and with its own internal consistency.

That atoms are not inferences from sensations means, of course, that from the existence of sensations we cannot validly infer the existence of atoms. And this means that we cannot have any reason at all to believe that they exist. And that is why I propose to argue that they do not exist — or at any rate that no one could know it if they did, and that we have absolutely no evidence of their existence.

What status have they, then? Is it meant that they are false and worthless, merely untrue? Certainly not. No one supposes that the entries in the nautical almanac "exist" anywhere except on the pages of that book and in the brains of its compilers and readers. Yet they are "true," inasmuch as they enable us to predict certain sensations, namely, the positions and times of certain perceived objects which we call the stars. And so the formulae of the atomic theory are true in the same sense, and perform a similar function.

I suggest that they are nothing but shorthand formulae, ingeniously worked out by the human mind, to enable it to predict its experience, i.e. to predict what sensations will be given to it. By "predict" here I do not mean to refer solely to the future. To calculate that there was an eclipse of the sun visible in Asia Minor in the year 585 B.C. is, in the sense in which I am using the term, to predict.

In order to see more clearly what is meant, let us apply the same idea to another case, that of gravitation. Newton formulated a law of gravitation in terms of "forces." It was supposed that this law—which was nothing but a mathematical formula—governed the operation of these existent forces. Nowadays it is no longer believed that these forces exist at all. And yet the law can be applied just as well without them to the prediction of astronomical phenomena. It is a matter of no importance to the scientific man whether the forces exist or not. That may be said to be a purely philosophical question. And I think the philosopher should pronounce them fictions. But that would not make the law useless or untrue. If it could still be used to predict phenomena, it would be just as true as it was.

It is true that fault is now found with Newton's law, and that another law, that of Einstein, has been substituted for it. And it is sometimes supposed that the reason for this is that forces are no longer believed in. But this is not the case. Whether forces exist or not simply does not matter. What matters is the discovery that Newton's law does *not* enable us accurately to predict certain astronomical facts such as the exact position of the planet Mercury. Therefore another formula, that of Einstein, has been substituted for it which permits correct predictions. This new law, as it happens, is a formula in terms of geometry. It is pure mathematics and nothing else. It does not contain anything about forces. In its pure form it does not even contain, so I am informed, anything about "humps and hills in space-time." And it does not matter whether any such humps and hills exist. It is truer than Newton's law, not because it substitutes humps and hills for forces, but solely because it is a more accurate formula of prediction.

Not only may it be said that forces do not exist. It may with equal truth be said that "gravitation" does not exist. Gravitation is not a "thing," but a mathematical formula, which exists only in the heads of mathematicians. And as a mathematical formula cannot cause a body to fall, so gravitation cannot cause a body to fall. Ordinary language misleads us here. We speak of the law "of" gravitation, and suppose that this law "applies to" the heavenly bodies. We are thereby misled into supposing that there are *two* things, namely, the gravitation and the heavenly bodies, and that one of these things, the gravitation, causes changes in the other. In reality nothing exists except the moving bodies. And neither Newton's law nor Einstein's law is, strictly speaking, a law of gravitation. They are both laws of moving bodies, that is to say, formulae which tell us how these bodies will move.

Now, just as in the past "forces" were foisted into Newton's law (by himself, be it said), so now certain popularizers of relativity foisted "humps and hills in space-time" into Einstein's law. We hear that the reason why the planets move in curved courses is that they cannot go through these humps and hills, but have to go round them! The planets just get "shoved about," not by forces, but by the humps and hills! But these humps and hills are pure metaphors. And anyone who takes them for "existences" gets asked awkward questions as to what "curved space" is curved "in."

It is not irrelevant to our topic to consider *why* human beings invent these metaphysical monsters of forces and bumps in space-time. The reason is that they have never emancipated themselves from the absurd idea that science "explains" things. They were not content to have laws which merely told them *that* the planets will, as a matter of fact, move in such and such ways. They wanted to know "why" the planets move in those ways. So Newton replied, "Forces." "Oh," said humanity, "that explains it. We understand forces. We feel them every time someone pushes or pulls us." Thus the movements were supposed to be "explained" by entities familiar because analogous to the muscular sensations which human beings feel. The humps and hills were introduced for exactly the same reason. They seem so familiar. If there is a bump in the billiard table, the rolling billiard ball is diverted from a straight to a curved course. Just the same with the planets. "Oh, I see!" says humanity, "that's quite simple. That *explains* everything."

But scientific laws, properly formulated, never "explain" anything. They simply state, in an abbreviated and generalized form, *what happens*. No scientist, and in my opinion no philosopher, knows *why* anything happens, or can "explain" anything. Scientific laws do nothing except state the brute fact that "when *A* happens, *B* always happens too." And laws of this kind obviously enable us to predict. If certain scientists substituted humps and hills for forces, then they have just substituted one superstition for another. For my part I do not believe that *science* has done this, though some *scientists* may have. For scientists, after all, are human beings with the same craving for "explanations" as other people.

I think that atoms are in exactly the same position as forces and the humps and hills of space-time. In reality the mathematical formulae which are the scientific ways of stating the atomic theory are simply formulae for calculating what sensations will appear in given conditions. But just as the weakness of the human mind demanded that there should correspond to the formula of gravitation a real "thing" which could be called "gravitation itself" or "force," so the same weakness demands that there should be a real thing corresponding to the atomic formulae, and this real thing is called the atom. In reality the atoms no more cause sensations than gravitation causes apples to fall. The only causes of sensations are other sensations. And the relation of atoms to sensations to be felt is not the relation of cause to effect, but the relation of a mathematical formula to the facts and happenings which it enables the mathematician to calculate.

Some writers have said that the physical world has no color, no sound, no taste, no smell. It has no spatiality. Probably it has not even number. We must not suppose that it is in any way like our world, or that we can understand it by attributing to it the characters of our world. Why not carry this progress to its logical conclusion? Why not give up the idea that it has even the character of "existence" which our familiar world has? We have given up smell, color, taste. We have given up even space and shape. We have given up number. Surely, after all that, mere existence is but a little thing to give up. No? Then is it that the idea of existence conveys "a sort of halo"? I suspect so. The "existence" of atoms is but the expiring ghost of the pellet and billiard-ball atoms of our forefathers. They, of course, had size, shape, weight, hardness. These have gone. But thinkers still cling to their existence, just as their fathers clung to the existence of forces, and for the same reason. Their reason is not in the slightest that science has any use for the existent atom. But the *imagination* has. It seems somehow to explain things, to make them homely and familiar.

It will not be out of place to give one more example to show how common fictitious existences are in science, and how little it matters whether they really exist or not. This example has no strange and annoying talk of "bent spaces" about it. One of the foundations of physics is, or used to be, the law of the conservation of energy. I do not know how far, if at all, this has been affected by the theory that matter sometimes turns into energy. But that does not affect the lesson it has for us. The law states, or used to state, that the amount of energy in the universe is always constant, that energy is never either created or destroyed. This was highly convenient, but it seemed to have obvious exceptions. If you throw a stone up into the air, you are told that it exerts in its fall the same amount of energy which it took to throw it up. But suppose it does not fall. Suppose it lodges on the roof of your house and stays there. What has happened to the energy which you can nowhere perceive as being exerted? It seems to have disappeared out of the universe. No, says the scientist, it still exists as *potential* energy. Now what does this blessed word "potential"—which is thus brought in to save the situation—mean as applied to energy? It means, of course, that the energy does not exist in any of its regular "forms," heat, light, electricity, etc. But this is merely negative. What positive meaning has the term? Strictly speaking, none whatever. Either the energy exists or it does not exist. There is no realm of the "potential" half-way between existence and non-existence. And the existence of energy can only consist in its being exerted. If the energy is not being exerted, then it is not energy and does not exist. Energy can no more exist without energizing than heat can exist without being hot. The "potential" existence of the energy is, then, a fiction. The actual empirically verifiable facts are that if a certain quantity of energy e exists in the universe and then disappears out of the universe (as happens when the stone lodges on the roof), the same amount of energy e will always reappear, begin to exist again, in certain known conditions. That is the fact which the law of the conservation of energy actually expresses. And the fiction of potential energy is introduced simply because it is convenient and makes the equations easier to work. They could be worked quite well without it, but would be slightly more complicated. In either case the function of the law is the same. Its object is to apprise us that if in certain conditions we have certain perceptions (throwing up the stone), then in certain other conditions we shall get certain other perceptions (heat, light, stone hitting skull, or other such). But there will always be a temptation to hypostatize the potential energy as an "existence," and to believe that it is a "cause" which "explains" the phenomena.

If the views which I have been expressing are followed out, they will lead to the conclusion that, strictly speaking, *nothing exists except sensations* (and the minds which perceive them). The rest is mental construction or fiction. But this does not mean that the conception of a star or the conception of an electron are worthless or untrue. Their truth and value consist in their capacity for helping us to organize our experience and predict our sensations.

13

The Ontological Status
of Theoretical Entities

Grover Maxwell

That anyone today should seriously contend that the entities referred to by scientific theories are only convenient fictions, or that talk about such entities is translatable without remainder into talk about sense contents or everyday physical objects, or that such talk should be regarded as belonging to a mere calculating device and, thus, without cognitive content—such contentions strike me as so incongruous with the scientific and rational attitude and practice that I feel this paper *should* turn out to be a demolition of straw men. But the instrumentalist views of outstanding physicists such as Bohr and Heisenberg are too well known to be cited, and in a recent book of great competence, Professor Ernest Nagel concludes that "the opposition between [the realist and the instrumentalist] views [of theories] is a conflict over preferred modes of speech" and "the question as to which of them is the 'correct position' has only terminological interest."[1] The phoenix, it seems, will not be laid to rest.

The literature on the subject is, of course, voluminous, and a comprehensive treatment of the problem is far beyond the scope of one essay. I shall limit myself to a small number of constructive arguments (for a radically realistic interpretation of theories) and to a critical examination of some of the more crucial assumptions (sometimes tacit, sometimes explicit) that seem to have generated most of the problems in this area.[2]

The Problem

Although this essay is not comprehensive, it aspires to be fairly self-contained. Let me, therefore, give a pseudohistorical introduction to the problem with a piece of science fiction (or fictional science).

In the days before the advent of microscopes, there lived a Pasteur-like scientist whom, following the usual custom, I shall call Jones. Reflecting on the fact that certain diseases seemed to be transmitted from one person to another by means of bodily contact or by contact with articles handled previously by an afflicted person, Jones began to speculate about the mechanism of the transmission. As a "heuristic crutch," he recalled that there is an obvious *observable* mechanism for transmission of certain afflictions (such as body lice), and he postulated that all, or most, infectious diseases were spread in a similar manner but that in most cases the corresponding "bugs" were too small to be seen and, possibly, that some of them lived inside the bodies of their hosts. Jones proceeded to develop his theory and to examine its testable consequences. Some of these seemed to be of great importance for preventing the spread of disease.

After years of struggle with incredulous recalcitrance, Jones managed to get some of his preventative measures adopted. Contact with or proximity to diseased persons was avoided when possible, and articles which they handled were "disinfected" (a word coined by Jones) either by means of high temperatures or by treating them with certain toxic preparations which Jones termed "disinfectants." The results were spectacular: within ten years the death rate had declined 40 per cent. Jones and his theory received their well-deserved recognition.

However, the "crobes" (the theoretical term coined by Jones to refer to the disease-producing organisms) aroused considerable anxiety among many of the philosophers and philosophically inclined scientists of the day. The expression of this anxiety usually began something like this: "In order to account for the facts, Jones must assume that his crobes are too small to be seen. Thus the very postulates of his theory preclude their being observed; they are *unobservable in principle*." (Recall that no one had envisaged such a thing as a microscope.) This common prefatory remark was then followed by a number of different "analyses" and "interpretations" of Jones' theory. According to one of these, the tiny organisms were merely convenient fictions—*façons de parler*—extremely useful as heuristic devices for facilitating (in the "context of discovery") the thinking of scientists but not to be taken seriously in the sphere of cognitive knowledge (in the "context of justification"). A closely related view was that Jones' theory was merely an instrument, useful for organizing observation statements and (thus) for producing desired results, and that, therefore, it made no more sense to ask what was the nature of the entities to which it referred than it did to ask what was the nature of the entities to which a hammer or any other tool referred.[3] "Yes," a philosopher might have said, "Jones' theoretical expressions are just meaningless sounds or marks on paper which, when correlated with observation sentences by appropriate syntactical rules, enable us to predict successfully and otherwise organize data in a convenient fashion." These philosophers called themselves "instrumentalists."

According to another view (which, however, soon became unfashionable), although expressions containing Jones' theoretical terms were genuine sentences, they were translatable without remainder into a set (perhaps infinite) of observation sentences. For example, 'There are crobes of disease X on this article' was said to translate into something like this: 'If a person handles this article without taking certain precautions, he will (probably) contact disease X; and if this article is first raised to a high temperature, then if a person handles it at any time afterward, before it comes into contact with another person with disease X, he will (probably) not contract disease X; and . . .'

Now virtually all who held any of the views so far noted granted, even insisted, that theories played a useful and legitimate role in the scientific enterprise. Their concern was the elimination of "pseudo problems" which might arise, say, when one began wondering about the "reality of supraempirical entities," etc. However, there was also a school of thought, founded by a psychologist named Pelter, which differed in an interesting manner from such positions as these. Its members held that while Jones' crobes might very well exist and enjoy "full-blown reality," they should not be the concern of medical research at all. They insisted that if Jones had employed the correct methodology, he would have discovered, even sooner and with much less effort, all of the observation laws related to disease contraction, transmission, etc. without introducing superfluous links (the crobes) into the causal chain.

Now, lest any reader find himself waxing impatient, let me hasten to emphasize that this crude parody is not intended to convince anyone, or even to cast serious doubt upon sophisticated varieties of any of the reductionistic positions caricatured (some of them not too severely, I would contend) above. I am well aware that there are theoretical entities and theoretical entities, some of whose conceptual and theoretical statuses differ in important respects from Jones' crobes. (I shall discuss some of these later.) Allow me, then, to bring the Jonesean prelude to our examination of observability to a hasty conclusion.

Now Jones had the good fortune to live to see the invention of the compound microscope. His crobes were "observed" in great detail, and it became possible to identify the specific kind of *microbe* (for so they began to be called) which was responsible for each different disease. Some philosophers freely admitted error and were converted to realist positions concerning theories. Others resorted to subjective idealism or to a thoroughgoing phenomenalism, of which there were two principal varieties. According to one, the one "legitimate" observation language had for its descriptive terms only those which referred to sense data. The other maintained the stronger thesis that *all* "factual" statements were *translatable* without remainder into the sense-datum language. In either case, any two non-sense data (e.g., a theoretical entity and what would ordinarily be called an "observable physical object") had virtually the same status. Others contrived means of modifying their views much less drastically. One group maintained that Jones' crobes actually never had been unobservable in principle, for, they said, the theory did not imply the impossibility of finding a means (e.g., the microscope) of observing them. A more radical contention was that the crobes were not observed at all; it was argued that what was seen by means of the microscope was just a shadow or an image rather than a corporeal organism.

The Observational-Theoretical Dichotomy

Let us turn from these fictional philosophical positions and consider some of the actual ones to which they roughly correspond. Taking the last one first, it is interesting to note the following passage from Bergmann: "But it is only fair to point out that if this . . . methodological and terminological analysis [for the thesis that there are no atoms] . . . is strictly adhered to, even stars and microscopic objects are not physical things in a literal sense, but merely by courtesy of language and pictorial imagination. This might seem awkward. But when I look through a microscope, all I see is a patch of color which creeps through the field like a shadow over a wall. And a shadow, though real, is certainly not a physical thing."[4]

  I should like to point out that it is also the case that if this analysis is strictly adhered to, we cannot observe physical things through opera glasses, or even through ordinary spectacles, and one begins to wonder about the status of what we see through an ordinary windowpane. And what about distortions due to temperature gradients—however small and, thus, always present—in the ambient air? It really *does* "seem awkward" to say that when people who wear glasses describe what they see they are talking about shadows, while those who employ unaided vision talk about physical things—or that when we look through a windowpane, we can only *infer* that it is raining, while if we raise the window, we may "observe directly" that it is. The point I am making is that there is, in principle, a continuous series beginning with looking through a vacuum and containing these as members: looking through a windowpane, looking through glasses, looking through binoculars, looking through a low-power microscope, looking through a high-power microscope, etc., in the order given. The important consequence is that, so far, we are left without criteria which would enable us to draw a nonarbitrary line between "observation" and "theory." Certainly, we will often find it convenient to draw such a to-some-extent-arbitrary line; but its position will vary widely from context to context. (For example, if we are determining the resolving characteristics of a certain microscope, we would certainly draw the line beyond ordinary spectacles, probably beyond simple magnifying glasses, and possibly beyond another microscope with a lower power of resolution.) But what ontological ice does a mere methodologically convenient observational-theoretical dichotomy cut? Does an entity attain physical thinghood and/or "real existence" in one context only to lose it in another? Or, we may ask, recalling the continuity from observable to unobservable, is what is seen through spectacles a "little bit less real" or does it "exist to a slightly less extent" than what is observed by unaided vision?[5]

However, it might be argued that things seen through spectacles and binoculars look like ordinary physical objects, while those seen through microscopes and telescopes look like shadows and patches of light. I can only reply that this does not seem to me to be the case, particularly when looking at the moon, or even Saturn, through a telescope or when looking at a small, though "directly observable," physical object through a low-power microscope. Thus, again, a continuity appears.

"But," it might be objected, "theory tells us that what we see by means of a microscope is a real image, which is certainly distinct from the object on the stage." Now first of all, it should be remarked that it seems odd that one who is espousing an austere empiricism which requires a sharp observational-language/theoretical-language distinction (and one in which the former language has a privileged status) should need a theory in order to tell him what is observable. But, letting this pass, what is to prevent us from saying that we still observe the object on the stage, even though a "real image" may be involved? Otherwise, we shall be strongly tempted by phenomenalistic demons, and at this point we are considering a physical-object observation language rather than a sense-datum one. (Compare the traditional puzzles: Do I see one physical object or two when I punch my eyeball? Does one object split into two? Or do I see one object and one image? Etc.)

Another argument for the continuous transition from the observable to the unobservable (theoretical) may be adduced from theoretical considerations themselves. For example, contemporary valency theory tells us that there is a virtually continuous transition from very small molecules (such as those of hydrogen) through "medium-sized" ones (such as those of the fatty acids, polypeptides, proteins, and viruses) to extremely large ones (such as crystals of the salts, diamonds, and lumps of polymeric plastic). The molecules in the last-mentioned group are macro, "directly observable" physical objects but are, nevertheless, genuine, single molecules; on the other hand, those in the first mentioned group have the same perplexing properties as subatomic particles (de Broglie waves, Heisenberg indeterminacy, etc.). Are we to say that a large protein molecule (e.g., a virus) which can be "seen" only with an electron microscope is a little less real or exists to somewhat less an extent than does a molecule of a polymer which can be seen with an optical microscope? And does a hydrogen molecule partake of only an infinitesimal portion of existence or reality? Although there certainly *is* a continuous transition from observability to unobservability, any talk of such a continuity from full-blown existence to nonexistence is, clearly, nonsense.

Let us now consider the next to last modified position which was adopted by our fictional philosophers. According to them, it is only those entities which are *in principle* impossible to observe that present special problems. What kind of impossibility is meant here? Without going into a detailed discussion of the various types of impossibility, about which there is abundant literature with which the reader is no doubt familiar, I shall assume what usually seems to be granted by most philosophers who talk of entities which are unobservable in principle—i.e., that the theory(s) itself (coupled with a physiological theory of perception, I would add) entails that such entities are unobservable.

We should immediately note that if this analysis of the notion of unobservability (and, hence, of observability) is accepted, then its use as a means of delimiting the observation language seems to be precluded for those philosophers who regard theoretical expressions as elements of a calculating device—as meaningless strings of symbols. For suppose they wished to determine whether or not 'electron' was a theoretical term. First, they must see whether the theory entails the sentence 'Electrons are unobservable.' So far, so good, for their calculating devices are said to be able to select genuine sentences, provided they contain no theoretical terms. But what about the selected "sentence" itself? Suppose that 'electron' is an observation term. It follows that the expression is a genuine sentence and asserts that electrons are unobservable. But this entails that 'electron' is *not* an observation term. Thus if 'electron' is an observation term, then it is *not* an observation term. Therefore it is not an observation term. But then it follows that 'Electrons are unobservable' is not a genuine sentence and does not assert that electrons are unobservable, since it is a meaningless string of marks and does not assert anything whatever. Of course, it could be stipulated that when a theory "selects" a meaningless expression of the form 'Xs are unobservable,' then 'X' is to be taken as a theoretical term. But this seems rather arbitrary.

But, assuming that well-formed theoretical expressions are genuine sentences, what shall we say about unobservability in principle? I shall begin by putting my head on the block and argue that the present-day status of, say, electrons is in many ways similar to that of Jones' crobes before microscopes were invented. I am well aware of the numerous theoretical arguments for the impossibility of observing electrons. But suppose new entities are discovered which interact with electrons in such a mild manner that if an electron is, say, in an eigenstate of position, then, in certain circumstances, the interaction does not disturb it. Suppose also that a drug is discovered which vastly alters the human perceptual apparatus—perhaps even activates latent capacities so that a new sense modality emerges. Finally, suppose that in our altered state we are able to perceive (not necessarily visually) by means of these new entities in a manner roughly analogous to that by which we now see by means of photons. To make this a little more plausible, suppose that the energy eigenstates of the electrons in some of the compounds present in the relevant perceptual organ are such that even the weak interaction with the new entities alters them and also that the cross sections, relative to the new entities, of the electrons and other particles of the gases of the air are so small that the chance of any interaction here is negligible. Then we might be able to "observe directly" the position and possibly the approximate diameter and other properties of some electrons. It would follow, of course, that quantum theory would have to be altered in some respects, since the new entities do not conform to all its principles. But however improbable this may be, it does not, I maintain, involve any logical or conceptual absurdity. Furthermore, the modification necessary for the inclusion of the new entities would not necessarily change the meaning of the term 'electron.'[6]

Consider a somewhat less fantastic example, and one which does not involve any change in physical theory. Suppose a human mutant is born who is able to "observe" ultraviolet radiation, or even X rays, in the same way we "observe" visible light.

Now I think that it is extremely improbable that we will ever observe electrons directly (i.e., that it will ever be reasonable to assert that we have so observed them). But this is neither here nor there; it is not the purpose of this essay to predict the future development of scientific theories, and, hence, it is not its business to decide what actually is observable or what will become observable (in the more or less intuitive sense of 'observable' with which we are now working). After all, we are operating, here, under the assumption that it is theory, and thus science itself, which tells us what is or is not, in this sense, observable (the 'in principle' seems to have become superfluous). And this is the heart of the matter; for it follows that, at least for this sense of 'observable,' there are no a priori or philosophical criteria for separating the observable from the unobservable. By trying to show that we can talk about the *possibility* of observing electrons without committing logical or conceptual blunders, I have been trying to support the thesis that any (nonlogical) term is a possible candidate for an observation term.

There is another line which may be taken in regard to delimitation of the observation language. According to it, the proper term with which to work is not 'observable' but, rather 'observed.' There immediately comes to mind the tradition beginning with Locke and Hume (No idea without a preceding impression!), running through Logical Atomism and the Principle of Acquaintance, and ending (perhaps) in contemporary positivism. Since the numerous facets of this tradition have been extensively examined and criticized in the literature, I shall limit myself here to a few summary remarks.

Again, let us consider at this point only observation languages which contain ordinary physical-object terms (along with observation predicates, etc., of course). Now, according to this view, all descriptive terms of the observation language must refer to that which has been observed. How is this to be interpreted? Not too narrowly, presumably, otherwise each language user would have a different observation language. The name of my Aunt Mamie, of California, whom I have never seen, would not be in my observation language, nor would 'snow' be an observation term for many Floridians. One could, of course, set off the observation language by means of this awkward restriction, but then, obviously, not being the referent of an observation term would have no bearing on the ontological status of Aunt Mamie or that of snow.

Perhaps it is intended that the referents of observation terms must be members of a *kind* some of whose members have been observed or instances of a *property* some of whose instances have been observed. But there are familiar difficulties here. For example, given any entity, we can always find a kind whose only member is the entity in question; and surely expressions such as 'men over 14 feet tall' should be counted as observational even though no instances of the "property" of being a man over 14 feet tall have been observed. It would seem that this approach must soon fall back upon some notion of simples or determinables vs. determinates. But is it thereby saved? If it is held that only those terms which refer to observed simples or observed determinates are observation terms, we need only remind ourselves of such instances as Hume's notorious missing shade of blue. And if it is contended that in order to be an observation term an expression must at least refer to an observed determinable, then we can always find such a determinable which is broad enough in scope to embrace any entity whatever. But even if these difficulties can be circumvented, we see (as we knew all along) that this approach leads inevitably into phenomenalism, which is a view with which we have not been concerning ourselves.

Now it is not the purpose of this essay to give a detailed critique of phenomenalism. For the most part, I simply assume that it is untenable, at least in any of its translatability varieties.[7] However, if there are any unreconstructed phenomenalists among the readers, my purpose, insofar as they are concerned, will have been largely achieved if they will grant what I suppose most of them would stoutly maintain anyway, i.e., that theoretical entities are no worse off than so-called observable physical objects.

Nevertheless, a few considerations concerning phenomenalism and related matters may cast some light upon the observational-theoretical dichotomy and, perhaps, upon the nature of the "observation language." As a preface, allow me some overdue remarks on the latter. Although I have contended that the line between the observable and the unobservable is diffuse, that it shifts from one scientific problem to another, and that it is constantly being pushed toward the "unobservable" end of the spectrum as we develop better means of observation—better instruments—it would, nevertheless, be fatuous to minimize the importance of the observation base, for it is absolutely necessary as a confirmation base for statements which do refer to entities which are unobservable at a given time. But we should take as its basis and its unit not the "observational term" but, rather, the quickly decidable sentence. (I am indebted to Feyerabend, *loc. cit.,* for this terminology.) A quickly decidable sentence (in the technical sense employed here) may be defined as a singular, nonanalytic sentence such that a reliable, reasonably sophisticated language user can very quickly decide[8] whether to assert it or deny it when he is reporting on an occurrent situation. 'Observation term' may now be defined as a 'descriptive (nonlogical) term which may occur in a quickly decidable sentence,' and 'observation sentence' as a 'sentence whose only descriptive terms are observation terms.'

Returning to phenomenalism, let me emphasize that I am not among those philosophers who hold that there are no such things as sense contents (even sense data), nor do I believe that they play no important role in our perception of "reality." But the fact remains that the referents of most (not all) of the statements of the linguistic framework used in everyday life and in science are *not* sense contents but, rather, physical objects and other publicly observable entities. Except for pains, odors, "inner states," etc., *we do not usually observe sense contents*; and although there is good reason to believe that they play an indispensable role in observation, *we are usually not aware of them when we visually* (or tactilely) observe physical objects. For example, when I observe a distorted, obliquely reflected image in a mirror, I may seem to be seeing a baby elephant standing on its head; later I discover it is an image of Uncle Charles taking a nap with his mouth open and his hand in a peculiar position. Or, passing my neighbor's home at a high rate of speed, I observe that he is washing a car. If asked to report these observations I could quickly and easily report a baby elephant and a washing of a car; I probably would not, without subsequent observations, be able to report what colors, shapes, etc. (i.e., what sense data) were involved.

Two questions naturally arise at this point. How is it that we can (sometimes) quickly decide the truth or falsity of a pertinent observation sentence? and, What role do sense contents play in the appropriate tokening of such sentences? The heart of the matter is that these are primarily scientific-theoretical questions rather than "purely logical," "purely conceptual," or "purely epistemological." If theoretical physics, psychology, neuro-physiology, etc., were sufficiently advanced, we could give satisfactory answers to these questions, using, in all likelihood, the physical-thing language as our observation language and *treating sensations, sense contents, sense data, and "inner states" as theoretical* (yes, theoretical!) *entities.*[9]

It is interesting and important to note that, even before we give completely satisfactory answers to the two questions considered above, we can, with due effort and reflection, train ourselves to "observe directly" what were once theoretical entities—the sense contents (color sensations, etc.)—involved in our perception of physical things. As has been pointed out before, we can also come to observe other kinds of entities which were once theoretical. Those which most readily come to mind involve the use of instruments as aids to observation. Indeed, using our painfully acquired theoretical knowledge of the world, we come to see that we "directly observe" many kinds of so-called theoretical things. After listening to a dull speech while sitting on a hard bench, we begin to become poignantly aware of the presence of a considerably strong gravitational field, and as Professor Feyerabend is fond of pointing out, if we were carrying a heavy suitcase in a changing gravitational field, we could observe the changes of the $G\mu\nu$ of the metric tensor.

I conclude that our drawing of the observational-theoretical line at any given point is an accident and a function of our physiological makeup, our current state of knowledge, and the instruments we happen to have available and, therefore, that it has no ontological significance whatever. . . .

NOTES

1. E. Nagel, *The Structure of Science* (New York: Harcourt, Brace, and World, 1961), Ch. 6.

2. For the genesis and part of the content of some of the ideas expressed herein, I am indebted to a number of sources; some of the more influential are H. Feigl, "Existential Hypotheses," *Philosophy of Science,* 17: 35–62 (1950); P. K. Feyerabend, "An Attempt at a Realistic Interpretation of Experience," *Proceedings of the Aristotelian Society,* 58: 144–170 (1958); N. R. Hanson, *Patterns of Discovery* (Cambridge: Cambridge University Press, 1958); E. Nagel, *loc. cit*; Karl Popper, *The Logic of Scientific Discovery* (London: Hutchinson, 1959); M. Scriven, "Definitions, Explanations, and Theories," in *Minnesota Studies in the Philosophy of Science,* Vol. II, H. Feigl, M. Scriven, and G. Maxwell, eds. (Minneapolis: University of Minnesota Press, 1958); Wilfrid Sellars, "Empiricism and the Philosophy of Mind," in *Minnesota Studies in the Philosophy of Science,* Vol. I, H. Feigl and M. Scriven, eds. (Minneapolis: University of Minnesota Press, 1956), and "The Language of Theories," in *Current Issues in the Philosophy of Science,* H. Feigl and G. Maxwell, eds. (New York: Holt, Rinehart, and Winston, 1961).

3. I have borrowed the hammer analogy from E. Nagel, "Science and [Feigl's] Semantic Realism," *Philosophy of Science,* 17: 174–181 (1950), but it should be pointed out that Professor Nagel makes it clear that he does not necessarily subscribe to the view which he is explaining.

4. G. Bergmann, "Outline of an Empiricist Philosophy of Physics," *American Journal of Physics,* 11: 248–258; 335–342 (1943), reprinted in *Readings in the Philosophy of Science,* H. Feigl and M. Brodbeck, eds. (New York: Appleton-Century-Crofts, 1953), pp. 262–287.

5. I am not attributing to Professor Bergmann the absurd views suggested by these questions. He seems to take a sense-datum language as his observation language (the base of what he called "the empirical hierarchy"), and, in some ways, such a position is more difficult to refute than one which purports to take an "observable-physical-object" view. However, I believe that demolishing the straw men with which I am now dealing amounts to desirable preliminary "therapy." Some nonrealist interpretations of theories which embody the presupposition that the observable-theoretical distinction is sharp and ontologically crucial seem to me to entail positions which correspond to such straw men rather closely.

6. For arguments that it is possible to alter a theory without altering the meanings of its terms, see my "Meaning Postulates in Scientific Theories," in *Current Issues in the Philosophy of Science,* Feigl and Maxwell, eds.

7. The reader is no doubt familiar with the abundant literature concerned with this issue. See, for example, Sellars' "Empiricism and the Philosophy of Mind," which also contains references to other pertinent works.

8. We may say "noninferentially" decide, provided this is interpreted liberally enough to avoid starting the entire controversy about observability all over again.

9. Cf. Sellars, "Empiricism and the Philosophy of Mind." As Professor Sellars points out, this is the crux of the "other-minds" problem. Sensations and inner states (relative to an intersubjective observation language, I would add) are theoretical entities (and they "really exist") and *not* merely actual and/or possible behavior. Surely it is the unwillingness to countenance theoretical entities—the hope that every sentence is translatable not only into some observation language but into the physical-thing language—which is responsible for the "logical behaviorism" of the neo-Wittgensteinians.

STUDY QUESTIONS FOR PART THREE

"Theories" — J. Hospers

1. Evaluate Hospers's contention that we *construct* theories and *discover* laws. Are laws and theories of the same ontological type? Put crudely, are they made of the same type of "stuff"?

2. After reading Stace's paper, consider Hospers's claim that if atoms were only convenient fictions there usefulness in explaining and predicting would be mysterious.

3. Evaluate Hospers's distinctions among theories, laws, and hypotheses. It might help to keep the following in mind: he claims that the statement "there are protons and electrons" is a theory. He also claims that the distinctive feature of hypotheses is that they express particular facts (or presumed facts).

"Observation" — N. Hanson

1. Suppose one tried to defend the claim that there is a sense in which two scientists see the same thing by citing the fact that they have the same retinal images. What does Hanson find wrong with this?

2. Here is another defense of the claim above. Try to state precisely what Hanson's objections are to it. "Surely all normal observers, when looking at the same thing, see the same thing. They have the same visual sense-data. Their differences are not in what they see but in the way they interpret what they see."

3. Hanson admits that "wherever it makes sense to say that two scientists looking at X do not see the same thing, there must always be a prior sense in which they do see the same thing." Sometimes he spells out the prior sense by saying that they must be *visually aware* of the same object. Why isn't this just to grant the point of those Hanson is arguing against?

4. Try to state some of the consequences of Hanson's view for the standard observation/theoretical distinction.

"The Theory-Ladenness of Observation" — C. Kordig

1. Why does Kordig believe that Hanson's analogy undercuts Hanson's own position?

2. Is Kordig's explanation of why Hanson finds 'sees that' rather than a neutral sense of 'see' central to science correct?

3. Your selection from Kordig does not include his treatment of 'believes that' as a reading of 'sees that.' Plug 'believes that' into Kordig's reconstruction of Hanson's argument and determine where the argument breaks down.

4. Given Kordig's work, can you think of a sympathetic reading of what Hanson means when he says that observers with different theories don't see the same things?

"Science and the Physical World" — W. Stace

1. Recently, when driving along the Interstate, I heard a thumping sound and made a causal inference to the car having a flat tire. I was right, unfortunately. Notice that initially I did not directly perceive the flat tire but made an inference to it from something I did directly perceive. What is the difference between this example and the kind of inference scientists supposedly make to electrons, and so on? Surely my inference is innocent enough but, according to Stace, the scientist's is not. How come?

2. Stace does not believe that theoretical entities exist. But he seems to believe that theoretical statements are true. Why? What does he think it is for a statement to be true?

3. Why do scientists, according to Stace, insist on thinking that forces, electrons, and so on, exist? Try to evaluate his argument. You might wish to refer back to Hospers's discussion of 'describing' and 'explaining.'

4. What does Stace mean by 'directly perceive'? Do we directly perceive our hands, desks, or distant airplanes?

"The Ontological Status of Theoretical Entities" — G. Maxwell

1. Compare the type of reasoning that supports the existence of "crobes" in Maxwell's story with the type of reasoning Stace believes is used to support the existence of theoretical entities.

2. Reconstruct Maxwell's argument(s) for his view that where the observational/theoretical line is drawn is a consequence of historical contingencies — our anatomy, our instruments, our knowledge.

3. Why would the view expressed in 2 count against the observational/theoretical distinction having the ontological significance Stace gives to it?

SELECTED BIBLIOGRAPHY

[1] Dretske, F. *Seeing and Knowing.* Chicago: University of Chicago Press, 1969. Relevant to understanding observation; intermediate level.

[2] Eddington, A. *The Nature of the Physical World.* New York: Cambridge University Press, 1929, pp. ix–xii, 282–289. A super-realist.

[3] Gutting, G. "Philosophy of Science." In *The Synoptic Vision,* ed. by F. Delaney et al. Notre Dame, Ind.: University of Notre Dame Press, 1977, pp. 73–86. Overview of W. Sellars's super-realism; will guide one to the Sellars literature.

[4] Harre, R. *The Philosophies of Science.* Oxford: Oxford University Press, 1972, chap. 3. Nice examples of theories; elementary.

[5] Hempel, C. *Philosophy of Natural Science.* Englewood Cliffs, N.J.: Prentice-Hall, 1966, chap. 6.

[6] Holton, G., and Roller, D. *Foundations of Modern Physical Science.* Reading, Mass.: Addison-Wesley, 1958. An introduction to physics written by a physicist and a historian of science; only requires high school algebra; good examples of theories.

[7] Kuhn, T. *The Structure of Scientific Revolutions.* Chicago: University of Chicago Press, 1962. An exciting work that has received enormous attention; presents an account of theories, observation (similar to Hanson's) and theory change; views are developed by way of historical examples.

[8] Nagel, E. *The Structure of Science.* New York: Harcourt, Brace and World, 1961, chaps. 5 and 6. An excellent discussion of the status of unobservables.

[9] Suppe, F. *The Structure of Theories.* Urbana, Ill.: University of Illinois Press, 1977. Detailed description of the standard view of theories, plus a review of all the work critical of the standard view; excellent bibliography; intermediate level.

[10] Toulmin, S. *The Philosophy of Science.* New York: Harper and Bros., 1953, chap. 4. Argues for a version of instrumentalism.

Part Four

Confirmation and Acceptance

Part 4: Confirmation and Acceptance

Introduction

In Parts 2 and 3, we addressed the structure of explanations and theories. In this part, the concern will be with how we justify our belief in a particular explanation or theory. Since hypotheses or lawlike statements are at the heart of both explanations and theories, the discussion will center on their justifications.

We are often told that scientists justify their hypotheses by comparing them to the facts or seeing how they fare in the face of the facts. There is a certain truth in this, though as with most slogans it is a misleading oversimplification. A purpose of the orienting remarks that follow is to make and motivate some of the bold distinctions to which a sensitive theory of justification must attend.

I. Confirmation

The following are examples of hypotheses:
- (a) Lead melts at 327° C.
- (b) Myelinated A-fibers transmit signals at 100 meters/second.
- (c) The price of goods is inversely proportional to the demand for the goods.
- (d) The pressure of a gas times the volume of the gas is a constant. ($PV = c$ or $P_1/P_2 = V_2/V_1$)

How do scientists justify their belief (or disbelief in (a)-(d)? Of course, they give reasons for their belief.

A place to begin in order to understand this reason-giving process is a distinction between two basic kinds of reasons: (i) epistemic reasons, (ii) practical reasons. Epistemic reasons count toward the truth or falsity of a hypothesis. Practical reasons do not bear on truth or falsity but rather on the usefulness of believing the hypothesis.

Consider the hypothesis that your next-door neighbor is madly in love with you. That she gives you gifts and winks at you as you leave for classes is an epistemic reason for believing the hypothesis.

Suppose further that you have an incredible inferiority complex. Unless you believe that people will greet you with open arms, you simply cannot face them at all. Furthermore, suppose that you wish very much to meet your neighbor. Now these facts about your psychology and goals are reasons for believing the hypothesis, but not epistemic reasons. They do not count toward the truthfulness of the hypothesis, but they make it useful for you to believe the hypothesis.

Later in this essay, the distinction above will be put to work. But, at least at the outset, it is not unreasonable to think that scientists offer epistemic reasons for (a)–(d). But what reasons?

In example (a), intuitively 'that this piece of lead melted at 327° C' is an epistemic reason for believing (a). Typically it is called a confirming instance. A hypothesis that has the form of our (a), viz., everything that is A is B, has four types of instances: this thing is A and B, this thing is A and not B, this thing is not A and B, this thing is not A and not B. Members of the first type are confirming instances; members of the second type are disconfirming instances.

So far then, a hypothesis is justified by appeal to epistemic reasons where epistemic reasons are confirming instances of the hypothesis. Even though a certain amount of jargon has been introduced, what has been said is fairly commonsensical. In fact it seems to be what is behind the slogans that initiated the discussion. But notice that, even supposing what has been said is beyond doubt, it is vague at some crucial points.

We spelled out the meaning of 'instance' for a hypothesis of the form 'Everything that is A is B.' But what about hypotheses of other forms such as (c) or (d)? A similar account of 'instance' is needed for these and hypotheses of other forms. Though this will require a little formal sophistication there are no obvious theoretical problems in the way.

The second vague spot belies a deeper issue. Though we gave examples of *confirming* and *disconfirming* instances we did not define these notions. A careful reader might have noticed that we were conspicuously silent on how to understand 'this thing is not A and B' and 'this thing is not A and not B.'

One worry can be simply put. A hypothesis of the form 'Everything that is A is B' is logically equivalent to a hypothesis of the form 'Everything that is not B is not A.' If you think that 'this A is a B' confirms the first hypothesis, as surely you must, then by parity of reasoning 'this thing is not B and not A' ought to confirm the latter. Since the two hypotheses are logically equivalent it seems reasonable that 'This thing is not B and not A' should also confirm 'Everything that is A is B.' But then 'that this wax melts at 100° C' would confirm 'all lead melts at 327° C.'[1]

There are further paradoxes but we need not pursue any of these.[2] The purpose of our reconnaissance is to spot pockets of resistance not necessarily to eliminate them.

Suppose we could, with a little philosophical tenacity, define confirming (disconfirming) instance. Our theory of justification is far from complete. Just as one birdie does not a good round of golf make, though it surely helps, one confirming instance does not a hypothesis justify, though it counts toward the justification. We have been discussing what philosophers call qualitative features of confirmation—that is defining *a* confirming instance. There are also quantitative features of confirmation.

An example will illustrate these features. The eighteenth-century chemist Robert Boyle is credited with discovering the law expressed in (d). He justified the hypothesis by providing the reader with the following data. In the first column, *A*, are values for the volume of a gas. In the second column, *B*, are the corresponding experimentally determined values for the pressure.

A	B	C
48	29 2/16	29 2/16
46	30 9/16	30 6/16
44	31 15/16	31 12/16
42	33 8/16	33 1/7
40	35 5/16	35
38	37	36 15/19
36	39 5/16	38 7/8
34	41 10/16	41 2/17
32	44 3/16	43 11/16
30	47 1/16	46 3/5
28	50 5/16	50
26	54 5/16	53 10/13
24	58 13/16	58 2/8
23	61 5/16	60 18/23
22	64 1/16	63 6/11
21	67 1/16	66 4/7
20	70 11/16	70
19	74 2/16	73 11/19
18	77 14/16	77 2/3
17	82 12/16	82 4/17
16	87 14/16	87 3/8
15	93 1/16	93 1/5
14	100 7/16	99 6/7
13	107 13/16	107 7/13
12	117 9/16	116 4/8

Boyle justifies his hypothesis not by a single confirming instance but by a set of confirming instances. Furthermore, he does not give us repeated measurements corresponding to a row in the chart but measures the pressure and volume over a range of values. Column *C* is also significant. If one plugs in V_1

and V_2 and P_1 and computes P_2 according to (d), he will discover a difference between the computed values for P_2 and the experimentally determined values. *C* gives the expected values. The differences are so slight that Boyle was willing to chalk them up to human error and defects in the experimental apparatus.[3]

This example brings out clearly that certain quantitative features of confirmation are relevant to the justification of a hypothesis. As a rule of thumb—the larger the quantity, the more diverse; and the more accurate the confirming instances, the greater the justification for the hypothesis.

Though this rule is surely correct, at least as a rough rule, it does not pass without inducing philosophical perplexity. How accurately must the "confirming instances" conform to the expected data in order to be *confirming* rather than disconfirming? How many instances are required before one can reasonably assert the truth of the hypothesis?

So far we have seen that a theory of justification must contain a theory of confirmation. Furthermore, one must distinguish qualitative aspects of confirmation from the quantitative aspects. There is one final feature of confirmation that needs to be sketched.

The distinction between observational and theoretical hypotheses, putting aside the difficulties in precisely drawing the distinction, is well known. The sample lawlike statements (a)–(d) are a subclass of observation statements. One important consequence of this is that the instances of these hypotheses can be known by observation in a fairly direct way. One can simply see if this piece of lead melts at 327° C. But with theoretical hypotheses this will not be true.

Some simple examples of theoretical hypotheses follow:

(e) All electrons have a charge of minus one.

(f) The weight of the "sea of air" surrounding the earth exerts a pressure on objects within the sea.

(g) The current density for any point within a conductor equals the reciprocal of the resistance of the conductor times the applied electrical field or $j = \frac{l}{p}E$.

The problem presented by theoretical hypotheses is not that they do not have instances, but rather since their instances cannot be directly known by observation one cannot in any straightforward way "compare the hypothesis to the facts."

Theoretical hypotheses are tested by deducing a test implication from the hypothesis and observing whether the test implication is true or not. If it is true, it is said to confirm the hypothesis; otherwise disconfirm. To distinguish these "instances" from the confirming and disconfirming instances let us call them confirming or disconfirming test implications.

Continuing our pneumatic examples, consider (f). Toricelli, a precursor of Boyle, deduced a number of test implications from the hypothesis. In one particularly impressive case he reasoned as follows: If (f) is true, then the pressure on objects at the top of a mountain ought to be less than on objects in the mountain's valley, since the column of air pushing on them would be less. So if (f) is true, then a barometer should read less on the top of a mountain than on the bottom. He tried this out and found a spectacular confirming test implication. The general structure of Torricelli's reasoning or the successful justification of a theoretical hypothesis can be represented by the following schema: (H – hypothesis; TI – test implication)

    (s) if H then TI

       TI

    Therefore, H is confirmed.

The disconfirmation schema looks like this:

    (t) if H then TI

       not TI

    Therefore, H is disconfirmed.

Just as quantitative considerations are important in justifying observational hypotheses, so are they in justifying theoretical hypotheses. The quantity and diversity and accuracy of test implications are again relevant. Torricelli strengthened his case by just such considerations.

One final point about confirmation must be made. There is a dissimilarity between (s) and (t). (s) is not a valid argument form. Hence from the premises of (s) it is fallacious to conclude that H is true. This point is marked by claiming only that H is supported or confirmed. H *could* turn out to be false, though in so far as TI is the case, one has some reason to believe that H is true.

Since schema (t) is valid, it looks like one ought to be able to conclude not just that H is disconfirmed but that H is false. Given what has been said that would be the correct conclusion. But the assumption that (t) captures the structure of disconfirmation is an oversimplification.

The actual form of the first premise of (t), and for that matter (s), is

    (n) if H and A, and A_2 and . . . A_n, then TI.

A_1, A_2, etc. are auxiliary hypotheses—hypotheses assumed when it is claimed that H implies TI. For example, Torricelli was at least assuming that the pressure on the surface of a liquid is transferred through the liquid and that one can veridically observe the height of mercury through a glass barrier, i.e., that glass does not have some systematic distorting effects that will make it look as though the mercury has not risen when it really has. Contemporary philosophers believe that the first premise will always contain auxiliary hypotheses.

If one is more realistic about the first premise of (t), then formally the conclusion is: Therefore, H is false or A_1 is false or A_2 is false or . . . A_n is false. In other words, if we substitute (n) for the first premise of (t), then as a matter of pure logic alone one has a reason to conclude that at least one of H-A_n is false, but he has no reason to pick out a particular false statement from this set.

The point of this discussion of schema (t) is to show that just as confirmation does not prove that H is true, disconfirmation does not prove that H is false. On the assumption that the needed auxiliary hypotheses are highly confirmed the appropriate conclusion from (t) is that *H* has been disconfirmed, not found false.

II. Acceptance

There is a difference between having reasons that count toward the truth or falsity of a hypothesis and the acceptance of a hypothesis as true or the rejection of a hypothesis as false. During the Watergate hearings, most people would admit that testimony was given that counted toward the truth of the statement, "Nixon is a liar." Yet many of these people would not accept the statement as true.

Acceptance as usually understood is a broader notion than confirmation. Part of the reason for the acceptance or rejection of a hypothesis is the extent to which it has been confirmed or disconfirmed. But many philosophers have thought that there are additional criteria for acceptance or rejection. The following are often mentioned candidates: (a) the simplicity of a hypothesis, (b) the theoretical fecundity of a hypothesis, (c) the compatibility of a hypothesis with certain moral, political, or religious views, (d) the harm incurred if one's decision to accept or reject a hypothesis is mistaken.

We have already seen some hints as to what motivates the need for additional criteria. By way of confirmation and disconfirmation one cannot *prove* that a hypothesis is true or false. Of course, quantitative features are relevant, but when has one observed enough test implications or when are the test results accurate enough? There are further problems. When evaluating a given hypothesis, various quantitative features may be at odds, for example, as the variety of test implications increases the accuracy may decrease. It is also often the case that for a given hypothesis that is to explain certain phenomena there will be alternative hypotheses that are equally well confirmed and explain the same phenomena.

We shall briefly summarize the readings momentarily, but first some general questions to keep in mind when evaluating the importance of additional criteria such as (a)-(d). (1) From the scientific point of view, should *practical reasons* have anything to do with the acceptance of a hypothesis? (2) Of the additional criteria suggested, which are epistemic and which are practical? (3) Is it really the business of the scientist as a scientist to *accept* hypotheses? (The paper by R. Rudner in Part 5 will help here.)

The essays "Reflections On My Critics," by T. Kuhn, and "The Variety of Reasons for the Acceptance of Scientific Theories," by P. Frank, deal exclusively with what we have called additional criteria of acceptance. Kuhn wishes to emphasize the difference between argumentation in mathematics and argumentation in science. A consequence of the differences is, according to Kuhn, that two scientists can disagree about which of two competing hypotheses to accept, after a "complete" effort at confirmation, yet both scientists have good reasons for their views. Supposing Kuhn is correct, what does it tell us about the nature of the additional criteria?

Frank insists that, historically, such factors as compatibility with religious views have been decisive in which hypothesis is accepted. Furthermore, given how he understands theoretical talk, viz., instrumentally, he does not find such factors improper, or unscientific. In fact, Frank seems to support the use of these criteria.

In the essay "Hypothesis," W. V. Quine and J. S. Ullian list six virtues of a hypothesis. It is not clear whether some of these are fancy ways of talking about confirmation or whether they are additional criteria. You will have to evaluate each one individually.

Quine and Ullian have a lengthy discussion of simplicity. Simplicity is the additional criterion most often mentioned by philosophers. Try to decide whether it is an epistemic or practical reason for acceptance. This essay has some arguments on the matter.

A.D.K.

NOTES

1. See Hempel's "Studies in the Logic of Confirmation" in [1] of the bibliography for Part 4.
2. See Goodman's "The New Riddle of Induction" in [2] of the bibliography for Part 4.
3. For the historical and logical details of Boyle's work, see "Robert Boyle's Experiments in Pneumatics" in *Harvard Case Histories in Experimental Science,* ed. by J. Conant and L. Nash. Vol. 1. Cambridge, Massachusetts: Harvard University Press, 1948. pp. 1–64.

14

Hypothesis

W. V. Quine and J. S. Ullian

Some philosophers once held that whatever was true could in principle be proved from self-evident beginnings by self-evident steps. The trait of absolute demonstrability, which we attributed to the truths of logic in a narrow sense and to relatively little else, was believed by those philosophers to pervade all truth. They thought that but for our intellectual limitations we could find proofs for any truths, and so, in particular, predict the future to any desired extent. These philosophers were the rationalists. Other philosophers, a little less sanguine, had it that whatever was true could be proved by self-evident steps from two-fold beginnings: self-evident truths and observations. Philosophers of both schools, the rationalists and the somewhat less sanguine ones as well, strained toward their ideals by construing self-evidence every bit as broadly as they in conscience might, or somewhat more so.

Actually even the truths of elementary number theory are presumably not in general derivable, we noted, by self-evident steps from self-evident truths. We owe this insight to Godel's theorem, which was not known to the old-time philosophers.

What then of the truths of nature? Might these be derivable still by self-evident steps from self-evident truths together with observations? Surely not. Take the humblest generalization from observation: that giraffes are mute, that sea water tastes of salt. We infer these from our observations of giraffes and sea water because we expect instinctively that what is true of all observed samples is true of the rest. The principle involved here, far from being self-evident, does not always lead to true generalizations. It worked for the giraffes and the sea water, but it would have let us down if we had inferred from a hundred observations of swans that all swans are white.

Such generalizations already exceed what can be proved from observations and self-evident truths by self-evident steps. Yet such generalizations are still only a small part of natural science. Theories of molecules and atoms are not related to any observations in the direct way in which the generalizations about giraffes and sea water are related to observations of mute giraffes and salty sea water.

It is now recognized that deduction from self-evident truths and observation is not the sole avenue to truth nor even to reasonable belief. A dominant further factor, in solid science as in daily life, is *hypothesis*. In a word, hypothesis is guesswork; but it can be enlightened guesswork:

It is the part of scientific rigor to recognize hypothesis as hypothesis and then to make the most of it. Having accepted the fact that our observations and our self-evident truths do not together suffice to predict the future, we frame hypotheses to make up the shortage.

Calling a belief a hypothesis says nothing as to what the belief is about, how firmly it is held, or how well founded it is. Calling it a hypothesis suggests rather what sort of reason we have for adopting or entertaining it. People adopt or entertain a hypothesis because it would explain, if it were true, some things that they already believe. Its evidence is seen in its consequences. . . .

Hypothesis, where successful, is a two-way street, extending back to explain the past and forward to predict the future. What we try to do in framing hypotheses is to explain some otherwise unexplained happenings by inventing a plausible story, a plausible description or history of relevant portions of the world. What counts in favor of a hypothesis is a question not to be lightly answered. We may note five virtues that a hypothesis may enjoy in varying degrees.

Virtue I is *conservatism*. In order to explain the happenings that we are inventing it to explain, the hypothesis may have to conflict with some of our previous beliefs; but the fewer the better. Acceptance of a hypothesis is of course like acceptance of any belief in that it demands rejection of whatever conflicts with it. The less rejection of prior beliefs required, the more plausible the hypothesis—other things being equal.

Often some hypothesis is available that conflicts with no prior beliefs. Thus we may attribute a click at the door to arrival of mail through the slot. Conservatism usually prevails in such a case; one is not apt to be tempted by a hypothesis that upsets prior beliefs when there is no need to resort to one. When the virtue of conservatism deserves notice, rather, is when something happens that cannot evidently be reconciled with our prior beliefs.

There could be such a case when our friend the amateur magician tells us what card we have drawn. How did he do it? Perhaps by luck, one chance in fifty-two; but this conflicts with our reasonable belief, if all unstated, that he would not have volunteered a performance that depended on that kind of luck. Perhaps the cards were marked; but this conflicts with our belief that he had had no access to them, they being ours. Perhaps he peeked or pushed, with help of a sleight-of-hand; but this conflicts with our belief in our perceptiveness. Perhaps he resorted to telepathy or clairvoyance; but this would wreak havoc with our whole web of belief. The counsel of conservatism is the sleight-of-hand.

Conservatism is rather effortless on the whole, having inertia in its favor. But it is sound strategy too, since at each step it sacrifices as little as possible of the evidential support, whatever that may have been, that our overall system of beliefs has hitherto been enjoying. The truth may indeed be radically remote from our present system of beliefs, so that we may need a long series of conservative steps to attain what might have been attained in one rash leap. The longer the leap, however, the more serious an angular error in the direction. For a leap in the dark the likelihood of a happy landing is severely limited. Conservatism holds out the advantages of limited liability and a maximum of live options for each next move.

Virtue II, closely akin to conservatism, is *modesty*. One hypothesis is more modest than another if it is weaker in a logical sense: if it is implied by the other, without implying it. A hypothesis *A* is more modest than *A* and *B* as a joint hypothesis. Also, one hypothesis is more modest than another if it is more humdrum: that is, if the events that it assumes to have happened are of a more usual and familiar sort, hence more to be expected.

Thus suppose a man rings our telephone and ends by apologizing for dialing the wrong number. We will guess that he slipped, rather than that he was a burglar checking to see if anyone was home. It is the more modest of the two hypotheses, butterfingers being rife. We could be wrong, for crime is rife too. But still the butterfingers hypothesis scores better on modesty than the burglar hypothesis, butterfingers being rifer.

We habitually practice modesty, all unawares, when we identify recurrent objects. Unhesitatingly we recognize our car off there where we parked it, though it may have been towed away and another car of the same model may have happened to pull in at that spot. Ours is the more modest hypothesis, because staying put is a more usual and familiar phenomenon than the alternative combination.

It tends to be the counsel of modesty that the lazy world is the likely world. We are to assume as little activity as will suffice to account for appearances. This is not all there is to modesty. It does not apply to the preferred hypothesis in the telephone example, since Mr. Butterfingers is not assumed to be a less active man than one who might have plotted burglary. Modesty figured there merely in keeping the assumptions down, rather than in actually assuming inactivity. In the example of the parked car, however, the modest hypothesis does expressly assume there to be less activity than otherwise. This is a policy that guides science as well as common sense. It is even erected into an explicit principle of mechanics under the name of the law of least action.

Between modesty and conservatism there is no call to draw a sharp line. But by Virtue I we meant conservatism only in a literal sense—conservation of past beliefs. Thus there remain grades of modesty still to choose among even when Virtue I—compatibility with previous beliefs—is achieved to perfection; for both a slight hypothesis and an extravagant one might be compatible with all previous beliefs.

Modesty grades off in turn into Virtue III, *simplicity*. Where simplicity considerations become especially vivid is in drawing curves through plotted points on a graph. Consider the familiar practice of plotting measurements. Distance up the page represents altitude above sea level, for instance, and distance across represents the temperature of boiling water. We plot our measurements on the graph, one dot for each pair. However many points we plot, there remain infinitely many curves that may be drawn through them. Whatever curve we draw represents our generalization from the data, our prediction of what boiling temperatures would be found at altitudes as yet untested. And the curve we will choose to draw is the simplest curve that passes through or reasonably close to all the plotted points.

There is a premium on simplicity in any hypothesis, but the highest premium is on simplicity in the giant joint hypothesis that is science, or the particular science, as a whole. We cheerfully sacrifice simplicity of a part for greater simplicity of the whole when we see a way of doing so. Thus consider gravity. Heavy objects tend downward: here is an exceedingly simple hypothesis, or even a mere definition. However, we complicate matters by accepting rather the hypothesis that the heavy objects around us are slightly attracted also by one another, and by the neighboring mountains, and by the moon, and that all these competing forces detract slightly from the downward one. Newton propounded this more complicated hypothesis even though, aside from tidal effects of the moon, he had no means of detecting the competing forces; for it meant a great gain in the simplicity of physics as a whole. His hypothesis of universal gravitation, which has each body attracting each in proportion to mass and inversely as the square of the distance, was what enabled him to make a single neat system of celestial and terrestrial mechanics.

A modest hypothesis that was long supported both by theoretical considerations and by observation is that the trajectory of a projectile is a parabola. A contrary hypothesis is that the trajectory deviates imperceptibly from a parabola, constituting rather one end of an ellipse whose other end extends beyond the center of the earth. This hypothesis is less modest, but again it conduces to a higher simplicity: Newton's laws of motion and, again, of gravitation. The trajectories are brought into harmony with Kepler's law of the elliptical orbits of the planets.

Another famous triumph of this kind was achieved by Count Rumford and later physicists when they showed how the relation of gas pressure to temperature could be accounted for by the impact of oscillating particles, for in this way they reduced the theory of gases to the general laws of motion. Such was the kinetic theory of gases. In order to achieve it they had to add the hypothesis, by no means a modest one, that gas consists of oscillating particles or molecules; but the addition is made up for, and much more, by the gain in simplicity accruing to physics as a whole.

What is simplicity? For curves we can make good sense of it in geometrical terms. A simple curve is continuous, and among continuous curves the simplest are perhaps those whose curvature changes most gradually from point to point. When scientific laws are expressed in equations, as they so often are, we can make good sense of simplicity in terms of what mathematicians call the degree of an equation, or the order of a differential equation. This line was taken by Sir Harold Jeffreys. The lower the degree, the lower the order, and the fewer the terms, the simpler the equation. Such simplicity ratings of equations agree with the simplicity ratings of curves when the equations are plotted as in analytical geometry.

Simplicity is hard to define when we turn away from curves and equations. Sometimes in such cases it is not to be distinguished from modesty. Commonly a hypothesis *A* will count as simpler than *A* and *B* together; thus far simplicity and modesty coincide. On the other hand the simplicity gained by Newton's hypothesis of universal gravitation was not modesty, in the sense that we have assigned to that term; for the hypothesis was not logically implied by its predecessors, nor was it more humdrum in respect of the events that it assumed. Newton's hypothesis was simpler than its predecessors in that it covered in a brief unified story what had previously been covered only by two unrelated accounts. Similar remarks apply to the kinetic theory of gases.

In the notion of simplicity there is a nagging subjectivity. What makes for a brief unified story depends on the structure of our language, after all, and on our available vocabulary, which need not reflect the structure of nature. This subjectivity of simplicity is puzzling, if simplicity in hypotheses is to make for plausibility. Why should the subjectively simpler of two hypotheses stand a better chance of predicting objective events? Why should we expect nature to submit to our subjective standard of simplicity?

That would be too much to expect. Physicists and others are continually finding that they have to complicate their theories to accommodate new data. At each stage, however, when choosing a hypothesis subject to subsequent correction, it is still best to choose the simplest that is not yet excluded. This strategy recommends itself on much the same grounds as the strategies of conservatism and modesty. The longer the leap, we reflected, the more and wilder ways of going wrong. But likewise, the more complex the hypothesis, the more and wilder ways of going wrong; for how can we tell which complexities to adopt? Simplicity, like conservatism and modesty, limits liability. Conservatism can be good strategy even though one's present theory be ever so far from the truth, and simplicity can be good strategy even though the world be ever so complicated. Our steps toward the complicated truth can usually be laid out most dependably if the simplest hypothesis that is still tenable is chosen at each step. It has even been argued that this policy will lead us at least asymptotically toward a theory that is true.

There is more, however, to be said for simplicity: the simplest hypothesis often just is the likeliest, apparently, quite apart from questions of cagy strategy. Why should this be? There is a partial explanation in our ways of keeping score on predictions. The predictions based on the simpler hypotheses tend to be scored more leniently. Thus consider curves, where simplicity comparisons are so clear. If a curve is kinky and complex, and if some measurement predicted from the curve turns out to miss the mark by a distance as sizable as some of the kinks of the curve itself, we will count the prediction a failure. We will feel that so kinky a curve, if correct, would have had a kink to catch this wayward point. On the other hand, a miss of the same magnitude might be excused if the curve were smooth and simple. It might be excused as due to inaccuracy of measurement or to some unexplained local interference. This cynical doctrine of selective leniency is very plausible in the case of the curves. And we may reasonably expect a somewhat similar but less easily pictured selectivity to be at work in the interest of the simple hypotheses where curves are not concerned.

Considering how subjective our standards of simplicity are, we wondered why we should expect nature to submit to them. Our first answer was that we need not expect it; the strategy of favoring the simple at each step is good anyway. Now we have noted further that some of nature's seeming simplicity is an effect of our bookkeeping. Are we to conclude that the favoring of simplicity is entirely our doing, and that nature is neutral in the matter? Not quite. Darwin's theory of natural selection offers a causal connection between subjective simplicity and objective truth in the following way. Innate subjective standards of simplicity that make people prefer some hypotheses to others will have survival value insofar as they favor successful prediction. Those who predict best are likeliest to survive and reproduce their kind, in a state of nature anyway, and so their innate standards of simplicity are handed down. Such standards will also change in the light of experience, becoming still better adapted to the growing body of science in the course of the individual's lifetime. (But these improvements do not get handed down genetically.)

Virtue IV is *generality*. The wider the range of application of a hypothesis, the more general it is. When we find electricity conducted by a piece of copper wire, we leap to the hypothesis that all copper, not just long thin copper, conducts electricity.

The plausibility of a hypothesis depends largely on how compatible the hypothesis is with our being observers placed at random in the world. Funny coincidences often occur, but they are not the stuff that plausible hypotheses are made of. The more general the hypothesis is by which we account for our present observation, the less of a coincidence it is that our present observation should fall under it. Hence, in part, the power of Virtue IV to confer plausibility.

The possibility of testing a hypothesis by repeatable experiment presupposes that the hypothesis has at least some share of Virtue IV. For in a repetition of an experiment the test situation can never be exactly what it was for the earlier run of the experiment; and so, if both runs are to be relevant to the hypothesis, the hypothesis must be at least general enough to apply to both test situations.[1] One would of course like to have it much more general still.

Virtues I, II, and III made for plausibility. So does Virtue IV to some degree, we see, but that is not its main claim; indeed generality conflicts with modesty. But generality is desirable in that it makes a hypothesis interesting and important if true.

We lately noted a celebrated example of generality in Newton's hypothesis of universal gravitation, and another in the kinetic theory of gases. It is no accident that the same illustrations should serve for both simplicity and generality. Generality without simplicity is cold comfort. Thus take celestial mechanics with its elliptical orbits, and take also terrestrial mechanics with its parabolic trajectories, just take them in tandem as a bipartite theory of motion. If the two together cover everything covered by Newton's unified laws of motion, then generality is no ground for preferring Newton's theory to the two taken together. But Virtue II, simplicity, is. When a way is seen of gaining great generality with little loss of simplicity, or great simplicity with no loss of generality, the conservatism and modesty give way to scientific revolution.

The aftermath of the famous Michelson-Morley experiment of 1887 is a case in point. The purpose of this delicate and ingenious experiment was to measure the speed with which the earth travels through the ether. For two centuries, from Newton onward, it had been a well entrenched tenet that something called the ether pervaded all of what we think of as empty space. The great physicist Lorentz (1853–1928) had hypothesized that the ether itself was stationary. What the experiment revealed was that the method that was expected to enable measurement of the earth's speed through the ether was totally inadequate to that task. Supplementary hypotheses multiplied in an attempt to explain the failure without seriously disrupting the accepted physics. Lorentz, in an effort to save the hypothesis of stationary ether, shifted to a new and more complicated set of formulas in his mathematical physics. Einstein soon cut through all this, propounding what is called the special theory of relativity.

This was a simplification of physical theory. Not that Einstein's theory is as simple as Newton's had been; but Newton's physics had been shown untenable by the Michelson-Morley experiment. The point is that Einstein's theory is simpler than Newton's as corrected and supplemented and complicated by Lorentz and others. It was a glorious case of gaining simplicity at the sacrifice of conservatism; for the time-honored ether went by the board, and far older and more fundamental tenets went by the board too. Drastic changes were made in our conception of the very structure of space and time. . . .

Yet let the glory not blind us to Virtue I. When our estrangement from the past is excessive, the imagination boggles; genius is needed to devise the new theory, and high talent is needed to find one's way about in it. Even Einstein's revolution, moreover, had its conservative strain; Virtue I was not wholly sacrificed. The old physics of Newton's classical mechanics is, in a way, preserved after all. For the situations in which the old and the new theories would predict contrary observations are situations that we are not apt to encounter without sophisticated experiment—because of their dependence on exorbitant velocities or exorbitant distances. This is why classical mechanics held the field so long. Whenever, even having switched to Einstein's relativity theory, we dismiss those exorbitant velocities and distances for the purpose of some practical problem, promptly the discrepancy between Einstein's theory and Newton's becomes too small to matter. Looked at from this angle, Einstein's theory takes on the aspect not of a simplification but a generalization. We might say that the sphere of applicability of Newtonian mechanics in its original simplicity was shown, by the Michelson-Morley experiment and related results, to be less than universal; and then Einstein's theory comes as a generalization, presumed to hold universally. Within its newly limited sphere, Newtonian mechanics retains its old utility. What is more, the evidence of past centuries for Newtonian mechanics even carries over, within these limits, as evidence for Einstein's physics; for, as far as it goes, it fits both.

What is thus illustrated by Einstein's relativity is more modestly exemplified elsewhere, and generally aspired to: the retention, in some sense, of old theories in new ones. If the new theory can be so fashioned as to diverge from the old only in ways that are undetectable in most ordinary circumstances, then it inherits the evidence of the old theory rather than having to overcome it. Such is the force of conservatism even in the context of revolution.

Virtues I through IV may be further illustrated by considering Neptune. That Neptune is among the planets is readily checked by anyone with reference material; indeed it passes as common knowledge, and there is for most of us no need to check it. But only through extensive application of optics and geometry was it possible to determine, in the first instance, that the body we call Neptune exists, and that it revolves around the sun. This required not only much accumulated science and mathematics, but also powerful telescopes and cooperation among scientists.

In fact it happens that Neptune's existence and planethood were strongly suspected even before that planet was observed. Physical theory made possible the calculation of what the orbit of the planet Uranus should be, but Uranus' path differed measurably from its calculated course. Now the theory on which the calculations were based was, like all theories, open to revision or refutation. But here conservatism operates: one is loath to revise extensively a well-established set of beliefs, especially a set so deeply entrenched as a basic portion of physics. And one is even more loath to abandon as spurious immense numbers of observation reports made by serious scientists. Given that Uranus had been observed to be as much as two minutes of arc from its calculated position, what was sought was a discovery that would render this deviation explicable within the framework of accepted theory. Then the theory and its generality would be unimpaired, and the new complexity would be minimal.

It would have been possible in principle to speculate that some special characteristic of Uranus exempted that planet from the physical laws that are followed by other planets. If such a hypothesis had been resorted to, Neptune would not have been discovered; not then, at any rate. There was a reason, however, for not resorting to such a hypothesis. It would have been what is called an *ad hoc hypothesis,* and ad hoc hypotheses are bad air; for they are wanting in Virtues III and IV. Ad hoc hypotheses are hypotheses that purport to account for some particular observations by supposing some very special forces to be at work in the particular cases at hand, and not generalizing sufficiently beyond those cases. The vice of an ad hoc hypothesis admits of degrees. The extreme case is where the hypothesis covers only the observations it was invented to account for, so that it is totally useless in prediction. Then also it is insusceptible of confirmation, which would come of our verifying its predictions.

Another example that has something of the implausibility of an ad hoc hypothesis is the water-diviner's belief that a willow wand held above the ground can be attracted by underground water. The force alleged is too special. One feels, most decidedly, the lack of an intelligible mechanism to explain the attraction. And what counts as intelligible mechanism? A hypothesis strikes us as giving an intelligible mechanism when the hypothesis rates well in familiarity, generality, simplicity. We attain the ultimate in intelligibility of mechanism, no doubt, when we see how to explain something in terms of physical impact or the familiar and general laws of motion.

There is an especially notorious sort of hypothesis which, whether or not properly classified also as ad hoc, shares the traits of insusceptibility of confirmation and uselessness in prediction. This is the sort of hypothesis that seeks to save some other hypothesis from refutation by systematically excusing the failures of its predictions. When the Voice from Beyond is silent despite the incantations of the medium, we may be urged to suppose that "someone in the room is interfering with the communication." In an effort to save the prior hypothesis that certain incantations will summon forth the Voice, the auxiliary hypothesis that untoward thoughts can thwart audible signals is advanced. This auxiliary hypothesis is no wilder than the hypothesis that it was invoked to save, and thus an uncritical person may find the newly wrinkled theory no harder to accept than its predecessor had been. On the other hand the critical observer sees that evidence has ceased altogether to figure. Experimental failure is being milked to fatten up theory.

These reflections bring a fifth virtue to the fore: *refutability,* Virtue V. It seems faint praise of a hypothesis to call it refutable. But the point, we have now seen, is approximately this: some imaginable event, recognizable if it occurs, must suffice to refute the hypothesis. Otherwise the hypothesis predicts nothing, is confirmed by nothing, and confers upon us no earthly good beyond perhaps a mistaken peace of mind.

This is too simple a statement of the matter. Just about any hypothesis, after all, can be held unrefuted no matter what, by making enough adjustments in other beliefs—though sometimes doing so requires madness. We think loosely of a hypothesis as implying predictions when, strictly speaking, the implying is done by the hypothesis together with a supporting chorus of ill-distinguished background beliefs. It is done by the whole relevant theory taken together.

Properly viewed, therefore, Virtue V is a matter of degree, as are its four predecessors. The degree to which a hypothesis partakes of Virtue V is measured by the cost of retaining the hypothesis in the face of imaginable events. The degree is measured by how dearly we cherish the previous beliefs that would have to be sacrificed to save the hypothesis. The greater the sacrifice, the more refutable the hypothesis.

A prime example of deficiency in respect of Virtue V is astrology. Astrologers can so hedge their predictions that they are devoid of genuine content. We may be told that a person will "tend to be creative" or "tend to be outgoing," where the evasiveness of a verb and the fuzziness of adjectives serve to insulate the claim from repudiation. But even if a prediction should be regarded as a failure, astrological devotees can go on believing that the stars rule our destinies; for there is always some item of information, perhaps as to a planet's location at a long gone time, that may be alleged to have been overlooked. Conflict with other beliefs thus need not arise.

All our contemplating of special virtues of hypotheses will not, we trust, becloud the fact that the heart of the matter is observation. Virtues I through V are guides to the framing of hypotheses that, besides conforming to past observations, may plausibly be expected to conform to future ones. When they fail on the latter score, questions are reopened. Thus it was that the Michelson-Morley experiment led to modifications, however inelegant, of Newton's physics at the hands of Lorentz. When Einstein came out with a simpler way of accommodating past observations, moreover, his theory was no mere reformulation of the Newton-Lorentz system; it was yet a third theory, different in some of its predicted observations and answerable to them. Its superior simplicity brought plausibility to its distinctive consequences.

Hypotheses were to serve two purposes: to explain the past and predict the future. Roughly and elliptically speaking, the hypothesis serves these purposes by implying the past events that it was supposed to explain, and by implying future ones. More accurately speaking, as we saw, what does the implying is the whole relevant theory taken together, as newly revised by adoption of the hypothesis in question. Moreover, the predictions that are implied are mostly not just simple predictions of future observations or other events; more often they are conditional predictions. The hypothesis will imply that we will make these further observations if we look in such and such a place, or take other feasible steps. If the predictions come out right, we can win bets or gain other practical advantages. Also, when they come out right, we gain confirmatory evidence for our hypotheses. When they come out wrong, we go back and tinker with our hypotheses and try to make them better.

What we called limiting principles in Chapter IV are, when intelligible, best seen as hypotheses—some good, some bad. Similarly, of course, for scientific laws generally. And similarly for laws of geometry, set theory, and other parts of mathematics. All these laws—those of physics and those of mathematics equally—are among the component hypotheses that fit together to constitute our inclusive scientific theory of the world. The most general hypotheses tend to be the least answerable to any particular observation, since subsidiary hypotheses can commonly be juggled and adjusted to accommodate conflicts; and on this score of aloofness there is no clear boundary between theoretical physics and mathematics. Of course hypotheses in various fields of inquiry may tend to receive their confirmation from different kinds of investigation, but this should in no way conflict with our seeing them all as hypotheses.

We talk of framing hypotheses. Actually we inherit the main ones, growing up as we do in a going culture. The continuity of belief is due to the retention, at each particular time, of most beliefs. In this retentiveness science even at its most progressive is notably conservative. Virtue I looms large. A reasonable person will look upon some of his or her retained beliefs as self-evident, on others as common knowledge though not self-evident, on others as vouched for by authority in varying degree, and on others as hypotheses that have worked all right so far.

But the going culture goes on, and each of us participates in adding and dropping hypotheses. Continuity makes the changes manageable. Disruptions that are at all sizable are the work of scientists, but we all modify the fabric in our small way, as when we conclude on indirect evidence that the schools will be closed and the planes grounded or that an umbrella thought to have been forgotten by one person was really forgotten by another.

NOTE

1. We are indebted to Nell E. Scroggins for suggesting this point.

15

Theory-Choice

T. S. Kuhn

. . . .

In my *Scientific Revolutions* normal science is at one point described as "a strenuous and devoted attempt to force nature into the conceptual boxes supplied by professional education."[1] Later, discussing the problems which surround the choice between competing sets of boxes, theories, or paradigms, I described them as[2]:

> about techniques of persuasion, or about argument and counter argument in a situation in which . . . neither proof nor error is at issue. The transfer of allegiance from paradigm to paradigm is a conversion experience that cannot be forced. Lifelong resistance . . . is not a violation of scientific standards but an index to the nature of scientific research itself. . . . Though the historian can always find men—Priestley, for instance—who were unreasonable to resist for as long as they did, he will not find a point at which resistance becomes illogical or unscientific. At most he may wish to say that the man who continues to resist after his whole profession has been converted has *ipso facto* ceased to be a scientist.

Not surprisingly (though I have myself been very much surprised), passages like these are in some quarters read as implying that, in the developed sciences, might makes right. Members of a scientific community can, I am held to have claimed, believe anything they please if only they will first decide what they agree about and then enforce it both on their colleagues and on nature. The factors which determine what they do choose to believe are fundamentally irrational, matters of accident and personal taste. Neither logic nor observation nor good reason is implicated in theory-choice. Whatever scientific truth may be, it is through-and-through relativistic.

These are all damaging misinterpretations, whatever my responsibility may be for making them possible. Though their elimination will still leave a deep divide between my critics and me, it is prerequisite even to discovering our disagreement. Before treating them individually, however, one general remark should be helpful. The sorts of misinterpretations just outlined are voiced only by philosophers, a group already familiar with the points at which I aim in passages like the above. Unlike readers to whom the point is less familiar, they sometimes suppose that I intend more than I do. What I mean to be saying, however, is only the following.

In a debate over choice of theory, neither party has access to an argument which resembles a proof in logic or formal mathematics. In the latter, both premises and rules of inference are stipulated in advance. If there is disagreement about conclusions, the parties to the debate can retrace their steps one by one, checking each against prior stipulation. At the end of that process, one or the other must concede that at an isolable point in the argument he has made a mistake, violated or misapplied a previously accepted rule. After that concession he has no recourse, and his opponent's proof is then compelling. Only if the two discover instead that they differ about the meaning or applicability of a stipulated rule, that their prior agreement does not provide a sufficient basis for proof, does the ensuing debate resemble what inevitably occurs in science.

Nothing about this relatively familiar thesis should suggest that scientists do not *use* logic (and mathematics) in their arguments, including those which aim to persuade a colleague to renounce a favoured theory and embrace another. I am dumbfounded by Sir Karl's attempt to convict me of self-contradiction because I employ logical arguments myself. What might better be said is that I do not expect that, merely because my arguments are logical, they will be compelling. Sir Karl underscores my point, not his, when he describes them as logical but mistaken, and then makes no attempt to isolate the mistake or to display its logical character. What he means is that, though my arguments are logical, he disagrees with my conclusion. Our disagreement must be about premises or the manner in which they are to be applied, a situation which is standard among scientists debating theory-choice. When it occurs, their recourse is to persuasion as a prelude to the possibility of proof.

To name persuasion as the scientist's recourse is not to suggest that there are not many good reasons for choosing one theory rather than another.[3] It is emphatically *not* my view that "adoption of a new scientific theory is an intuitive or mystical affair, a matter for psychological description rather than logical or methodological codification.[4] On the contrary, the chapter of my *Scientific Revolutions* from which the preceding quotation was abstracted explicitly denies "that new paradigms triumph ultimately through some mystical aesthetic," and the pages which precede that denial contain a preliminary codification of good reasons for theory choice.[5] These are furthermore, reasons of exactly the kind standard in philosophy of science: accuracy, scope, simplicity, fruitfulness, and the like. It is vitally important that scientists be taught to value these characteristics and that they be provided with examples that illustrate them in practice. If they did not hold values like these, their disciplines would develop very differently. Note, for example, that the periods in which the history of art was a history of progress were also the periods in which the artist's aim was accuracy of representation. With the abandonment of that value, the developmental pattern changed drastically though very significant development continued.[6]

What I am denying then is neither the existence of good reasons nor that these reasons are of the sort usually described. I am, however, insisting that such reasons constitute values to be used in making choices rather than rules of choice. Scientists who share them may nevertheless make different choices in the same concrete situation. Two factors are deeply involved. First, in many concrete situations, different values, though all constitutive of good reasons, dictate different conclusions, different choices. In such cases of value-conflict (e.g. one theory is simpler but the other is more accurate) the relative weight placed on different values by different individuals can play a decisive role in individual choice. More important, though scientists share these values and must continue to do so if science is to survive, they do not all apply them in the same way. Simplicity, scope, fruitfulness, and even accuracy can be judged quite differently (which is not to say they may be judged arbitrarily) by different people. Again, they may differ in their conclusions without violating any accepted rule. . . .

NOTES

1. Cf. my [1962], p. 5.
2. *Op cit.* p. 151.
3. For one version of the view that Kuhn insists that "the decision of a scientific group to adopt a new paradigm cannot be based on good reasons of any kind, factual or otherwise," see Shapere [1966], especially p. 67.
4. Cf. Scheffler [1967], p. 18
5. Cf. my [1962], p. 157.
6. Gombrich, [1960], pp. 11 f.

REFERENCES

Gombrich [1960]: *Art and Illusion,* 1960.
Kuhn [1962]: *The Structure of Scientific Revolutions,* 1962.
Scheffler [1967]: *Science and Subjectivity,* 1967.
Shapere [1966]: "Meaning and Scientific Change," in Colodny (ed.): *Mind and Cosmos: Essays in Contemporary Science and Philosophy,* 1966, pp. 41–85.

16

The Variety of Reasons
for the Acceptance of Scientific Theories

Philipp G. Frank

Among scientists it is taken for granted that a theory "should be" accepted if and only if it is "true"; to be true means in this context to be in agreement with the observable facts that can be logically derived from the theory. Every influence of moral, religious, or political considerations upon the acceptance of a theory is regarded as "illegitimate" by the so-called "community of scientists." This view certainly has had a highly salutary effect upon the evolution of science as a human activity. It tells the truth—but not the whole truth. It has never happened that all the conclusions drawn from a theory have agreed with the observable facts. The scientific community has accepted theories only when a vast number of facts has been derived from few and simple principles. A familiar example is the derivation of the immensely complex motions of celestial bodies from the simple Newtonian formula of gravitation, or the large variety of electromagnetic phenomena from Maxwell's field equations.

If we restrict our attention to the two criterions that are called "agreement with observations" and "simplicity," we remain completely within the domain of activities that are cultivated and approved by the community of scientists. But, if we have to choose a certain theory for acceptance, we do not know what respective weight should be attributed to these two criterions. There is obviously no theory that agrees with *all* observations and no theory that has "perfect" simplicity. Therefore, in every individual case, one has to make a choice of a theory by a compromise between both criterions. However, when we try to specify the degree of "simplicity" in different theories, we soon notice that attempts of this kind lead us far beyond the limits of physical science. Everybody would agree that a linear function is simpler than a function of the second or higher degree; everybody would also admit that a circle is simpler than an ellipse. For this reason, physics is filled with laws that express proportionality, such as Hooke's law in elasticity or Ohm's law in electrodynamics. In all these cases, there is no doubt that a nonlinear relationship would describe the facts in a more accurate way, but one tries to get along with a linear law as much as possible.

There was a time when, in physics, laws that could be expressed without using differential calculus were preferred, and in the long struggle between the corpuscular and the wave theories of light, the argument was rife that the corpuscular theory was mathematically simpler, while the wave theory required the solution of boundary problems of partial differential equations, a highly complex matter. We note that even a purely mathematical estimation of simplicity depends upon the state of culture of a certain period. People who have grown up in a mathematical atmosphere—that is, saturated with ideas about invariants—will find that Einstein's theory of gravitation is of incredible beauty and simplicity; but to people for whom ordinary calculus is the center of interest, Einstein's theory will be of immense complexity, and this low degree of simplicity will not be compensated by a great number of observed facts.

However, the situation becomes much more complex, if we mean by *simplicity* not only simplicity of the mathematical scheme but also simplicity of the whole discourse by which the theory is formulated. We may start from the most familiar instance, the decision between the Copernican (heliocentric) and the Ptolemaic (geocentric) theories. Both parties, the Roman Church and the followers of Copernicus, agreed that Copernicus' system, from the purely mathematical angle, was simpler than Ptolemy's. In the first one, the orbits of planets were plotted as a system of concentric circles with the sun as center, whereas in the geocentric system, the planetary orbits were sequences of loops. The observed facts covered by these systems were approximately the same ones. The criterions of acceptance that are applied in the community of scientists today are, according to the usual way of speaking, in agreement with observed facts and mathematical simplicity. According to them, the Copernican system had to be accepted unhesitatingly. Since this acceptance did not happen before a long period of doubt, we see clearly that the criterions "agreement with observed facts" and "mathematical simplicity" were not the only criterions that were considered as reasons for the acceptance of a theory.

As a matter of fact, there were three types of reasons against the acceptance of the Copernican theory that remained unchallenged at the time when all "scientific" reasons were in favor of that theory. First, there was the incompatibility of the Copernican system with the traditional interpretation of the Bible. Second, there was the disagreement between the Copernican system and the prevailing philosophy of that period, the philosophy of Aristotle as it was interpreted by the Catholic schoolmen. Third, there was the objection that the mobility of the earth, as a real physical fact, is incompatible with the common-sense interpretation of nature. Let us consider these three types of reasons more closely. In the Book of Joshua this leader prays to God to stop the sun in its motion in order to prolong the day and to enable the people of Israel to win a decisive victory. God indeed "stopped the sun." If interpreted verbally, according to the usage of words in our daily language, this means that the sun is moving, in flagrant contradiction with the Copernican theory. One could, of course, give a more sophisticated interpretation and say that "God stopped the sun" means that he stopped it in its motion relative to the earth. This is no longer contradictory to the Copernican system. But now the question arises: Should we adopt a simple mathematical description and a complicated, rather "unnatural" interpretation of the Bible or a more complicated mathematical description (motion in loops) and a simple "natural" interpretation of the biblical text? The decision certainly cannot be achieved by any argument taken from physical science.

If one believes that all questions raised by science must be solved by the "methods" of this special science, one must say: Every astronomer who lived in the period between Copernicus and Galileo was "free" to accept either the Copernican or the Ptolemaic doctrine; he could make an "arbitrary" decision. However, the situation is quite different if one takes into consideration that physical science is only a part of science in general. Building up astronomical theories is a particular act of human behavior. If we consider human behavior in general, we look at physical science as a part of a much more general endeavor that embraces also psychology and sociology. It is called by some authors "behavioristics." From this more general angle, the effect of a simplification in the mathematical formula and the simplification in biblical interpretation are quite comparable with each other. There is meaning in asking by which act the happiness of human individuals and groups is more favorably influenced. This means that, from the viewpoint of a general science of human behavior, the decision between the Copernican and Ptolemaic systems was never a matter of arbitrary decision.

The compatibility of a physical theory with a certain interpretation of the Bible is a special case of a much more general criterion: the compatibility of a physical theory with theories that have been advanced to account for observable phenomena outside the domain of a physical science. The most important reason for the acceptance of a theory beyond the "scientific criterions" in the narrower sense (agreement with observation and simplicity of the mathematical pattern) is the fitness of a theory to be generalized, to be the basis of a new theory that does not logically follow from the original one, and to allow prediction of more observable facts. This property is often called the "dynamical" character or the "fertility" of a theory. In this sense, the Copernican theory is much superior to the geocentric one. Newtonian laws of motion have a simple form only if the sun is taken as a system of reference and not the earth. But the decision in favor of the Copernican theory on this basis could be made only when Newton's laws came into existence. This act requires, however, creative imagination or, to speak more flippantly, a happy guessing that leads far beyond the Copernican and Ptolemaic systems.

However, long before the "dynamical" character of the Copernican system was recognized, the objection was raised that the system was incompatible with "the general laws of motion" that could be derived from principles regarded as "immediately intelligible" or, in other words, "self-evident" without physical experiment or observations. From such "self-evident" principles there followed, for example, the physical law that only celestial bodies (like sun or moon) moved "naturally" in circular orbits, while terrestrial bodies (like our earth) moved naturally along straight lines as a falling stone does. Copernicus' theory of a "motion of the earth in a circular orbit" was, therefore, incompatible with "self-evident" laws of nature.

Medieval scientists were faced with the alternatives: Should they accept the Copernican theory with its simple mathematical formulas and drop the self-evident laws of motion, or should they accept the complicated mathematics of the Ptolemaic system along with the intelligible and self-evident general laws of motion. Acceptance of Copernicus' theory would imply dropping the laws of motion that had been regarded as self-evident and looking for radically new laws. This would also mean dropping the contention that a physical law can be derived from "intelligible" principles. Again, from the viewpoint of physical science, this decision cannot be made. Although an arbitrary decision may seem to be required, if one looks at the situation from the viewpoint of human behavior it is clear that the decision, by which the derivation of physical laws from self-evident principles is abandoned, would alter the situation of man in the world fundamentally. For example, an important argument for the existence of spiritual beings would lose its validity. Thus, social science had to decide whether the life of man would become happier or unhappier by the acceptance of the Copernican system.

The objections to this system, on the basis of self-evident principles, have also been formulated in a way that looks quite different but may eventually, when the chips are down, not be so very different. Francis Bacon, the most conspicuous adversary of Aristotelianism in the period of Galileo, fought the acceptance of the Copernican theory on the basis of common-sense experience. He took it for granted that the principles of science should be as analogous as possible to the experience of our daily life. Then, the principles could be presented in the language that has been shaped for the purpose of describing, in the most convenient way, the experience of our daily existence—the language that everyone has learned as a child and that is called "common-sense language." From this daily experience, we have learned that the behavior of the sun and the planets is very different from that of the earth. While the earth does not emit any light, the sun and the planets are brilliant; while every earthly object that becomes separated from the main body will tend to fall back toward the center and stop there, the celestial objects undergo circular motion eternally around the center.

To separate the sun from the company of the planets and put the earth among these brilliant and mobile creatures, as Copernicus suggested, would have been not only unnatural but a serious violation of the rule to keep the principles of science as close to common sense as possible. We see by this example that one of the reasons for the acceptance of a theory has frequently been the compatibility of this theory with daily life experience or, in other words, the possibility of expressing the theory in common-sense language. Here is, of course, the source of another conflict between the "scientific" reasons for the acceptance of a theory and other requirements that are not "scientific" in the narrower sense. Francis Bacon rejected the Copernican system because it was not compatible with common sense.

In the eighteenth and nineteenth centuries, Newton's mechanics not only had become compatible with common sense but had even been identified with common-sense judgment. As a result, in twentieth century physics, the theory of relativity and the quantum theory were regarded by many as incompatible with common sense. These theories were regarded as "absurd" or, at least, "unnatural." Lenard in Germany, Bouasse in France, O'Rahilly in Ireland, and Timiryaseff in Russia rejected the theory of relativity, as Francis Bacon had rejected the Copernican system. Looking at the historical record, we notice that the requirement of compatibility with common sense and the rejection of "unnatural theories" have been advocated with a highly emotional undertone, and it is reasonable to raise the question: What was the source of heat in those fights against new and absurd theories? Surveying these battles, we easily find one common feature, the apprehension that a disagreement with common sense may deprive scientific theories of their value as incentives for a desirable human behavior. In other words, by becoming incompatible with common sense, scientific theories lose their fitness to support desirable attitudes in the domain of ethics, politics, and religion.

Examples are abundant from all periods of theory-building. According to an old theory that was prevalent in ancient Greece and was accepted by such men as Plato and Aristotle, the sun, planets, and other celestial bodies were made of a material that was completely different from the material of which our earth consists. The great gap between the celestial and the terrestrial bodies was regarded as required by our common-sense experience. There were men—for example, the followers of Epicurus—who rejected this view and assumed that all bodies in the universe, earth and stars, consist of the same material. Nevertheless, many educators and political leaders were afraid that denial of the exceptional status of the celestial bodies in physical science would make it more difficult to teach the belief in the existence of spiritual beings as distinct from material bodies; and since it was their general conviction that the belief in spiritual beings is a powerful instrument to bring about a desirable conduct among citizens, a physical theory that supported this belief seemed to be highly desirable.

Plato, in his famous book *Laws,* suggested that people in his ideal state who taught the "materialistic" doctrine about the constitution of sun and stars should be jailed. He even suggested that people who knew about teachers of that theory and did not report them to the authorities should also be jailed. We learn from this ancient example how scientific theories have served as instruments of indoctrination. Obviously, fitness to support a desirable conduct of citizens or, briefly, to support moral behavior, has served through the ages as a reason for acceptance of a theory. When the "scientific criterions" did not uniquely determine a theory, its fitness to support moral or political indoctrination became an important factor for its acceptance. It is important to learn that the interpretation of a scientific theory as a support of moral rules is not a rare case but has played a role in all periods of history.

This role probably can be traced back to a fact that is well known to modern students of anthropology and sociology. The conduct of man has always been shaped according to the example of an ideal society; on the other hand, this ideal has been represented by the "behavior" of the universe, which is, in turn, determined by the laws of nature, in particular, by the physical laws. In this sense, the physical laws have always been interpreted as examples for the conduct of man or, briefly speaking, as moral laws. Ralph Waldo Emerson wrote in his essay *Nature* that "the laws of physics are also the laws of ethics." He used as an example the law of the lever, according to which "the smallest weight may be made to lift the greatest, the difference of weight being compensated by time."

We see the connection of the laws of desirable human conduct with the physical laws of the universe when we glance at the Book of Genesis. The first chapter presents a physical theory about the creation of the world. But the story of the creation serves also as an example for the moral behavior of men; for instance, because the creation took 7 days, we all feel obliged to rest on each seventh day. Perhaps the history of the Great Flood is even more instructive. When the Flood abated, God established a Covenant with the human race: "Never again shall all flesh be cut off by the waters of a flood; neither shall there any more be a flood to destroy the earth." As a sign of the Covenant the rainbow appeared: "When I bring clouds over the earth and the bow is seen in the clouds, I will remember the Covenant which is between me and you, and the waters shall never again become a flood to destroy all flesh." If we read the biblical text carefully, we understand that what God actually pledged was to maintain, without exception, the validity of the physical laws or, in other words, of the causal law. God pledged: "While the earth remains, seedtime and harvest, cold and heat, summer and winter, day and night shall not cease."

All the physical laws, including the law of causality, were given to mankind as a reward for moral behavior and can be canceled if mankind does not behave well. So even the belief in the validity of causal laws in the physical world has supported the belief in God as the supreme moral authority who would punish every departure from moral behavior by abolishing causality. We have seen that Epicurean physics and Copernican astronomy were rejected on moral grounds. We know that Newton's physics was accepted as supporting the belief in a God who was an extremely able engineer and who created the world as a machine that performed its motions according to its plans. Even the generalization of Newtonian science that was advanced by 18th century materialism claimed to serve as a support for the moral behavior of man. In his famous book *Man a Machine,* which has often been called an "infamous book," La Mettrie stresses the point that by regarding men, animals, and planets as beings of the same kind, man is taught to regard them all as his brothers and to treat them kindly.

It would be a great mistake to believe that this situation has changed in the nineteenth and twentieth centuries. A great many authors have rejected the biological theory that organisms have arisen from inanimate matter (spontaneous generation), because such a theory would weaken the belief in the dignity of man and in the existence of a soul and would, therefore, be harmful to moral conduct. In twentieth century physics, we have observed that Einstein's theory of relativity has been interpreted as advocating an "idealistic" philosophy, which, in turn, would be useful as a support of moral conduct. Similarly, the quantum theory is interpreted as supporting a weakening of mechanical determinism and, along with it, the introduction of "indeterminism" into physics. In turn, a great many educators, theologians, and politicians have enthusiastically acclaimed this "new physics" as providing a strong argument for the acceptance of "indeterminism" as a basic principle of science.

The special mechanism by which social powers bring about a tendency to accept or reject a certain theory depends upon the structure of the society within which the scientist operates. It may vary from a mild influence on the scientist by friendly reviews in political or educational dailies to promotion of his book as a best seller, to ostracism as an author and as a person, to loss of his job, or, under some social circumstances, even to imprisonment, torture, and execution. The honest scientist who works hard in his laboratory or computation-room would obviously be inclined to say that all this is nonsense—that his energy should be directed toward finding out whether, say, a certain theory is "true" and that he "should not" pay any attention to the fitness of a theory to serve as an instrument in the fight for educational or political goals. This is certainly the way in which the situation presents itself to most active scientists. However, scientists are also human beings and are definitely inclined toward some moral, religious, or political creed. Those who deny emphatically that there is any connection between scientific theories and religious or political creeds believe in these creeds on the basis of indoctrination that has been provided by organizations such as churches or political parties. This attitude leads to the conception of a "double truth" that is not only logically confusing but morally dangerous. It can lead to the practice of serving God on Sunday and the devil on weekdays.

The conviction that science is independent of all moral and political influences arises when we regard science either as a collection of facts or as a picture of objective reality. But today, everyone who has attentively studied the logic of science will know that science actually is an instrument that serves the purpose of connecting present events with future events and deliberately utilizes this knowledge to shape future physical events as they are desired. This instrument consists of a system of propositions—principles—and the operational definitions of their terms. These propositions certainly cannot be derived from the facts of our experience and are not uniquely determined by these facts. Rather they are hypotheses from which the facts can be logically derived. If the principles or hypotheses are not determined by the physical facts, by what are they determined? We have learned by now that, besides the agreement with observed facts, there are other reasons for the acceptance of a theory: simplicity, agreement with common sense, fitness for supporting a desirable human conduct, and so forth. All these factors participate in the making of a scientific theory. We remember, however, that according to the opinion of the majority of active scientists, these extrascientific factors "should not" have any influence on the acceptance of a scientific theory. But who has claimed and who can claim that they "should not"?

This firm conviction of the scientists comes from the philosophy that they have absorbed since their early childhood. The theories that are built up by "scientific" methods, in the narrower sense, are "pictures" of physical reality. Presumably they tell us the "truth" about the world. If a theory built up exclusively on the ground of its agreement with observable facts tells the "truth" about the world, it would be nonsense to assume seriously that a scientific theory can be influenced by moral or political reasons. However, we learned that "agreement with observed facts" does not single out one individual theory. We never have one theory that is in full agreement but several theories that are in partial agreement, and we have to determine the final theory by a compromise. The final theory has to be in fair agreement with observations and also has to be sufficiently simple to be usable. If we consider this point, it is obvious that such a theory cannot be "the truth." In modern science, a theory is regarded as an instrument that serves toward some definite purpose. It has to be helpful in predicting future observable facts on the basis of facts that have been observed in the past and the present. It should also be helpful in the construction of machines and devices that can save us time and labor. A scientific theory is, in a sense, a tool that produces other tools according to a practical scheme.

In the same way that we enjoy the beauty and elegance of an airplane, we also enjoy the "elegance" of the theory that makes the construction of the plane possible. In speaking about any actual machine, it is meaningless to ask whether the machine is "true" in the sense of its being "perfect." We can ask only whether it is "good" or sufficiently "perfect" for a certain purpose. If we require speed as our purpose, the "perfect" airplane will differ from one that is "perfect" for the purpose of endurance. The result will be different again if we choose safety, or fun, or convenience for reading and sleeping as our purpose. It is impossible to design an airplane that fulfills all these purposes in a maximal way. We have to make some compromises. But then, there is the question: Which is more important, speed or safety, or fun or endurance? These questions cannot be answered by any agreement taken from physical science. From the viewpoint of "science proper" the purpose is arbitrary, and science can teach us only how to construct a plane that will achieve a specified speed with a specified degree of safety. There will be a debate, according to moral, political, and even religious lines, by which it will be determined how to produce the compromise. The policymaking authorities are, from the logical viewpoint, "free" to make their choice of which type of plane should be put into production. However, if we look at the situation from the viewpoint of a unified science that includes both physical and social science, we shall understand how the compromise between speed and safety, between fun and endurance is determined by the social conditions that produce the conditioned reflexes of the policymakers. The conditioning may be achieved, for example, by letters written to congressmen. As a matter of fact, the building of a scientific theory is not essentially different from the building of an airplane.

If we look for an answer to the question of whether a certain theory, say the Copernican system or the theory of relativity, is perfect or true, we have to ask the preliminary questions: What purpose is the theory to serve? Is it only the purely technical purpose of predicting observable facts? Or is it to obtain a simple and elegant theory that allows us to derive a great many facts from simple principles? We choose the theory according to our purpose. For some groups, the main purpose of a theory may be to serve as a support in teaching a desirable way of life or to discourage an undesirable way of life. Then, we would prefer theories that give a rather clumsy picture of observed facts, provided that we can get from the theory a broad view of the universe in which man plays the role that we desire to give him. If we wish to speak in a more brief and general way, we may distinguish just two purposes of a theory: the usage for the construction of devices (technological purpose) and the usage for guiding human conduct (sociological purpose).

The actual acceptance of theories by man has always been a compromise between the technological and the sociological usage of science. Human conduct has been influenced directly by the latter, by supporting specific religious or political creeds, while the technological influence has been rather indirect. Technological changes have to produce social changes that manifest themselves in changing human conduct. Everybody knows of the Industrial Revolution of the nineteenth century and the accompanying changes in human life from a rural into an urban pattern. Probably the rise of atomic power will produce analogous changes in man's way of life.

The conflict between the technological and the sociological aims of science is the central factor in the history of science as a human enterprise. If thoroughly investigated, it will throw light upon a factor that some thinkers, Marxist as well as religious thinkers, regard as responsible for the social crisis of our time: the backwardness of social progress compared with technological progress. To cure this illness of our time, an English bishop recommended, some years ago, the establishment of a "truce" in the advancement in technology, in order to give social progress some time to keep up with technological advancement. We have seen examples of this conflict in Plato's indictment of astrophysical theories that could be used as a support of "materialism." We note the same purpose in the fight against the Copernican system and, in our own century, against the Darwinian theory of evolution, against Mendel's laws of heredity, and so forth.

A great many scientists and educators believe that such a conflict no longer exists in our time, because now it is completely resolved by "the scientific method," which theory is the only valid one. This opinion is certainly wrong if we consider theories of high generality. In twentieth century physics, we note clearly that a formulation of the general principles of subatomic physics (quantum theory) is accepted or rejected according to whether we believe that introduction of "indeterminism" into physics gives comfort to desirable ethical postulates or not. Some educators and politicians have been firmly convinced that the belief in "free will" is necessary for ethics and that "free will" is not compatible with Newtonian physics but is compatible with quantum physics. The situation in biology is similar. If we consider the attitude of biologists toward the question whether living organisms have developed from inanimate matter, we shall find that the conflict between the technological and the sociological purposes of theories is in full bloom. Some prominent biologists say that the existence of "spontaneous generation" is highly probable, while others of equal prominence claim that it is highly improbable. If we investigate the reasons for these conflicting attitudes, we shall easily discover that, for one group of scientists, a theory that claims the origin of man not merely from the "apes" but also from "dead matter" undermines their belief in the dignity of man, which is the indispensable basis for all human morality. We should note in turn that, for another group, desirable human behavior is based on the belief that there is a unity in nature that embraces all things.

In truth, many scientists would say that scientific theories "should" be based only on purely scientific grounds. But, exactly speaking, what does the word *should* mean in this context? With all the preceding arguments it can mean only: If we consider exclusively the technological purpose of scientific theories, we could exclude all criterions such as agreement with common sense or fitness for supporting desirable conduct. But even if we have firmly decided to do away with all "nonsense," there still remains the criterion of "simplicity," which is necessary for technological purposes and also contains, as we learned previously, a certain sociological judgment. Here, restriction to the purely technological purpose does not actually lead unambiguously to a scientific theory. The only way to include theory-building in the general science of human behavior is to refrain from ordering around scientists by telling them what they "should" do and to find how each special purpose can be achieved by a theory. Only in this way can science as a human activity be "scientifically" understood and the gap between the scientific and the humanistic aspect be bridged.

STUDY QUESTIONS FOR PART FOUR

"Hypothesis"—W. W. Quine and J. S. Ullian

1. How do Quine and Ullian dash the hope of absolute demonstrability in science?
2. What do Quine and Ullian mean by 'hypothesis'? You might compare their view with Hospers's view in "Theories."
3. Answer the critic who says that the so-called virtue of conservatism is just an expression of the reactionary attitude of the powerful scientists in the scientific institutions.
4. Do conservatism and modesty amount to more than features of quantitative confirmation?
5. How do Quine and Ullian define simplicity? Do they think considerations of simplicity provide epistemic or practical reasons? What is their argument?
6. Explain how modesty and generality conflict? How do Quine and Ullian reconcile them?
7. What are *ad hoc* hypotheses and what is wrong with them?
8. Why must hypotheses be refutable?
9. How can refutability and conservatism both be virtues of a hypothesis?

"Reflections On My Critics"—T. S. Kuhn

1. How does argument in formal mathematics differ from that found in science?
2. Explain how, on Kuhn's view, two scientists can both be reasonable yet disagree on which theory to accept?
3. Would Quine and Ullian agree with Kuhn's answer to the question above?

"The Variety of Reasons for the Acceptance of Scientific Theories"—P. G. Frank

1. How does Frank show that with the Copernican system other reasons than agreement with observed facts and simplicity were used in determining its acceptance? What were some of the additional criteria?
2. Is it unreasonable or unscientific to put much weight on the additional criteria? Discuss.
3. Why is it important that scientific theories agree with common sense?
4. What view, according to Frank, on the status of unobservables and theoretical statements fits best with the recognition of the importance of "additional criteria?" Evaluate his view.

SELECTED BIBLIOGRAPHY

[1] Brody, B., ed. *Readings in the Philosophy of Science.* Englewood Cliffs, N.J.: Prentice-Hall, 1970, part 3. Contains essays on confirmation and acceptance.

[2] Goodman, N. *Fact, Fiction, and Forecast.* New York: Bobbs-Merrill, 1965, part 3. Presents a serious paradox for a theory of confirmation.

[3] Kuhn, T. *The Structure of Scientific Revolutions.* Chicago: University of Chicago Press, 1962. A wealth of material on acceptance.

[4] Popper, K. *Conjectures and Refutations.* New York: Harper & Row, 1963. Contains many relevant essays; see especially 10.

[5] Scheffler, I. *The Anatomy of Inquiry.* New York: Knopf, 1969, part 3. Tight discussion of confirmation.

[6] ———. *Science and Subjectivity.* New York: Bobbs-Merrill, 1967. Tries to blunt what he sees as the relativistic consequences of Kuhn's view.

Part Five

Science and Values

Part 5: Science and Values

Introduction

I. The Problems

Problems about the role of value judgments within *science* and ethical disputes about *the uses of* science (e.g., in technology and social policy) are commonplace. Recent discussion of genetic engineering, behavior control, safety in nuclear power plants, human experimentation (e.g., deception and manipulation in research), medical ethics, and the IQ controversy are only the most dramatic instances of such problems. The basic problems are by no means new: just think of the controversy between Galileo and the Catholic Church about the nature of the solar system, or controversies between Darwin and his critics over the question of evolution.

However, underlying the many disputes covered by the phrase "science and values," there are several fundamental and persisting issues. Among the most obvious are these:

1. Can (is/should) science be "value free" or "neutral"? What do "value free" and "value neutral" mean in such contexts? What kinds of values are at issue? (This raises the question "What is value?" which must be bypassed here.) For instance, not all values are *moral* values, so that science may be *morally* neutral even though not *value* neutral in some other (or in a wider) sense. As we shall see, it is open to question whether science is or can be even morally neutral, much less value neutral.

2. If science *is* value free (in either the extended or the narrower sense of morally neutral: unless otherwise specified I shall mean the former whenever "value free" is used) what implications does this have for our conception of science, knowledge, and values, and for our views about the nature and aims of science and the social uses of science and technology? If science is *not* value free, what does *this* imply about the foregoing?

223

3. What are the best or most defensible concepts or theories about knowledge, values (moral or nonmoral), science, and the best ways of conceptualizing their interconnections (or the lack thereof)? Here we must eventually deal with the concepts of rationality, objectivity, subjectivity, pure and applied science, and so on. (See Part 1 of this volume.) And we must eventually decide what sorts of views will or will not be plausible candidates for helping us understand science, values, and their connections. For instance, will any theory according to which science and values are, and should be, totally unrelated be acceptable to us? This is a question that is discussed later on in this introduction.

It should come as no surprise that these and other issues have been given a variety of answers, and have generated a number of complex and often conflicting theories about science, values, and their interrelations. There is no point in attempting to even list, much less discuss, all or even many of these views in an introductory essay. Instead, some historical and philosophical backdrop for the selections in Part 5 will be provided. The emphasis will be on those developments which get to the very heart of the issues, and which bear most directly on the readings.

It will prove instructive to begin with the question of *why* there is, or even should be, any problems about the relationships between science and values at all. This question is by no means rhetorical, but rather gets to the nub of the issue, which involves the relationship between science (and, more generally, knowledge) and values which finds expression in modern science and philosophy. It is here, after all, that we must look for the ideas that continue to dominate our culture's general outlooks on these questions.

II. Why Is There a Problem About "Science and Values"?

The problems briefly outlined above come into existence with the advent of modern philosophy and science, especially the scientific revolution of the seventeenth century. In order to appreciate why, and how, this happens, it will be helpful to briefly sketch the views of the ancient Greeks—the molders of Western culture—on the issues of knowledge and values (or, at any rate, on the Greek approximations to these issues as they have since come to be understood). Then we can go on to sketch the development of the modern problems as they arise at the beginning of "modern history," which shall be dated here from the seventeenth century.

For the ancient Greeks, there are no distinctions between (a) science or knowledge and "values" or "the good" or between (b) science and philosophy or between (c) the objective and the subjective (as these concepts are understood within modern science and philosophy) or between (d) a "factual" or "descriptive" account of the world (e.g., in terms of the structural properties of things and the laws which govern them) and a "normative" or "evaluative" or (even) a "moral" interpretation of the world, as embodying a certain order, pattern, beauty, purpose, and even "goodness." (What is natural is also good in this view.) The most influential, and most forceful, presentation of the Greek view of the cosmos is articulated by Plato in his *Republic* (bks. iv–vii). For Plato, "objective" reality is characterized in terms of the idea or form of "the good." Reality is a unified, patterned, or ordered whole. In order to understand experience, we have to arrive at a knowledge of the laws and structures governing everything, as well as the order, patterning, or "purpose" which pervades all experience and which unifies it in a coherent and "meaningful" fashion. Such knowledge, which Plato calls "Dialectic," is not to be equated with what is today called "knowledge," since our term 'knowledge' is often used as a synonym for 'science' or 'scientific knowledge'. On Plato's view, the aims and methods of the empirical sciences are designed to give us at best only a limited insight into reality; more specifically, an insight into a certain kind of experience, a certain level of reality (the level of objects of experience like trees, rocks, and so on). They must be complemented by an insight into the more basic principles and patterns which govern everything. What for Plato is the "most real" is also the most abstract and the least accessible to ordinary experience. (Modern theoretical science, e.g., atomic theory, quantum physics, genetics, and chemistry, embody this Platonic ideal to some extent.) An "objective account" of things is not complete until everything is ordered into a unified picture; which involves the idea that purpose and norms (i.e., ordering principles) are not eliminable from such an account. It also involves the idea that science cannot give us either a complete account of everything, or an adequate account of even the objects of its legitimate concern, since these objects must be ultimately understood in terms of the principles which govern everything, including themselves and their place in the whole scheme.

One of the chief features of modern science and modern philosophy is the attempt to deny or else to truncate this platonic vision of the universe, and of the nature of knowledge and the good. In what follows, a brief sketch of these developments will be given.

III. Origin and Nature of the Problem

The issues concerning the relationship between science and values, including the role of values *in* science (the issue of value-neutrality) comes into modern Western history with the advent of the so-called mechanical picture of the world (especially classical Newtonian science), and the scientific revolution, most especially the epistemological and methodological revolution in science and philosophy inspired by its main architect, René Descartes (1596–1650).

According to this view, we must make a sharp distinction between what is objective and what is subjective in order to acquire reliable (i.e., "objective") knowledge of the world (including knowledge of human beings) by the use of reliable ("rational") methods of inquiry. Since, according to the mechanical

world picture, nature is a vast machine governed by quantitative laws and relationships (nature is written in the language of mathematics), the objective
features of the world turn out to be those features—matter, motion, and
physical magnitudes—which constitute the nuts and bolts of the machine,
together with the laws governing it. Only such features of experience are truly
objective. Thus a rational methodology for inquiring about the machine's
working—i.e., for acquiring knowledge—must take into account only those
features which can be quantified, i.e., written in nature's language. The very
essence of the world is given by the objective properties just mentioned, together
with the mechanical laws which govern them. (These essential features of the
world are dubbed "primary qualities" by Galileo and Locke.) Everything else,
e.g., colors, values, interpretations, purpose, and theories, are not "objective"
and thus do not belong in an objective account of the world, unless they can be
"reduced" to objective terms, or explained away as illusory phenomena by such
an account. (Later thinkers expanded the province of science to include those
"subjective" items—called "secondary qualities"—that had an autonomous
status for Descartes (e.g., mental phenomena and values) so that these came to
be "reduced" to objective features or explained away entirely, as in modern
behaviorism and materialism.)

In sum, objectivity means both: (a) objective in the sense of being about
what is objective, and (b) objective in the sense of arriving at objective truths
by methods which themselves take no account of anything "subjective," i.e.,
which are unbiased. (The search for mechanical "fool-proof" methods, e.g.,
computer algorithms, cost benefit decisions, is the ultimate outcome of this
ideal of rational, objective method.)

IV. Initial Objections to "Objectivism"

This view already has insuperable difficulties: at least *we* can now see this
(which is not to make the anachronistic claim that its classical proponents were
flawed for not seeing it: it is just as easy to be a "Monday-morning quarter-
back" in history as in football).

First of all, to paraphrase Woody Allen (*Love and Death*): Objectivity is
subjective, and subjectivity is objective; at least the latter point is certainly
true: the fact that I am (say) in pain is no less objective a fact about me than
the fact that I weigh 160 pounds.

Second, the view being considered is, paradoxically, rooted in the notion
that "objective knowledge" is a "rational reconstruction" of the private, i.e.,
"subjective," experiences of a collection of knowers (viz., out of those subjective experiences which represent the essence of the world). Modern
epistemology is rooted in this conundrum.

Third, as M. Polanyi [6] and others show, if we merely wanted objective
truths, we (as a species) would devote virtually all of our intellectual energy to
studying interstellar dust, and only a fraction of a microsecond studying

ourselves (or anything else, for that matter) since, objectively speaking, human beings are of no cosmic significance in the objective order of things! Obviously no one would take this requirement on objectivity seriously (this concept of objectivity is theological—God sees the world objectively as an outside omniscient observer). What we are seeking are truths which are interesting, which are useful and valuable to us. In a word, knowledge, truth, and objectivity are (or are rooted in) values and human purposes.

Fourth, not only do knowledge, objectivity, and truth—and thus methodology—turn out to be, or at least be grounded in, values; on some views they are, or are grounded in, moral ideals. In any event, the search for knowledge expresses a value; and thus distinctions between reliable and unreliable knowledge claims, between good and bad methods, and so on, are partly normative judgments. (See the articles by Rescher and Scriven; also [5].)

(The question as to whether knowledge is an *intrinsic* value, as it is for those, such as Plato and Aristotle, for whom the life of contemplation is the highest good or "*summum bonum,*" or whether it is an *instrumental* value (Bacon's maxim that knowledge is power), as well as the issue whether such values are (ultimately) moral values, will be taken up in the introduction to Part 6. At this point it may be worth citing the words of N. I. Bukharin, who says: "The idea of the self-sufficient character of science . . . is naive; it confuses the *subjective passions* of the professional scientist . . . with the *objective social role* of this kind of activity, as an activity of vast practical importance.")

V. Some Implications of "Objectivism"

Despite these "obvious" difficulties, the fact remains that the distinction between objectivity and subjectivity, as regards claims governing methodology and the content of an objective world picture, dominate modern science, philosophy, and Western culture from Descartes and Galileo up to the present. The ideas of value neutrality, the uses of cost-benefit analyses to arrive at "rational" decisions in politics, science, and technology, the attempt to use knowledge for social and political ends (e.g., behavior control techniques), and so on, are just sophisticated outgrowths of this cluster of ideas. So, too, is the idea that scientific method affords the only rational methods for solving problems, so that value judgments are either not rational, or else are concerned merely with problems about calculating efficiency, or about decision-making under conditions of uncertainty. Both proponents and opponents of the classical picture of the world and the attendant ideas of objectivity and rationality (e.g., behaviorists, on the one hand, and so-called "neo-romantics" and existentialists, on the other) share this conception that values are essentially subjective and irrational if they are anything more than predictions or calculations about means to an end.

But it is becoming glaringly obvious that these assumptions are connected with an inadequate conception of *both* science *and* values, and of the interrelationships between the two areas. (See especially the Scriven article in this part.)

As Gerwirth [5] points out, no one is reluctant to distinguish between "good" and "bad" science, or between science and nonscience or pseudoscience (e.g., astronomy vs. astrology). (See Part 1.) Yet on the view being discussed we are not supposed to be able to distinguish between good and bad moralities, or between acceptable vs. unacceptable value judgments. But this combination of "normative" science and "positive" ethics is internally incoherent and grossly inadequate.

VI. Some Corollaries of "Objectivism"

Two clusters of ideas conspire to produce this result. (i) Many advocates of the view being discussed either attempt to turn ethics into a science, or to explain value judgments scientifically (e.g., in terms of conditioning or historical or economic determinism). The latter kind of strategy usually leads to some form of ethical nihilism or extreme ethical relativism: All we can do is explain the origins of ethical behavior in terms of some objective scientific theory. On this view the point or content (i.e., the "autonomy") of value judgments is either lost or obscured. But this strategy explains why, for an advocate of this approach, "positive ethics" is the inevitable result. Moreover, it is just because "normative science" shows that objectivism is the only correct scientific approach that "positive ethics" turns out to be compatible, indeed, required by, objectivism. (ii) The idea of value neutrality, especially as this idea shows up in the social and policy sciences, is greatly influenced by views which attempt to reconcile "objective" science and "subjective" morality, by drawing theoretical limits to science, in order to save morality and human freedom. On this view, (a) science and morality cannot conflict (they are "complementary") since (b) they have nothing to do with each other: they govern different spheres of experience (these relate to the differences between "man as object" and as actor). But the price to be paid for this move is just the idea that science is, and must be, value neutral and that value judgments are merely subjective and irrational acts of the will.

Connected with this view is the idea that rational justification is essentially (hypothetico) deductive. The ultimate principles of a system, whether a moral system, a scientific theory, or a formal system, such as geometry, or the "brute facts" or "data" are beyond rational dispute. They must either be arbitrarily stipulated and accepted or be taken as self-evident, and then used to define what a rational proof or justification is *within* the system. The "ultimates" must be accepted as given. Relativism is the view that there are different, equally rational or acceptable (incompatible) ultimates. When one reaches these ultimates, be it a body of "hard facts," or axioms or moral principles, one has reached bedrock. One can then either accept or reject them. If the former, one can then show that the principles which follow from them are rational. In the latter case, one is free to adopt different "ultimates." At the same time, finding rational principles, or making rational decisions within the system, becomes a matter of, say, finding the best means of optimizing the ultimate principles or ends postulated by the system.

As P. Diesing [3] puts it:

> Such a conception of rationality limits its scope rather severely. The criterion of efficiency is applicable to means and not to ends. . . . Thus ultimate ends, the basic aims of life, cannot be selected or evaluated by rational procedures; they must be dealt with by arbitrary preference, or intuition, or by cultural and biological determinism. And yet it seems unfortunate to have rational procedures available for the relatively less important decisions of life and to have none for dealing with the most important decisions. Nor is this all. Several rather common types of activity can only with great difficulty be assimilated to the category of efficient action and be evaluated by standards of efficiency. ([3], p. 1)

Ultimately, this theory of justification is part and parcel of the idea that science is value free, that value judgments are really objective judgments about the best means of optimizing goals which cannot be rationally assessed, and of the view that value judgments can and must be explained objectively (e.g., by deterministic explanations) or else are merely arbitrary fiats of individuals or cultures (which view amounts to nihilism or relativism). This is the most dramatic way in which the theory of objectivity we are discussing is already pregnant with modern nihilism; for this view already structures our view of values as either just objective items of a social system or an individual's behavioral repertoire or else as just "subjective" reactions to the objective facts, which do not belong in an objective scientific picture of the world.

It turns out, as shall be seen in the part on science and culture, that *both* of these approaches to values amount to a kind of relativism, which often embodies a very conservative ideology, i.e., supports the idea that existing views of morality cannot be challenged, and thus that value judgments are really judgments about the best means to those ends and values which are in existence. (This is why cost-benefit analyses embody the view that the optimization of a given end is the only standard for making value judgments.) It thus turns out, on the "C–B" view, that the idea of value neutrality really amounts to the idea that value judgments (to the degree that they are rational and objective) are just judgments of efficiency concerning the best means to a given end. The ends (e.g., purposes) are determined scientifically, and this means (ultimately) they are either given or explained by some deterministic theory as being inevitable, ultimate facts. In any event, objectivism certainly is not "value-free."

VII. The Readings

In the essays in this part, these and related issues mentioned are discussed by the authors. The main contention of Rudner's classic paper is that scientists must make value judgments, "since no scientific hypothesis is ever completely verified, in accepting a hypothesis the scientist must make the decision that the evidence is *sufficiently* strong or that the probability is *sufficiently* high to warrant the acceptance of the hypothesis." Such a decision, he claims, is a function of the importance of the acceptance or rejection of a hypothesis, and is thus an ethical judgment. Rescher and Scriven develop some of Rudner's seminal paper in their own defense of the view that science is value-laden, although they also differ from Rudner in several important respects. For example, they do not hold the view that all value judgments in science are ethical. Nor do they seem to base their argument on a strong version of the verificationist principle espoused here by Rudner. It is also not clear to what extent they want to talk about "acceptance" conditions. (See here the introduction to the part on "Confirmation and Acceptance," to which Rudner's article also makes important contributions.) Finally, it is interesting to compare Rudner's remarks about value judgments, decisions, and the insufficient nature of the evidence (i.e., confirmation or epistemic principles) with Kuhn's analysis of theory choice in Part 4. Rescher takes up the role of value judgments within science, concentrating on those values relating to the nature and aims of science and attendant upon the uses of scientific method. Hempel discusses the relation between scientific evidence, goals, and value judgments, and shows that the idea of absolutely ultimate ends is an illusion. Scriven, in a forceful and provocative essay, questions the distinction between objective science and subjective values and gives detailed arguments and illustrations of the ways in which scientific and value judgments are interwoven. Together, these four essays provide fruitful suggestions and developments for more adequate conceptions of science, values, and their interconnections.

R.H.

17

The Scientist *Qua* Scientist
Makes Value Judgments

Richard Rudner

The question of the relationship of the making of value judgments in a typically ethical sense to the methods and procedures of science has been discussed in the literature at least to that point which e. e. cummings somewhere refers to as "The Mystical Moment of Dullness." Nevertheless, albeit with some trepidation, I feel that something more may fruitfully be said on the subject.

In particular the problem has once more been raised in an interesting and poignant fashion by recently published discussions between Carnap[1] and Quine[2] on the question of the ontological commitments which one may make in the choosing of language systems.

I shall refer to this discussion in more detail in the sequel; for the present, however, let us briefly examine the current status of what is somewhat loosely called the "fact-value dichotomy."

I have not found the arguments which are usually offered, by those who believe that scientists do essentially make value judgments, satisfactory. On the other hand the rebuttals of some of those with opposing viewpoints seem to have had at least a *prima facie* cogency although they too may in the final analysis prove to have been subtly perverse.

Those who contend that scientists do essentially make value judgments generally support their contentions by either

(a) pointing to the fact that our having a science at all somehow "involves" a value judgment, or
(b) by pointing out that in order to select, say among alternative problems, the scientist must make a value judgment; or (perhaps most frequently)
(c) by pointing to the fact that the scientist cannot escape his quite human self—he is a "mass of predilections," and these predilections must inevitably influence all of his activities not excepting his scientific ones.

To such arguments, a great many empirically oriented philosophers and scientists have responded that the value judgments involved in our decisions to have a science, or to select problem (a) for attention rather than problem (b) are, *of course,* extra-scientific. If (they say) it is necessary to make a decision to have a science before we can have one, then this decision is literally pre-scientific and the act has thereby certainly not been shown to be any part of the *procedures* of science. Similarly the decision to focus attention on one problem rather than another is extra-problematic and forms no part of the procedures involved in dealing with the problem *decided* upon. Since it is *these* procedures which constitute the method of science, value judgments, so they respond, have not been shown to be involved in the scientific method as such. Again, with respect to the inevitable presence of our predilections in the laboratory, most empirically oriented philosophers and scientists agree that this is "unfortunately" the case; but, they hasten to add, if science is to progress toward objectivity the influence of our personal feelings or biases on experimental results must be minimized. We must try not to let our personal idiosyncrasies affect our scientific work. The perfect scientist—the scientist *qua* scientist does not allow this kind of value judgment to influence his work. However much he may find doing so unavoidable *qua* father, *qua* lover, *qua* member of society, *qua* grouch, *when* he does so he is not behaving *qua* scientist.

As I indicated at the outset, the arguments of neither of the protagonists in this issue appear quite satisfactory to me. The empiricists' rebuttals, telling prima facie as they may against the specific arguments that evoke them, nonetheless do not appear ultimately to stand up, but perhaps even more importantly, *the original arguments* seem utterly too frail.

I believe that a much stronger case may be made for the contention that value judgments are essentially involved in the procedures of science. And what I now propose to show is that scientists as scientists *do* make value judgments.

Now I take it that no analysis of what constitutes the method of science would be satisfactory unless it comprised some assertion to the effect that the scientist as scientist accepts or rejects hypotheses.

But if this is so then clearly the scientist as scientist does make value judgments. For, since no scientific hypothesis is ever completely verified, in accepting a hypothesis the scientist must make the decision that the evidence is *sufficiently* strong or that the probability is *sufficiently* high to warrant the acceptance of the hypothesis. Obviously our decision regarding the evidence and respecting how strong is "strong enough," is going to be a function of the *importance,* in the typically ethical sense, of making a mistake in accepting or rejecting the hypothesis. Thus, to take a crude but easily manageable example, if the hypothesis under consideration were to the effect that a toxic ingredient of a drug was not present in lethal quantity, we would require a relatively high degree of confirmation or confidence before accepting the hypothesis—for the consequences of making a mistake here are exceedingly grave by our moral standards. On the other hand, if, say, our hypothesis stated that, on the basis of a sample, a certain lot of machine stamped belt buckles was not defective, the degree of confidence we should require would be relatively not so high. *How sure we need to be before we accept a hypothesis will depend on how serious a mistake would be.*

The examples I have chosen are from scientific inferences in industrial quality control. But the point is clearly quite general in application. It would be interesting and instructive, for example, to know just how high a degree of probability the Manhattan Project scientists demanded for the hypothesis that no uncontrollable pervasive chain reaction would occur, before they proceeded with the first atomic bomb detonation or first activated the Chicago pile above a critical level. It would be equally interesting and instructive to know why they decided that *that* probability value (if one was decided upon) was high enough rather than one which was higher; and perhaps most interesting of all to learn whether the problem in this form was brought to consciousness at all.

In general then, before we can accept any hypothesis, the value decision must be made in the light of the seriousness of a mistake, that the probability is *high enough* or that, the evidence is *strong enough,* to warrant its acceptance.

Before going further, it will perhaps be well to clear up two points which might otherwise prove troublesome below. First I have obviously used the term "probability" up to this point in a quite loose and pre-analytic sense. But my point can be given a more rigorous formulation in terms of a description of the process of making statistical inference and of the acceptance or rejection of hypotheses in statistics. As is well known, the acceptance or rejection of such a hypothesis presupposes that a certain level of significance or level of confidence or critical region be selected.[3]

It is with respect at least to the *necessary* selection of a confidence level or interval that the necessary value judgment in the inquiry occurs. For, "the size of the critical region (one selects) is related to *the risk one wants to accept* in testing a statistical hypothesis."[3*, p. 435]

And clearly how great a risk one is willing to take of being wrong in accepting or rejecting the hypothesis will depend upon how seriously in the typically ethical sense one views the consequences of making a mistake.

I believe, of course, that an adequate rational reconstruction of the procedures of science would show that every scientific inference is properly constructable as a statistical inference (i.e., as an inference from a set of characteristics of a sample of a population to a set of characteristics of the total population) and that such an inference would be scientifically in control only in so far as it is statistically in control. But it is not necessary to argue this point, for even if one believes that what is involved in some scientific inferences is not statistical probability but rather a concept like strength of evidence or degree of confirmation, one would still be concerned with making the decision that the evidence was *strong enough* or the degree of confirmation *high enough* to warrant acceptance of the hypothesis. Now, many empiricists who reflect on the foregoing considerations agree that acceptances or rejections of hypotheses do essentially involve value judgments, but they are nonetheless loathe to accept the conclusion. And one objection which has been raised against this line of argument by those of them who are suspicious of the intrusion of value questions into the "objective realm of science," is that actually the scientist's task is only to *determine* the degree of confirmation or the strength of the evidence which *exists* for a hypothesis. In short, they object that while it may be a function of the scientist *qua member of society* to decide whether a degree of probability associated with the hypothesis is high enough to warrant its acceptance, *still* the task of the scientist *qua* scientist is *just the determination* of the degree of probability or the strength of the evidence for a hypothesis and not the acceptance or rejection of that hypothesis.

But a little reflection will show that the plausibility of this objection is merely apparent. For the determination that the degree of confirmation is say, *p*, or that the strength of evidence is such and such, which is on this view being held to be the indispensable task of the scientist *qua* scientist, is clearly nothing more than *the acceptance by the scientist of the hypothesis that the degree of confidence is p or that the strength of the evidence is such and such*; and as these men have conceded, acceptance of hypotheses does require value decisions. The second point which it may be well to consider before finally turning our attention to the Quine-Carnap discussion has to do with the nature of the suggestions which have thus far been made in this essay. In this connection, it is important to point out that the preceding remarks do *not* have as their import that an empirical description of every present day scientist ostensibly going about his business would include the statement that he made a value judgment at such and such a juncture. This is no doubt the case; but it is a hypothesis which can only be confirmed by a discipline which cannot be said to have gotten extremely far along as yet; namely, the Sociology and Psychology of Science, whether such an empirical description is warranted, cannot be settled from the armchair.

My remarks have, rather, amounted to this: Any adequate analysis or (if I may use the term) rational reconstruction of the method of science must comprise the statement that the scientist *qua* scientist accepts or rejects hypotheses; and further that an analysis of that statement would reveal it to entail that the scientist *qua* scientist makes value judgments.

I think that it is in the light of the foregoing arguments, the substance of which has, in one form or another, been alluded to in past years by a number of inquirers (notably C. W. Churchman, R. L. Ackoff, and A. Wald), that the Quine-Carnap discussion takes on heightened interest. For, if I understand that discussion and its outcome correctly, although it apparently begins a good distance away from any consideration of the fact-value dichotomy, and although all the way through it both men touch on the matter in a way which indicates that they believe that questions concerning the dichotomy are, if anything, merely tangential to their main issue, yet it eventuates with Quine by an independent argument apparently in agreement with at least the conclusion here reached and also apparently having forced Carnap to that conclusion. (Carnap, however, is expected to reply to Quine's article and I may be too sanguine here.)

The issue of ontological commitment between Carnap and Quine has been one of relatively long standing. In this recent article,[1] Carnap maintains that we are concerned with two kinds of questions of existence relative to a given language system. One is what *kinds* of entities it would be permissible to speak about as existing when that language system is used, i.e., what kind of *framework* for speaking of entities should our system comprise. This, according to Carnap, is an *external* question. It is the *practical* question of what sort of linguistic system we want to choose. Such questions as "Are there abstract entities?" or "Are there physical entities?" thus are held to belong to the category of external questions. On the other hand, having made the decision regarding which linguistic framework to adopt, we can then raise questions like "Are there any black swans?" "What are the factors of 544?" etc. Such questions are *internal* questions.

For our present purposes, the important thing about all of this is that while for Carnap *internal* questions are theoretical ones, i.e., ones whose answers have cognitive content, external questions are not theoretical at all. They are *practical questions*—they concern our decisions to employ one language structure or another. They are of the kind that face us when for example we have to decide whether we ought to have a Democratic or a Republican administration for the next four years. In short, though neither Carnap nor Quine employ the epithet, they are *value questions*.

Now if this dichotomy of existence questions is accepted Carnap can still deny the essential involvement of the making of value judgments in the procedures of science by insisting that concern with *external* questions, admittedly necessary and admittedly axiological, is nevertheless in some sense a pre-scientific concern. But most interestingly, what Quine then proceeds to do is to show that the dichotomy, as Carnap holds it, is untenable. This is not the appropriate place to repeat Quine's arguments which are brilliantly presented in the article referred to. They are in line with the views he has expressed in his "Two Dogmas of Empiricism" essay and especially with his introduction to his recent book, *Methods of Logic.* Nonetheless the final paragraph of the Quine article I'm presently considering sums up his conclusions neatly:

> Within natural science there is a continuum of gradations, from the statements which report observations to those which reflect basic features say of quantum theory or the theory of relativity. The view which I end up with, in the paper last cited, is that statements of ontology or even of mathematics and logic form a continuation of this continuum, a continuation which is perhaps yet more remote from observation than are the central principles of quantum theory or relativity. The differences here are in my view differences only in degree and not in kind. Science is a unified structure, and in principle it is the structure as a whole, and not its component statements one by one, that experience confirms or shows to be imperfect. Carnap maintains that ontological questions, and likewise questions of logical or mathematical principle, are questions not of fact but of

choosing a convenient conceptual scheme or framework for science; and with this I agree only if the same be conceded for every scientific hypothesis. (n. 2, pp. 71–72.)

In the light of all this I think that the statement that *Scientists qua Scientists* make value judgments is also a consequence of Quine's position.

Now, if the major point I have here undertaken to establish is correct, then clearly we are confronted with a first order crisis in science and methodology. The positive horror which most scientists and philosophers of science have of the intrusion of value considerations into science is wholly understandable. Memories of the (now diminished but a certain extent still continuing) conflict between science and, e.g., the dominant religions over the intrusion of religious value considerations into the domain of scientific inquiry, are strong in many reflective scientists. The traditional search for objectivity exemplifies science's pursuit of one of its most precious ideals. But for the scientist to close his eyes to the fact that scientific method *intrinsically* requires the making of value decisions, for him to push out of his consciousness the fact that he does make them, can in no way bring him closer to the ideal of objectivity. To refuse to pay attention to the value decisions which *must* be made, to make them intuitively, unconsciously, haphazardly, is to leave an essential aspect of scientific method scientifically out of control.

What seems called for (and here no more than the sketchiest indications of the problem can be given) is nothing less than a radical reworking of the ideal of scientific objectivity. The slightly juvenile conception of the coldblooded, emotionless, impersonal, passive scientist mirroring the world perfectly in the highly polished lenses of his steel rimmed glasses—this stereotype—is no longer, if it ever was, adequate.

What is being proposed here is that objectivity for science lies at least in becoming precise about what value judgments are being and might have been made in a given inquiry—and even, to put it in its most challenging form, what value decisions ought to be made; in short that a science of ethics is a necessary requirement if science's progress toward objectivity is to be continuous.

Of course the establishment of such a science of ethics is a task of stupendous magnitude and it will probably not even be well launched for many generations. But a first step is surely comprised of the reflective self awareness of the scientist in making the value judgments he must make.

NOTES

1. R. Carnap, "Empiricism, Semantics, and Ontology," *Revue Internationale de Philosophie,* XI, 1950, pp. 20–40.

2. W. V. Quine, "On Carnap's Views on Ontology," *Philosophical Studies,* II, No. 5, 1951.

3. "In practice three levels are commonly used: 1 per cent, 5 per cent and 0.3 of one per cent. There is nothing sacred about these three values; *they have become established in practice without*

any rigid theoretical justification." (my italics) (subnote 3*, p. 435). To establish significance at the 5 per cent level means that one is willing to take the risk of accepting a hypothesis as true when one will be thus making a mistake, one time in twenty. Or in other words, that one will be wrong (over the long run) once every twenty times if one employed an .05 level of significance. See also (subnote 3† Chap. v) for such statements as "which of these two errors is most *important* to avoid (it being necessary to make such a decision in order to accept or reject the given hypothesis) is a *subjective matter . . ."* (my italics) (subnote 3†, p. 262).

* A. C. Rosander, *Elementary Principles of Statistics* (New York: D. Van Nostrand Co., 1951).
† J. Neyman, *First Course in Probability and Statistics* (New York: Henry Holt & Co., 1950).

18

The Ethical Dimension of Scientific Research

Nicholas Rescher

It has been frequently asserted that the creative scientist is distinguished by his objectivity. The scientist—so it is said—goes about his work in a rigidly impersonal and unfeeling way, unmoved by any emotion other than the love of knowledge and the delights of discovering the secrets of nature.

This widely accepted image of scientific inquiry as a cold, detached, and unhumane affair is by no means confined to the scientifically uninformed and to scientific outsiders, but finds many of its most eloquent spokesmen within the scientific community itself. Social scientists in particular tend to be outspoken supporters of the view that the scientist does not engage in making value judgments, and that science, real science, deals only with what is, and has no concern with what ought to be. Any recitation of concrete instances in which the attitudes, values, and temperaments of scientists have influenced their work or affected their findings is dismissed with the scornful dichotomy that such matters may bear upon the psychology or sociology of scientific inquiry, but have no relevance whatever to the *logic* of science.

This point of view that science is "value free" has such wide acceptance as to have gained for itself the distinctive, if somewhat awesome, label as the thesis of the *value neutrality of science*.

Now the main thesis that I propose is simply that this supposed division between the evaluative disciplines on the one hand and the nonevaluative sciences on the other is based upon mistaken views regarding the nature of scientific research. In paying too much attention to the abstract logic of scientific inquiry, many students of scientific method have lost sight of the fact that science is a human enterprise, carried out by flesh and blood men, and that scientific research must therefore inevitably exhibit some normative complexion. It is my aim to examine the proposition that evaluative, and more specifically *ethical,* problems crop up at numerous points within the framework of scientific research. I shall attempt to argue that the scientist does not, and cannot, put aside his common humanity and his evaluative capabilities when he puts on his laboratory coat.

Ethical Issues and the Collectivization of Scientific Research

Before embarking on a consideration of the ethical dimension of scientific research, a number of preliminary points are in order.

In considering ethical issues within the sciences, I do not propose to take any notice at all of the various moral problems that arise in relation to what is *done with* scientific discoveries once they have been achieved. I want to concern myself with scientific work as such, and insofar as possible to ignore the various technological and economic applications of science. We shall not be concerned with the very obviously ethical issues that have to do with the use of scientific findings for the production of the instrumentalities of good or evil. The various questions about the morality of the *uses* to which scientific discoveries are put by men other than the scientists themselves—questions of the sort that greatly exercise such organizations as, for example, the Society for Social Responsibility in Science—are substantially beside the point. We all know that the findings of science can be used to manufacture wonder drugs to promote man's welfare, or bacteriological weapons to promote his extermination. Such questions of what is done with the fruits of the tree of science, both bitter and sweet, are not problems that arise *within* science, and are not ethical choices that confront the scientist himself. This fact puts them outside of my limited area of concern. They relate to the exploitation of scientific research, not to its pursuit, and thus they do not arise *within* science in the way that concerns us here.

Before turning to a description of some of the ethical issues that affect the conduct of research in the sciences, I should like to say a word about their reason for being. Ethical questions—that is, issues regarding the rightness and wrongness of conduct—arise out of people's dealings with each other, and pertain necessarily to the duties, rights, and obligations that exist in every kind of interpersonal relationship. For a Robinson Crusoe, few, if any, ethical problems present themselves. One of the most remarkable features of the science of our time is its joint tendency toward collectivization of effort and dispersion of social involvement.

The solitary scientist laboring in isolation in his study or laboratory has given way to the institutionalized laboratory, just as the scientific paper has become a thing of almost inevitably multiple authorship, and the scientific calculation has shifted from the back of an envelope to the electronic computer. Francis Bacon's vision of scientific research as a group effort has come to realization. The scientist nowadays usually functions not as a detached individual unit, but as part of a group, as a "member of the team."

This phenomenon of the collectivization of scientific research leads increasingly to more prominent emphasis upon ethical considerations within science itself. As the room gets more crowded, if I may use a simile, the more acute becomes the need for etiquette and manners; the more people involved in a given corner of scientific research, the more likely ethical issues are apt to arise. It seems that these phenomena of the collectivization and increasing social diffusion of modern science are the main forces that have resulted in making a good deal of room for ethical considerations within the operational framework of modern science.

Ethical Problems Regarding Research Goals

Perhaps the most basic and pervasive way in which ethical problems arise in connection with the prosecution of scientific research is in regard to the choice of research problems, the setting of research goals, and the allocation of resources (both human and material) to the prosecution of research efforts. This ethical problem of choices relating to research goals arises at all levels of aggregation—the national, the institutional, and the individual. I should like to touch upon each of these in turn.

The National Level. As regards the national level, it is commonplace that the United States government is heavily involved in the sponsorship of research. The current level of federal expenditure on research and development is 8.4 billion dollars, which is around 10 per cent of the federal budget, and 1.6 per cent of the gross national product. If this seems like a modest figure, one must consider the historical perspective. The rate of increase of this budget item over the past ten years has been 10 per cent per annum, which represents a doubling time of seven years. Since the doubling time of our GNP is around twenty years, at these present rates our government will be spending all of our money on science and technology in about sixty-five years. But even today, long before this awkward juncture of affairs is reached, our government, that is to say our collective selves, is heavily involved in the sponsorship of scientific work. And since the man who pays the piper inevitably gets to call at least some of the tune, our society is confronted with difficult choices of a squarely ethical nature regarding the direction of these research efforts. Let me cite a few instances.

In the Soviet Union, 35 per cent of all research and academic trained personnel is engaged in the engineering disciplines, compared with 10 per cent in medicine and pharmaceutical science. Does this 3.5 to 1 ratio of technology to medicine set a pattern to be adopted by the United States? Just how are we to "divide the pie" in allocating federal support funds among the various areas of scientific work?

In our country, the responsibility for such choices is, of course, localized. The President's Science Advisory Committee and the Federal Council for

Science and Technology give a mechanism for establishing an overall science budget and thereby for making the difficult decisions regarding resource allocation. These decisions, which require weighing space probes against biological experimentation and atomic energy against oceanography are among the most difficult choices that have to be made by, or on behalf of, the scientific community. The entrance of political considerations may complicate, but cannot remove, the ethical issues that are involved in such choices.

What is unquestionably the largest ethical problem of scientific public policy today is a question of exactly this type. I refer to the difficult choices posed by the fantastic costs of the gadgetry of space exploration. The costs entailed by a systematic program of manned space travel are such as to necessitate major sacrifices in the resources our society can commit both to the advancement of knowledge as such in areas other than space and to medicine, agriculture, and other fields of technology bearing directly upon human welfare. Given the fact, now a matter of common knowledge, that modern science affords the means for effecting an almost infinite improvement in the material conditions of life for at least half the population of our planet, are we morally justified in sacrificing this opportunity to the supposed necessity of producing cold war spectaculars? No other question could more clearly illustrate the ethical character of the problem of research goals at the national level.

The Institutional Level. Let me now turn to the institutional level—that of the laboratory or department or research institute. Here again the ethical issue regarding research goals arises in various ways connected with the investment of effort, or, to put this same matter the other way around, with the selection of research projects.

One very pervasive problem at this institutional level is the classical issue of pure, or basic, versus applied, or practical, research. This problem is always with us and is always difficult, for the more "applied" the research contribution, the more it can yield immediate benefits to man; the more "fundamental," the deeper is its scientific significance and the more can it contribute to the development of science itself. No doubt it is often the case unfortunately that the issue is not dealt with on this somewhat elevated plane, but is resolved in favor of the applied end of the spectrum by the mundane, but inescapable, fact that this is the easier to finance.

I need scarcely add that this ethical issue can also arise at the institutional level in far more subtle forms. For instance, the directorship of a virology laboratory may have to choose whether to commit its limited resources to developing a vaccine which protects against a type of virus that is harmless as a rule but deadly to a few people, as contrasted with a variant type of virus that, while deadly to none, is very bothersome to many.

The Individual Level. The most painful and keenly felt problems are often not the greatest in themselves, but those that touch closest to home. At the level of the individual, too, the ethical question of research goals and the allocation of effort—namely that of the individual himself—can arise and present difficulties of the most painful kind. To cite one example, a young scientist may well ponder the question of whether to devote himself to pure or to applied work. Either option may present its difficulties for him, and these can, although they need not necessarily, be of an ethical nature.

Speaking now just of applied science, it is perfectly clear that characteristically ethical problems can arise for the applied scientist in regard to the nature of the application in question. This is at its most obvious in the choice of a military over against a nonmilitary problem context—A-bombs versus X-rays, poison gas versus pain killers. On this matter of the pressure of ethical considerations upon the conscience of an individual, I cannot forbear giving a brief, but eloquent, autobiographical quotation from C. P. Snow:

> I was an official for twenty years. I went into official life at the beginning of the war, for the reason that prompted my scientific friends to begin to make weapons. I stayed in that life until a year ago, for the same reason that made my scientific friends turn into civilian soldiers. The official's life in England is not quite so disciplined as a soldier's, but it is very nearly so. I think I know the virtues, which are very great, of the men who live that disciplined life. I also know what for me was the moral trap. I, too, had got onto an escalator. I can put the result in a sentence: I was coming to hide behind the institution; I was losing the power to say no.[1]

How many scientists in our day are passengers riding along on Snow's escalator and have dulled their moral sensitivities to this question of personal goals? I have myself known more than one scientist who has forgone the chance of being a public benefactor in favor of the more immediate opportunity to be a public servant.

Ethical Problems Regarding the Staffing of Research Activities

The recruitment and assignment of research personnel to particular projects and activities poses a whole gamut of problems of an ethical nature. I will confine myself to two illustrations.

It is no doubt a truism that scientists become scientists because of their interest in science. Devotion to a scientific career means involvement with scientific work: *doing* science rather than *watching* science done. The collectivization of science creates a new species—the science administrator whose very existence poses both practical and ethical problems. Alvin Weinberg, director of the Oak Ridge National Laboratory, has put it this way:

> Where large sums of public money are being spent there must be many administrators who see to it that the money is spent wisely.

> Just as it is easier to spend money than to spend thought, so it is easier to tell other scientists how and what to do than to do it oneself. The big scientific community tends to acquire more and more bosses. The Indians with bellies to the bench are hard to discern for all the chiefs with bellies to the mahogany desks. Unfortunately, science dominated by administrators is science understood by administrators, and such science quickly becomes attenuated, if not meaningless.[2]

The facts adduced by Weinberg have several ethical aspects. For one thing there is Weinberg's concern with what administrationitis may be doing to science. And this is surely a problem with ethical implications derived from the fact that scientists have a certain obligation to the promotion of science itself as an ongoing human enterprise. On the other hand, there is the ethical problem of the scientist himself, for a scientist turned administrator is frequently a scientist lost to his first love.

My second example relates to the use of graduate students in university research. There seems to me to be a very real problem in the use of students in the staffing of research projects. We hear a great many pious platitudes about the value of such work for the training of students. The plain fact is that the kind of work needed to get the project done is simply not always the kind of work that is of optimum value for the basic training of a research scientist in a given field. Sometimes instead of doing the student a favor by awarding him a remunerative research fellowship, we may be doing him more harm than good. In some instances known to me, the project work that was supposedly the training ground of a graduate student in actuality derailed or stunted the development of a research scientist.

Ethical Problems Regarding Research Methods

Let me now take up a third set of ethical problems arising in scientific research— those having to do with the *methods* of the research itself. Problems of this kind arise perhaps most acutely in biological or medical or psychological experiments involving the use of experimental animals. They have to do with the measures of omission and commission for keeping experimental animals from needless pain and discomfort. In this connection, let me quote Margaret Mead:

> The growth of importance of the study of human behavior raises a host of new ethical problems, at the head of which I would place the need for consent to the research by both observer and subject. Studies of the behavior of animals other than man introduced a double set of problems: how to control the tendency of the human observer to anthropomorphize, and so distort his observations, and how to protect both the animal and the experimenter from the effects of cruelty. In debates on the issue of cruelty it is usually recognized that callousness toward a living thing may produce suffering in the experimental subject, but

it is less often recognized that it may produce moral deterioration in the experimenter.[3]

It goes without saying that problems of this sort arise in their most acute form in experiments that risk human life, limb, well-being, or comfort.

Problems of a somewhat similar character come up in psychological or social science experiments in which the possibility of a compromise of human dignity or integrity is present, so that due measures are needed to assure treatment based on justice and fair play.

Ethical Problems Regarding Standards of Proof

I turn now to a further set of ethical problems relating to scientific research—those that are bound up with what we may call the standards of proof. These have to do with the amount of evidence that a scientist accumulates before he deems it appropriate to announce his findings and put forward the claim that such-and-so may be regarded as an established fact. At what juncture should scientific evidence be reasonably regarded as strong enough to give warrant for a conclusion, and how should the uncertainties of this conclusion be presented?

This problem of standards of proof is ethical, and not merely theoretical or methodological in nature, because it bridges the gap between scientific understanding and action, between thinking and doing. The scientist cannot conveniently sidestep the whole of the ethical impact of such questions by saying to the layman, "I'll tell you the scientific facts and then *you* decide on the proper mode of action." These issues are usually so closely interconnected that it is the scientific expert alone who can properly adjudge the bearing of the general scientific considerations upon the particular case in hand.

Every trained scientist knows, of course, that "scientific knowledge" is a body of statements of varying degrees of certainty—including a great deal that is quite unsure as well as much that is reasonably certain. But in presenting particular scientific results, and especially in presenting his own results, a researcher may be under a strong temptation to fail to do justice to the precise degree of certainty and uncertainty involved.

On the one hand, there may be some room for play given to a natural human tendency to exaggerate the assurance of one's own findings. Moreover, when much money and effort have been expended, it can be embarrassing—especially when talking with the nonscientific sponsors who have footed the bill—to derogate from the significance or suggestiveness of one's results by dwelling on the insecurities in their basis. The multiple studies and restudies made over the past ten years in order to assess the pathological and genetic effects of radioactive fallout afford an illustration of a struggle to pinpoint the extent of our knowledge and our ignorance in this area.

On the other hand, it may in some instances be tempting for a researcher to underplay the certainty of his findings by adopting an unreasonably high standard of proof. This is especially possible in medical research, where life-risking actions may be based upon a research result. In this domain, a researcher may be tempted to "cover" himself by hedging his findings more elaborately than the realities of the situation may warrant.

Especially when communicating with the laity, this matter of indicating in a convincing way the exact degree of assurance that attaches to a scientific opinion may be a task of great complexity and difficulty. Let me illustrate this by a quotation from W. O. Baker, Vice President for Research at Bell Laboratories:

> I happened to be one of a task force that was gathered officially, with State Department sanction, at the very beginning of 1946, to prepare a detailed scientific estimate . . . about the probable duration of the United States nuclear monopoly. We found, of course, the engineering truth that another country, explicitly the Soviet Union, would have nuclear weapons in a certain number of years after 1946 — a number which we carefully estimated. Our estimate, which is a matter of record, was off by little more than a year, and it was, indeed, too conservative an estimate. But it was by no means trusted, and — an equally sorry circumstance — we lacked the skill to make people believe and heed it.[4]

Ethical Problems Regarding the Dissemination of Research Findings

A surprising variety of ethical problems revolve around the general topic of the dissemination of research findings. It is so basic a truth as to be almost axiomatic that, with the possible exception of a handful of unusual cases in the area of national security classification, a scientist has not only the right, but even the duty, to communicate his findings to the community of fellow scientists, so that his results may stand or fall in the play of the open market place of ideas. Modern science differs sharply in this respect from science in Renaissance times, when a scientist shared his discoveries only with trusted disciples, and announced his findings to the general public only in cryptogram form, if at all.

This ethical problem of favoritism in the sharing of scientific information has come to prominence again in our day. Although scientists do generally publish their findings, the processes of publication consume time, so that anything between six months and three years may elapse between a scientific discovery and its publication in the professional literature. It has become a widespread practice to make prepublication announcements of findings, or even pre-prepublication announcements. The ethical problem is posed by the extent and direction of such exchanges, for there is no doubt that in many cases favoritism comes into the picture, and that some workers and laboratories exchange findings in a preferential way that amounts to a conspiracy to maintain themselves ahead of the state of the art in the world at large.

There is, of course, nothing reprehensible in the natural wish to overcome publication lags or in the normal desire for exchanges of ideas with fellow workers. But when such practices tend to become systematized in a prejudicial way, a plainly ethical problem comes into being.

Let us consider yet another ethical problem regarding the dissemination of research findings. The extensive dependence of science upon educated public opinion, in connection with its support both by the government and the foundations, has already been touched upon. This factor has a tendency to turn the reporting of scientific findings and the discussion of issues relating to scientific research into a kind of journalism. There is a strong incentive to create a favorable climate of public opinion for certain pet projects or concepts. Questions regarding scientific or technical merits thus tend to get treated not only in the proper forum of the science journals, but also in the public press and in Congressional or foundation committee rooms. Not only does this create the danger of scientific pressure groups devoted to preconceived ideas and endowed with the power of retarding other lines of thought, but it also makes for an unhealthy emphasis on the spectacular and the novel, unhealthy, that is, from the standpoint of the development of science itself. For such factors create a type of control over the direction of scientific research that is disastrously unrelated to the proper issue of strictly scientific merits.

The fact is that science has itself become vulnerable in this regard through its increasing sensitivity to public relations matters. Let me cite just one illustration—that of the issue of the fluoridation of municipal water supplies. Some scientists appear to have chosen this issue as a barricade at which to fight for what (to use a political analogue) might be called the "grandeur" of science.

Not long ago, local referenda in the state of Massachusetts gave serious defeats to the proponents of fluoridation. Not only were proposals to introduce this practice defeated in Wellesley and Brookline, but Andover, where fluoridation had been in effect for five years, voted discontinuance of the program. These defeats in towns of the highest educational and socioeconomic levels caused considerable malaise in the scientific community, and wails of anguish found their way even into *Science,* official journal of the American Association for the Advancement of Science. This annoyance over what is clearly not a *scientific* setback, but merely a failure in public relations or political effectiveness, sharply illustrates the sensitivity that scientists have developed in this area.

Ethical Problems Regarding the Control of Scientific "Misinformation"

Closely bound up with the ethical problems regarding the dissemination of scientific information are what might be thought of as the other side of the coin—the control, censorship, and suppression of scientific misinformation. Scientists clearly have a duty to protect both their own colleagues in other specialties and the lay public against the dangers of supposed research findings that are strictly erroneous, particularly in regard to areas such as medicine and nutrition, where the public health and welfare are concerned. And quite generally, of course, a scientist has an obligation to maintain the professional literature of his field at a high level of content and quality. The editors and editorial reviewers in whose hands rests access to the media of scientific publication clearly have a duty to preserve their readership from errors of fact and trivia of thought. But these protective functions must always be balanced by respect for the free play of ideas and by a real sensitivity to the possible value of the unfamiliar.

To give just one illustration of the importance of such considerations, I will cite the example of the nineteenth-century English chemist J. J. Waterson. His groundbreaking papers on physical chemistry, anticipating the development of thermodynamics by more than a generation, were rejected by the referees of the Royal Society for publication in its *Proceedings,* with the comment (among others) that "the paper is nothing but of nonsense." As a result, Waterson's work lay forgotten in the archives of the Royal Society until rescued from oblivion by Rayleigh some forty-five years later. Let me quote J. S. Haldane, whose edition of Waterson's works in 1928 decisively rehabilitated this important researcher:

> It is probable that, in the long and honorable history of the Royal Society, no mistake more disastrous in its actual consequences for the progress of science and the reputation of British science than the rejection of Waterson's papers was ever made. The papers were foundation stones of a new branch of scientific knowledge, molecular physics, as Waterson called it, or physical chemistry and thermodynamics as it is now called. There is every reason for believing that, had the papers been published, physical chemistry and thermodynamics would have developed mainly in this country [i.e., England], and along much simpler, more correct, and more intelligible lines than those of their actual development.[5]

Many other examples could be cited to show that it is vitally important that the gatekeepers of our scientific publications be keenly alive to the possible but unobvious value of unfamiliar and strange seeming conceptions.

It is worth emphasizing that this matter of "controlling" the dissemination of scientific ideas poses special difficulties due to an important, but much underrated, phenomenon: *the resistivity to novelty and innovation by the scientific community itself.* No feature of the historical course of development of the sciences is more damaging to the theoreticians' idealized conception of science as perfectly objective—the work of almost disembodied intellects governed by purely rational considerations and actuated solely by an abstract love of truth. The mere assertion that scientists can resist, and indeed frequently have resisted, acceptance of scientific discoveries clashes sharply with the stereotyped concept of the scientist as the purely objective, wholly rational, and entirely open-minded man. Although opposition to scientific findings by social groups other than scientists has been examined by various investigators, the resistance to scientific discoveries by scientists themselves is just beginning to attract the attention of sociologists.[6]

The history of science is, in fact, littered with examples of this phenomenon. Lister, in a graduation address to medical students, bluntly warned against blindness to new ideas such as he had himself encountered in advancing his theory of antisepsis. Pasteur's discovery of the biological character of fermentation was long opposed by chemists, including the eminent Liebig, and his germ theory met with sharp resistance from the medical fraternity of his day. No doubt due in part to the very peculiar character of Mesmer himself, the phenomenon of hypnosis, or mesmerism, was rejected by the scientifically orthodox of his time as so much charlatanism. At the summit of the Age of Reason, the French Academy dismissed the numerous and well-attested reports of stones falling from the sky (meteorites, that is to say) as mere folk stories. And this list could be prolonged *ad nauseam.*

Lord Rayleigh, the rediscoverer of J. J. Waterson, who had also himself been burned by scientific opposition to his research findings, became so pessimistic about the difficulties that new conceptions encounter before becoming established in science that he wrote:

> Perhaps one may . . . say that a young author who believes himself capable of great things would usually do well to secure the favorable recognition of the scientific world by work whose scope is limited, and whose value is easily judged, before embarking on greater flights.[7]

(The value of Rayeligh's advice is, of course, very questionable, in view of the fact that it is more than likely that any young scientist of promise who fritters away the maximally creative years of youthful freshness and enthusiasm by doing work of routine drudgery will almost inevitably blunt the keen edge of his productive capacities to a point where "great things" are simply no longer within his grasp.)

Those scientists who have themselves fallen victim to the resistance to new ideas on the part of their colleagues have invariably felt this keenly, and have given eloquent testimony to the existence of this phenomenon. Oliver Heaviside, whose important contributions to mathematical physics were slighted for over twenty-five years, is reported to have exclaimed bitterly that "even men who are not Cambridge mathematicians deserve justice." And Max Planck, after encountering analogous difficulties, wrote:

> This experience gave me also an opportunity to learn a new fact—a remarkable one in my opinion: A new scientific truth does not triumph by convincing its opponents and making them see the light, but rather because its opponents eventually die, and a new generation grows up that is familiar with it.[8]

In summary, the prominence, even in scientific work, of the human psychological tendency to resist new ideas must temper the perspective of every scientist when enforcing what he conceives to be his duty to safeguard others against misinformation and error.

At no point, however, does the ethical problem of information control in science grow more difficult and vexatious than in respect to the boundary line between proper science on the one hand and pseudo-science on the other. The plain fact is that truth is to be found in odd places, and that scientifically valuable materials turn up in unexpected spots.

No one, of course, would for a moment deny the abstract thesis that there is such a thing as pseudo-science, and that it must be contested and controlled. The headache begins with the question of just what is pseudo-science and what is not. We can all readily agree on some of the absurd cases so interestingly described in Martin Gardner's wonderful book *Fads and Fallacies in the Name of Science* (New York: Dover Publications, Inc., 1957). But the question of exactly where science ends and where pseudo-science begins is at once important and far from simple. There is little difficulty indeed with Wilbur Glenn Voliva, Gardner's Exhibit No. 1, who during the first third of this century thundered out of Zion (Illinois) that "the earth is flat as a pancake." But parapsychology, for example, is another study and a much more complicated one. And the handful of United States geneticists who, working primarily with yeasts, feel that they have experimental warrant for Lamarckian conclusions, much to the discomfort of the great majority of their professional colleagues, exemplify the difficulties of a hard and fast compartmentalization of pseudo-science in a much more drastic way. Nobody in the scientific community wants to let pseudo-science make headway. But the trouble is that one man's interesting possibility may be another man's pseudo-science.

On the one hand, reputable scientists have often opposed genuine scientific findings as being pseudo-scientific. Lord Bacon, the high priest of early modern science, denounced Gilbert's treatise *On the Magnet* as "a work of inconclusive writing," and he spoke disparagingly of the "electric energy concerning which Gilbert told so many fables." A more recent, if less clearcut, example is the extensive opposition encountered by psychoanalysis, particularly in its early years.

But on the other hand, we have the equally disconcerting fact that reputable scientists have advanced, and their fellow scientists accepted, findings that were strictly fraudulent. One instructive case is that of the French physicist René Blondlot, which is interestingly described in Derek Price's book *Science Since Babylon* (New Haven: Yale University Press, 1961). Blondlot allegedly discovered "N-rays," which were supposed to be something like X-rays. His curious findings attracted a great deal of attention and earned for Blondlot himself a prize from the French government. But the American physicist Robert W. Wood was able to show by careful experimental work that Blondlot and all who concurred in his findings were deluded. It is thus to be recognized not only that pseudo-science exists, but that it sometimes even makes its way into the sacred precincts of highly orthodox science. This, of course, does not help to simplify the task of discriminating between *real* and *pseudo*-science.

But let us return to the ethical issues involved. These have to do not with the uncontroversial thesis that pseudo-science must be controlled, but with procedural questions of the *means* to be used for the achievement of this worthy purpose. It is with this problem of the means for its control that pseudo-science poses real ethical difficulties for the scientific community.

The handiest instrumentalities to this end and the most temptingly simple to use are the old standbys of thought control—censorship and suppression. But these are surely dire and desperate remedies. It is no doubt highly unpleasant for a scientist to see views that he regards as "preposterous" and "crackpot" to be disseminated and even to gain a considerable public following. But surely we should never lose sensitivity to the moral worth of the methods for achieving our ends or forget that good ends do not justify questionable means. It is undeniably true that scientists have the duty to prevent the propagation of error and misinformation. But this duty has to be acted on with thoughtful caution. It cannot be construed to fit the conveniences of the moment. And it surely cannot be stretched to give warrant to the suppression of views that might prove damaging to the public "image" of science or to justify the protection of one school of thought against its critics. Those scientists who pressured the publisher of Immanuel Velikovsky's fanciful *Worlds in Collision* by threatening to boycott the firm's textbooks unless this work were dropped from its list resorted to measures that I should not care to be called on to defend, but the case is doubtless an extreme one. However, the control exercised by editors and guardians of foundation purse-strings is more subtle, but no less effective and no less problematic.

The main point in this regard is one that needs little defense or argument in its support. Surely scientists, of all people, should have sufficient confidence in the ability of truth to win out over error in the market place of freely interchanged ideas as to be unwilling to forgo the techniques of rational persuasion in favor of the unsavory instrumentalities of pressure, censorship, and suppression.

Ethical Problems Regarding the Allocation of Credit for Scientific Research Achievements

The final set of ethical problems arising in relation to scientific research that I propose to mention relate to the allocation of credit for the achievements of research work. Moral philosophers as well as students of jurisprudence have long been aware of the difficulties in assigning to individuals the responsibility for corporate acts, and thus to allocate to individual wrongdoers the blame for group misdeeds. This problem now faces the scientific community in its inverse form—the allocation to individuals of credit for the research accomplishments resulting from conjoint, corporate, or combined effort.

Particularly in this day of collectivized research, this problem is apt to arise often and in serious forms.

Let no one be put off by stories about scientific detachment and disinterestedness. The issue of credit for their findings has for many centuries been of the greatest importance to scientists. Doubts on this head are readily dispelled by the prominence of priority disputes in the history of science. Their significance is illustrated by such notorious episodes as the bitter and long-continuing dispute between Newton and Leibniz and their followers regarding priority in the invention of the calculus—a dispute that made for an estrangement between English and continental mathematics which lasted through much of the eighteenth century, considerably to the detriment of the quality of British mathematics during that era.

But to return to the present, the problem of credit allocation can come up nowadays in forms so complex and intricate as to be almost inconceivable to any mind not trained in the law. For instance, following out the implications of an idea put forward as an idle guess by X, Y, working under W's direction in Z's laboratory, comes up with an important result. How is the total credit to be divided? It requires no great imagination to think up some of the kinds of problems and difficulties that can come about in saying who is to be credited with what in this day of corporate and collective research. This venture lends itself to clever literary exploitation in the hands of a master like C. P. Snow.

Retrospect on the Ethical Dimension of Scientific Research

Let us now pause for a moment to survey the road that we have traveled thus far. The discussion to this point has made a guided tour of a major part of the terrain constituting the ethical dimension of scientific research. In particular, we have seen that questions of a strictly ethical nature arise in connection with scientific research at the following crucial junctures:

1. the choice of research goals
2. the staffing of research activities
3. the selection of research methods
4. the specification of standards of proof
5. the dissemination of research findings
6. the control of scientific misinformation
7. the allocation of credit for research accomplishments

In short, it seems warranted to assert that, at virtually every junction of scientific research work, from initial inception of the work to the ultimate reporting of its completed findings, issues of a distinctively ethical character may present themselves for resolution.

It is a regrettable fact that too many persons, both scientists and students of scientific method, have had their attention focused so sharply upon the abstracted "logic" of an idealized "scientific method" that this ethical dimension of science has completely escaped their notice. This circumstance seems to me to be particularly regrettable because it has tended to foster a harmful myth that finds strong support in both the scientific and the humanistic camps—namely, the view that science is antiseptically devoid of any involvement with human values. Science, on this way of looking at the matter, is so purely objective and narrowly factual in its concerns that it can, and indeed should, be wholly insensitive to the emotional, artistic, and ethical values of human life.

I hope that my analysis of the role of ethical considerations within the framework of science has been sufficiently convincing to show that this dichotomy, with its resultant divorce between the sciences and the humanities, is based on a wholly untenable conception of the actual division of labor between these two areas of intellectual endeavor. It is my strong conviction that both parties to this unasked for divorce must recognize the spuriousness of its alleged reasons for being, if the interests of a wholesome unity of human understanding are to be served properly.

The humanist, for his part, must not be allowed to forget that in the whole course of the intellectual history of the West, from Aristotle and his predecessors to Descartes, Newton, Kant, James, and Einstein, science has been a part of the cultural tradition in its larger sense. Throughout the whole course of the development of our civilization, science has always merited the historic epithet of "natural philosophy." No matter how much our way of describing the facts may change, there is little doubt that this basic circumstance of the formative role of science in molding the *Weltanschauung* basic to all of our areas of thought will remain invariant.

On the other hand, the scientist, for his part, should realize science has worth and status enough in its own right that its devotees can dispense with claiming that, although the handiwork of imperfect humans, it is somehow mysteriously endowed with virtually superhuman powers, such as are implied by claims of actual achievement of what in fact are remote ideals, and in some instances unworkable idealizations, like pure open-mindedness, complete objectivity, and perfect rationality.

From the standpoint of a realistic appreciation of the nature of science as a human creation and activity, it seems to me that a heightened awareness of the humanistic dimension of science—which I have tried to illustrate on the ethical side—can serve the best interests of both of these two important working areas of the human intellect. Instead of being nearly separable, these domains are interpenetrating and interdependent. Rather than being strange bedfellows, the sciences and the humanities are ancient and mutually beneficial partners in that pre-eminently humane enterprise of leading man to a better understanding both of himself and of the world in which he lives.

NOTES

1. *Science,* Vol. CXXXIII (1961), 258–259.
2. *Ibid.,* Vol. CXXXIX (1961), 162.
3. *Ibid.,* Vol. CXXXIII (1961), 164.
4. *Ibid.,* Vol. CXXXIII (1961), 261.
5. Quoted by Stephen G. Brush in *American Scientist,* Vol. XLIX (1961), 211–212.
6. To anyone interested in this curious topic, I refer the eye-opening article by Bernard Barber, "Resistance by Scientists to Scientific Discovery," *Science,* Vol. CXXXIV (1961).
7. Quoted by Brush, *op. cit.,* 210.
8. Quoted by Barber, *op. cit.*

19

Science and Human Values

Carl G. Hempel

1. The Problem

Our age is often called an age of science and of scientific technology, and with good reason: the advances made during the past few centuries by the natural sciences, and more recently by the psychological and sociological disciplines, have enormously broadened our knowledge and deepened our understanding of the world we live in and of our fellow men; and the practical application of scientific insights is giving us an ever increasing measure of control over the forces of nature and the minds of men. As a result, we have grown quite accustomed, not only to the idea of a physico-chemical and biological technology based on the results of the natural sciences, but also to the concept, and indeed the practice, of a psychological and sociological technology that utilizes the theories and methods developed by behavioral research.

This growth of scientific knowledge and its applications has vastly reduced the threat of some of man's oldest and most formidable scourges, among them famine and pestilence; it has raised man's material level of living, and it has put within his reach the realization of visions which even a few decades ago would have appeared utterly fantastic, such as the active exploration of interplanetary space.

But in achieving these results, scientific technology has given rise to a host of new and profoundly disturbing problems: The control of nuclear fission has brought us not only the comforting prospect of a vast new reservoir of energy, but also the constant threat of the atom bomb and of grave damage, to the present and to future generations, from the radioactive by-products of the fission process, even in its peaceful uses. And the very progress in biological and medical knowledge and technology which has so strikingly reduced infant mortality and increased man's life expectancy in large areas of our globe has significantly contributed to the threat of the "population explosion," the rapid growth of the earth's population which we are facing today, and which, again, is a matter of grave concern to all those who have the welfare of future generations at heart.

Clearly, the advances of scientific technology on which we pride ourselves, and which have left their characteristic imprint on every aspect of this "age of science," have brought in their train many new and grave problems which urgently demand a solution. It is only natural that, in his desire to cope with these new issues, man should turn to science and scientific technology for further help. But a moment's reflection shows that the problems that need to be dealt with are not straightforward technological questions but intricate complexes of technological and moral issues. Take the case of the population explosion, for example, To be sure, it does pose specific technological problems. One of these is the task of satisfying at least the basic material needs of a rapidly growing population by means of limited resources; another is the question of means by which population growth itself may be kept under control. Yet these technical questions do not exhaust the problem. For after all, even now we have at our disposal various ways of counteracting population growth; but some of these, notably contraceptive methods, have been and continue to be the subject of intense controversy on moral and religious grounds, which shows that an adequate solution of the problem at hand requires, not only knowledge of technical means of control, but also standards for evaluating the alternative means at our disposal; and this second requirement clearly raises moral issues.

There is no need to extend the list of illustrations: any means of technical control that science makes available to us may be employed in many different ways, and a decision as to what use to make of it involves us in questions of moral valuation. And here arises a fundamental problem to which I would now like to turn: Can such valuational questions be answered by means of the objective methods of empirical science, which have been so successful in giving us reliable, and often practically applicable, knowledge of our world? Can those methods serve to establish objective criteria of right and wrong and thus to provide valid moral norms for the proper conduct of our individual and social affairs?

2. Scientific Testing

Let us approach this question by considering first, if only in brief and sketchy outline, the way in which objective scientific knowledge is arrived at. We may leave aside here the question of *ways of discovery*; i.e., the problem of how a new scientific idea arises, how a novel hypothesis or theory is first conceived; for our purposes it will suffice to consider the scientific *ways of validation*; i.e., the manner in which empirical science goes about examining a proposed new hypothesis and determines whether it is to be accepted or rejected. I will use the word 'hypothesis' here to refer quite broadly to any statements or set of statements in empirical science, no matter whether it deals with some particular event or purports to set forth a general law or perhaps a more or less complex theory.

As is well known, empirical science decides upon the acceptability of a proposed hypothesis by means of suitable tests. Sometimes such a test may involve nothing more than what might be called direct observation of pertinent facts. This procedure may be used, for example, in testing such statements as 'It is raining outside,' 'All the marbles in this urn are blue,' 'The needle of this ammeter will stop at the scale point marked 6,' and so forth. Here a few direct observations will usually suffice to decide whether the hypothesis at hand is to be accepted as true or to be rejected as false.

But most of the important hypotheses in empirical science cannot be tested in this simple manner. Direct observation does not suffice to decide, for example, whether to accept or to reject the hyeptheses that the earth is a sphere, that hereditary characteristics are transmitted by genes, that all Indo-European languages developed from one common ancestral language, that light is an electromagnetic wave process, and so forth. With hypotheses such as these, science resorts to indirect methods of test and validation. While these methods vary greatly in procedural detail, they all have the same basic structure and rationale. First, from the hypothesis under test, suitable other statements are inferred which describe certain directly observable phenomena that should be found to occur under specifiable circumstances if the hypothesis is true; then those inferred statements are tested directly; i.e., by checking whether the specified phenomena do in fact occur; finally, the proposed hypothesis is accepted or rejected in the light of the outcome of these tests. For example, the hypothesis that the earth is spherical in shape is not directly testable by observation, but it permits us to infer that a ship moving away from the observer should appear to be gradually dropping below the horizon; that circumnavigation of the earth should be possible by following a straight course; that high-altitude photographs should show the curving of the earth's surface; that certain geodetic and astronomical measurements should yield such and such results; and so forth. Inferred statements such as these can be tested more or less directly; and as an increasing number and variety of them are actually borne out, the hypothesis becomes increasingly confirmed. Eventually, a hypothesis may be so well confirmed by the available evidence that it is accepted as having been established beyond reasonable doubt. Yet no scientific hypothesis is ever proved completely and definitively; there is always at least the theoretical possibility that new evidence will be discovered which conflicts with some of the observational statements inferred from the hypothesis, and which thus leads to its rejection. The history of science records many instances in which a once accepted hypothesis was subsequently abandoned in the light of adverse evidence.

3. Instrumental Judgments of Value

We now turn to the question whether this method of test and validation may be used to establish moral judgments of value, and particularly judgments to the effect that a specified course of action is good or right or proper, or that it is better than certain alternative courses of action, or that we ought—or ought not—to act in certain specified ways.

By way of illustration, consider the view that it is good to raise children permissively and bad to bring them up in a restrictive manner. It might seem that, at least in principle, this view could be scientifically confirmed by appropriate empirical investigations. Suppose, for example, that careful research had established (1) that restrictive upbringing tends to generate resentment and aggression against parents and other persons exercising educational authority, and that this leads to guilt and anxiety and an eventual stunting of the child's initiative and creative potentialities; whereas (2) permissive upbringing avoids these consequences, makes for happier interpersonal relations, encourages resourcefulness and self-reliance, and enables the child to develop and enjoy his potentialities. These statements, especially when suitably amplified, come within the purview of scientific investigation; and though our knowledge in the matter is in fact quite limited, let us assume, for the sake of the argument, that they had actually been strongly confirmed by careful tests. Would not scientific research then have objectively shown that it is indeed better to raise children in a permissive rather than in a restrictive manner?

A moment's reflection shows that this is not so. What would have been established is rather a conditional statement; namely, that *if* our children are to become happy, emotionally secure, creative individuals rather than guilt-ridden and troubled souls *then* it is better to raise them in a permissive than in a restrictive fashion. A statement like this represents a *relative, or instrumental, judgment of value*. Generally, a relative judgment of value states that a certain kind of action, M, is good (or that it is better than a given alternative M_1) *if* a specified goal G is to be attained; or more accurately, that M is good, or appropriate, for the attainment of goal G. But to say that is tantamount to asserting either that, in the circumstances at hand, course of action M will definitely (or probably) lead to the attainment of G, or that failure to embark on course of action M will definitely (or probably) lead to the nonattainment of G. In other words, the instrumental value judgment asserts either that M is a (definitely or probably) sufficient means for attaining the end or goal G, or that it is a (definitely or probably) necessary means for attaining it. Thus, a relative, or instrumental, judgment of value can be reformulated as a statement which expresses a universal or a probabilistic kind of means-ends relationship, and which contains no terms of moral discourse—such as 'good,' 'better,' 'ought to'—at all. And a statement of this kind surely is an empirical assertion capable of scientific test.

4. Categorical Judgments of Value

Unfortunately, this does not completely solve our problem; for after a relative judgment of value referring to a certain goal *G* has been tested and, let us assume, well confirmed, we are still left with the question of whether the goal *G* ought to be pursued, or whether it would be better to aim at some alternative goal instead. Empirical science can establish the conditional statement, for example, that if we wish to deliver an incurably ill person from intolerable suffering, then a large dose of morphine affords a means of doing so; but it may also indicate ways of prolonging the patient's life, if also his suffering. This leaves us with the question whether it is right to give the goal of avoiding hopeless human suffering precedence over that of preserving human life. And this question calls, not for a relative but for an *absolute, or categorical, judgment of value* to the effect that a certain state of affairs (which may have been proposed as a goal or end) is good, or that it is better than some specified alternative. Are such categorical value judgments capable of empirical test and confirmation?

Consider, for example, the sentence 'Killing is evil'. It expresses a categorical judgment of value which, by implication, would also categorically qualify euthanasia as evil. Evidently, the sentence does not express an assertion that can be directly tested by observation; it does not purport to describe a directly observable fact. Can it be indirectly tested, then, by inferring from it statements to the effect that under specified test conditions such and such observable phenomena will occur? Again, the answer is clearly in the negative. Indeed, the sentence 'Killing is evil' does not have the function of expressing an assertion that can be qualified as true or false; rather, it serves to express a standard for moral appraisal or a norm for conduct. A categorical judgment of value may have other functions as well; for example, it may serve to convey the utterer's approval or disapproval of a certain kind of action, or his commitment to the standards of conduct expressed by the value judgment. Descriptive empirical import, however, is absent; in this respect a sentence such as 'Killing is evil' differs strongly from, say, 'Killing is condemned as evil by many religions', which expresses a factual assertion capable of empirical test.

Categorical judgments of value, then, are not amenable to scientific test and confirmation or disconfirmation; for they do not express assertions but rather standards or norms for conduct. It was Max Weber, I believe, who expressed essentially the same idea by remarking that science is like a map: it can tell us how to get to a given place, but it cannot tell us where to go. Gunnar Myrdal, in his book *An American Dilemma* (p. 1052), stresses in a similar vein that "factual or theoretical studies alone cannot logically lead to a practical recommendation. A practical or valuational conclusion can be derived only when there is at least one valuation among the premises."

Nevertheless, there have been many attempts to base systems of moral standards on the findings of empirical science; and it would be of interest to examine in some detail the reasoning which underlies those procedures. In the present context, however, there is room for only a few brief remarks on this subject.

It might seem promising, for example, to derive judgments of value from the results of an objective study of human needs. But no cogent derivation of this sort is possible. For this procedure would presuppose that it is right, or good, to satisfy human needs—and this presupposition is itself a categorical judgment of value: it would play the role of a valuational premise in the sense of Myrdal's statement. Furthermore, since there are a great many different, and partly conflicting, needs of individuals and of groups, we would require not just the general maxim that human needs ought to be satisfied, but a detailed set of rules as to the preferential order and degree in which different needs are to be met, and how conflicting claims are to be settled; thus, the valuational premise required for this undertaking would actually have to be a complex system of norms; hence, a derivation of valuational standards simply from a factual study of needs is out of the question.

Several systems of ethics have claimed the theory of evolution as their basis; but they are in serious conflict with each other even in regard to their most fundamental tenets. Some of the major variants are illuminatingly surveyed in a chapter of G. G. Simpson's book, *The Meaning of Evolution.* One type, which Simpson calls a "tooth-and-claw ethics," glorifies a struggle for existence that should lead to a survival of the fittest. A second urges the harmonious adjustment of groups or individuals to one another so as to enhance the probability of their survival, while still other systems hold up as an ultimate standard the increased aggregation of organic units into higher levels of organization, sometimes with the implication that the welfare of the state is to be placed above that of the individuals belonging to it. It is obvious that these conflicting principles could not have been validly inferred from the theory of evolution—unless indeed that theory were self-contradictory, which does not seem very likely.

But if science cannot provide us with categorical judgments of value, what then can serve as a source of unconditional valuations? This question may either be understood in a pragmatic sense, as concerned with the sources from which human beings do in fact obtain their basic values. Or it may be understood as concerned with a systematic aspect of valuation; namely, the question where a proper system of basic values is to be found on which all other valuations may then be grounded.

The pragmatic question comes within the purview of empirical science. Without entering into details, we may say here that a person's values—both those he professes to espouse and those he actually conforms to—are largely absorbed from the society in which he lives, and especially from certain influential subgroups to which he belongs, such as his family, his schoolmates, his associates on the job, his church, clubs, unions, and other groups. Indeed his values may vary from case to case depending on which of these groups dominates the situation in which he happens to find himself. In general, then, a person's basic valuations are no more the result of careful scrutiny and critical appraisal of possible alternatives than is his religious affiliation. Conformity to the standards of certain groups plays a very important role here, and only rarely are basic values seriously questioned. Indeed, in many situations, we decide and act unreflectively in an even stronger sense; namely, without any attempt to base our decisions on some set of explicit, consciously adopted, moral standards.

Now, it might be held that this answer to the pragmatic version of our question reflects a regrettable human inclination to intellectual and moral inertia; but that the really important side of our question is the systematic one: If we do want to justify our decisions, we need moral standards of conduct of the unconditional type—but how can such standards be established? If science cannot provide categorical value judgments, are there any other sources from which they might be obtained? Could we not, for example, validate a system of categorical judgments of value by pointing out that it represents the moral standards held up by the Bible, or by the Koran, or by some inspiring thinker or social leader? Clearly, this procedure must fail, for the factual information here adduced could serve to validate the value judgments in question only if we were to use, in addition, a valuational presupposition to the effect that the moral directives stemming from the source invoked *ought* to be complied with. Thus, if the process of justifying a given decision or a moral judgment is ever to be completed, certain judgments of value have to be accepted without any further justification, just as the proof of a theorem in geometry requires that some propositions be accepted as postulates, without proof. The quest for a justification of *all* our valuations overlooks this basic characteristic of the logic of validation and of justification. The value judgments accepted without further justification in a given context need not, however, be accepted once and for all, with a commitment never to question them again. This point will be elaborated further in the final section of this essay.

As will hardly be necessary to stress, in concluding the present phase of our discussion, the ideas set forth in the preceding pages do not imply or advocate moral anarchy; in particular, they do not imply that any system of values is just as good, or just as valid, as any other, or that everyone should adopt the moral principles that best suit his convenience. For all such maxims have the character of categorical value judgments and cannot, therefore, be implied by the preceding considerations, which are purely descriptive of certain logical, psychological, and social aspects of moral valuation.

5. Rational Choice: Empirical and Valuational Components

To gain further insight into the relevance of scientific inquiry for categorical valuation let us ask what help we might receive, in dealing with a moral problem, from science in an ideal state such as that represented by Laplace's conception of a superior scientific intelligence, sometimes referred to as

Laplace's demon. This fiction was used by Laplace, early in the nineteenth century, to give a vivid characterization of the idea of universal causal determinism. The demon is conceived as a perfect observer, capable of ascertaining with infinite speed and accuracy all that goes on in the universe at a given moment; he is also an ideal theoretician who knows all the laws of nature and has combined them into one universal formula; and finally, he is a perfect mathematician who, by means of that universal formula, is able to infer, from the observed state of the universe at the given moment, the total state of the universe at any other moment; thus past and future are present before his eyes. Surely, it is difficult to imagine that science could ever achieve a higher degree of perfection!

Let us assume, then, that, faced with a moral decision, we are able to call upon the Laplacean demon as a consultant. What help might we get from him? Suppose that we have to choose one of several alternative courses of action open to us, and that we want to know which of these we *ought* to follow. The demon would then be able to tell us, for any contemplated choice, what its consequences would be for the future course of the universe, down to the most minute detail, however remote in space and time. But, having done this for each of the alternative courses of action under consideration, the demon would have completed his task: he would have given us all the information that an ideal science might provide under the circumstances. And yet he would not have resolved our moral problem, for this requires a decision as to which of the several alternative sets of consequences mapped out by the demon as attainable to us is the best; which of them we ought to bring about. And the burden of this decision would still fall upon our shoulders: it is we who would have to commit ourselves to an unconditional judgment of value by singling out one of the sets of consequences as superior to its alternatives. Even Laplace's demon, or the ideal science he stands for, cannot relieve us of this responsibility.

In drawing this picture of the Laplacean demon as a consultant in decision-making, I have cheated a little; for if the world were as strictly deterministic as Laplace's fiction assumes, then the demon would know in advance what choice we were going to make, and he might disabuse us of the idea that there were several courses of action open to us. However that may be, contemporary physical theory has cast considerable doubt on the classical conception of the universe as a strictly deterministic system: the fundamental laws of nature are now assumed to have a statistical or probabilistic rather than a strictly universal, deterministic, character.

But whatever may be the form and the scope of the laws that hold in our universe, we will obviously never attain a perfect state of knowledge concerning them; confronted with a choice, we never have more than a very incomplete knowledge of the laws of nature and of the state of the world at the time when we must act. Our decisions must therefore always be made on the basis of incomplete information, a state which enables us to anticipate the consequences of alternative choices at best with probability. Science can render an indispensable service by providing us with increasingly extensive and reliable information relevant to our purpose; but again it remains for us to *evaluate* the various probable sets of consequences of the alternative choices under consideration. And this requires the adoption of pertinent valuational standards which are not objectively determined by the empirical facts.

This basic point is reflected also in the contemporary mathematical theories of decision-making. One of the objectives of these theories is the formulation of decision rules which will determine an optimal choice in situations where several courses of action are available. For the formulation of decision rules, these theories require that at least two conditions be met: (1) Factual information must be provided specifying the available courses of action and indicating for each of these its different possible outcomes—plus, if feasible, the probabilities of their occurrence; (2) there must be a specification of the values—often prosaically referred to as utilities—that are attached to the different possible outcomes. Only when these factual and valuational specifications have been provided does it make sense to ask which of the available choices is the best, considering the values attaching to their possible results.

In mathematical decision theory, several criteria of optimal choice have been proposed. In case the probabilities for the different outcomes of each action are given, one standard criterion qualifies a choice as optimal if the probabilistically expectable utility of its outcome is at least as great as that of any alternative choice. Other rules, such as the maximin and the maximax principles, provide criteria that are applicable even when the probabilities of the outcomes are not available. But interestingly, the various criteria conflict with each other in the sense that, for one and the same situation, they will often select different choices as optimal.

The policies expressed by the conflicting criteria may be regarded as reflecting different attitudes towards the world, different degrees of optimism or pessimism, of venturesomeness or caution. It may be said therefore that the analysis offered by current mathematical models indicates two points at which decision-making calls not solely for factual information, but for categorical valuation, namely, in the assignment of utilities to the different possible outcomes and in the adoption of one among many competing decision rules or criteria of optimal choice. . . .

6. Valuational "Presuppositions" of Science

The preceding three sections have been concerned mainly with the question whether, or to what extent, valuation and decision presuppose scientific investigation and scientific knowledge. This problem has a counterpart which deserves some attention in a discussion of science and valuation; namely, the question whether scientific knowledge and method presuppose valuation.

The word 'presuppose' may be understood in a number of different senses which require separate consideration here. First of all, when a person decides to devote himself to scientific work rather than to some other career, and again, when a scientist chooses some particular topic of investigation, these choices will presumably be determined to a large extent by his preferences, i.e., by how highly he values scientific research in comparison with the alternatives open to him, and by the importance he attaches to the problems he proposes to investigate. In this explanatory, quasi-causal sense the scientific *activities* of human beings may certainly be said to presuppose valuations.

Much more intriguing problems arise, however, when we ask whether judgments of value are presupposed by the body of scientific *knowledge,* which might be represented by a system of statements accepted in accordance with the rules of scientific inquiry. Here presupposing has to be understood in a systematic-logical sense. One such sense is invoked when we say, for example, that the statement 'Henry's brother-in-law is an engineer' presupposes that Henry has a wife or a sister: in this sense, a statement presupposes whatever can be logically inferred from it. But, as we noted earlier, no set of scientific statements logically implies an unconditional judgment of value; hence, scientific knowledge does not, in this sense, presuppose valuation.

There is another logical sense of presupposing, however. We might say, for example, that in Euclidean geometry the angle-sum theorem for triangles presupposes the postulate of the parallels in the sense that that postulate is an essential part of the basic assumptions from which the theorem is deduced. Now, the hypotheses and theories of empirical science are not normally validated by deduction from supporting evidence (though it may happen that a scientific statement, such as a prediction, is established by deduction from a previously ascertained, more inclusive set of statements); rather, as was mentioned in section 2, they are usually accepted on the basis of evidence that lends them only partial, or "inductive," support. But in any event it might be asked whether the statements representing scientific knowledge presuppose valuation in the sense that the grounds on which they are accepted include, sometimes or always, certain unconditional judgments of value. Again the answer is in the negative. The grounds on which scientific hypotheses are accepted or rejected are provided by empirical evidence, which may include observational findings as well as previously established laws and theories, but surely no value judgments. Suppose for example that, in support of the hypothesis that a radiation belt of a specified kind surrounds the earth, a scientist were to adduce, first, certain observational data, obtained perhaps by rocket-borne instruments; second, certain previously accepted theories invoked in the interpretation of those data; and finally, certain judgments of value, such as 'it is good to ascertain the truth'. Clearly, the judgments of value would then be dismissed as lacking all logical relevance to the proposed hypothesis since they can contribute neither to its support nor to its disconfirmation.

But the question whether science presupposes valuation in a logical sense can be raised, and recently has been raised, in yet another way, referring more specifically to valuational presuppositions of scientific *method*. In the preceding considerations, scientific knowledge was represented by a system of statements which are sufficiently supported by available evidence to be accepted in accordance with the principles of scientific test and validation. We noted that as a rule the observational evidence on which a scientific hypothesis is accepted is far from sufficient to establish that hypothesis conclusively. For example, Galileo's law refers not only to past instances of free fall near the earth, but also to all future ones; and the latter surely are not covered by our present evidence. Hence, Galileo's law, and similarly any other law in empirical science, is accepted on the basis of incomplete evidence. Such acceptance carries with it the "inductive risk" that the presumptive law may not hold in full generality, and that future evidence may lead scientists to modify or abandon it.

A precise statement of this conception of scientific knowledge would require, among other things, the formulation of rules of two kinds: First, *rules of confirmation*, which would specify what kind of evidence is confirmatory, what kind disconfirmatory for a given hypothesis. Perhaps they would also determine a numerical *degree* of evidential support (or confirmation, or inductive probability) which a given body of evidence could be said to confer upon a proposed hypothesis. Secondly, there would have to be *rules of acceptance*: these would specify how strong the evidential support for a given hypothesis has to be if the hypothesis is to be accepted into the system of scientific knowledge; or, more generally, under what conditions a proposed hypothesis is to be accepted, under what conditions it is to be rejected by science on the basis of a given body of evidence.

Recent studies of inductive inference and statistical testing have devoted a great deal of effort to the formulation of adequate rules of either kind. In particular, rules of acceptance have been treated in many of these investigations as special instances of decision rules of the sort mentioned in the preceding section. The decisions in question are here either to accept or to reject a proposed hypothesis on the basis of given evidence. As was noted earlier, the formulation of "adequate" decision rules requires, in any case, the antecedent specification of valuations that can then serve as standards of adequacy. The requisite valuations, as will be recalled, concern the different possible outcomes of the choices which the decision rules are to govern. Now, when a scientific rule of acceptance is applied to a specified hypothesis on the basis of a given body of evidence, the possible "outcomes" of the resulting decision may be divided into four major types: (1) the hypothesis is accepted (as presumably true) in accordance with the rule and is in fact true; (2) the hypothesis is rejected (as presumably false) in accordance with the rule and is in fact false; (3) the hypothesis is accepted in accordance with the rule, but is in fact false; (4) the hypothesis is rejected in accordance with the rule, but is in fact true. The former two cases are what science aims to achieve; the possibility of the latter two represents the inductive risk that any acceptance rule must involve. And the problem of formulating adequate rules of acceptance and rejection has no clear meaning unless standards of adequacy have been provided by assigning definite values or disvalues to those different possible "outcomes" of acceptance or rejection. It is in this sense that the method of establishing scientific hypotheses "presupposes" valuation: the justification of the rules of acceptance and rejection requires reference to value judgments.

In the cases where the hypothesis under test, if accepted, is to be made the basis of a specific course of action, the possible outcomes may lead to success or failure of the intended practical application; in these cases, the values and disvalues at stake may well be expressible in terms of monetary gains or losses; and for situations of this sort, the theory of decision functions has developed various decision rules for use in practical contexts such as industrial quality control. But when it comes to decision rules for the acceptance of hypotheses in pure scientific research, where no practical applications are contemplated, the question of how to assign values to the four types of outcome mentioned earlier becomes considerably more problematic. But in a general way, it seems clear that the standards governing the inductive procedures of pure science reflect the objective of obtaining a certain goal, which might be described somewhat vaguely as the attainment of an increasingly reliable, extensive, and theoretically systematized body of information about the world. Note that if we were concerned, instead, to form a system of beliefs or a world view that is emotionally reassuring or esthetically satisfying to us, then it would not be reasonable at all to insist, as science does, on a close accord between the beliefs we accept and our empirical evidence; and the standards of objective testability and confirmation by publicly ascertainable evidence would have to be replaced by acceptance standards of an entirely different kind. The standards of procedure must in each case be formed in consideration of the goals to be attained; their justification must be relative to those goals and must, in this sense, presuppose them.

7. Concluding Comparisons

If, as has been argued in section 4, science cannot provide a validation of categorical value judgments, can scientific method and knowledge play any role at all in clarifying and resolving problems of moral valuation and decision? The answer is emphatically in the affirmative. I will try to show this in a brief survey of the principal contributions science has to offer in this context.

First of all, science can provide factual information required for the resolution of moral issues. Such information will always be needed, for no matter what system of moral values we may espouse—whether it be egoistic or altruistic, hedonistic or utilitarian, or of any other kind—surely the specific course of action it enjoins us to follow in a given situation will depend upon the facts about that situation; and it is scientific knowledge and investigation that must provide the factual information which is needed for the application of our moral standards.

More specifically, factual information is needed, for example, to ascertain (a) whether a contemplated objective can be attained in a given situation; (b) if it can be attained, by what alternative means and with what probabilities; (c) what side effects and ulterior consequences the choice of a given means may have apart from probably yielding the desired end; (d) whether several proposed ends are jointly realizable, or whether they are incompatible in the sense that the realization of some of them will definitely or probably prevent the realization of others.

By thus giving us information which is indispensable as a factual basis for rational and responsible decision, scientific research may well motivate us to change some of our valuations. If we were to discover, for example, that a certain kind of goal which we had so far valued very highly could be attained only at the price of seriously undesirable side effects and ulterior consequences, we might well come to place a less high value upon that goal. Thus, more extensive scientific information may lead to a change in our basic valuations—not by "disconfirming" them, of course, but rather by motivating a change in our total appraisal of the issues in question.

Secondly, and in a quite different manner, science can illuminate certain problems of valuation by an objective psychological and sociological study of the factors that affect the values espoused by an individual or a group; of the ways in which such valuational commitments change; and perhaps of the manner in which the espousal of a given value system may contribute to the emotional security of an individual or the functional stability of a group.

Psychological, anthropological, and sociological studies of valuational behavior cannot, of course, "validate" any system of moral standards. But their results can psychologically effect changes in our outlook on moral issues by broadening our horizons, by making us aware of alternatives not envisaged, or not embraced, by our own group, and by thus providing some safeguard against moral dogmatism or parochialism.

Finally, a comparison with certain fundamental aspects of scientific knowledge may help to illuminate some further questions concerning valuation.

If we grant that scientific hypotheses and theories are always open to revision in the light of new empirical evidence, are we not obliged to assume that there is another class of scientific statements which cannot be open to doubt and reconsideration, namely, the observational statements describing experiential findings that serve to test scientific theories? Those simple, straightforward reports of what has been directly observed in the laboratory or in scientific field work, for example—must they not be regarded as immune from any conceivable revision, as irrevocable once they have been established by direct observation? Reports on directly observed phenomena have indeed often been considered as an unshakable bedrock foundation for all scientific hypotheses and theories. Yet this conception is untenable; even here, we find no definitive, unquestionable certainty.

For, first of all, accounts of what has been directly observed are subject to error that may spring from various physiological and psychological sources. Indeed, it is often possible to check on the accuracy of a given observation report by comparing it with the reports made by other observers, or with relevant data obtained by some indirect procedure, such as a motion picture taken of the finish of a horse race; and such comparison may lead to the rejection of what had previously been considered as a correct description of a directly observed phenomenon. We even have theories that enable us to explain and anticipate some types of observational error, and in such cases, there is no hesitation to question and to reject certain statements that purport simply to record what has been directly observed.

Sometimes relatively isolated experimental findings may conflict with a theory that is strongly supported by a large number and variety of other data; in this case, it may well happen that part of the conflicting data, rather than the theory, is refused admission into the system of accepted scientific statements—even if no satisfactory explanation of the presumptive error of observation is available. In such cases it is not the isolated observational finding which decides whether the theory is to remain in good standing, but it is the previously well-substantiated theory which determines whether a purported observation report is to be regarded as describing an actual empirical occurrence. For example, a report that during a spiritualistic séance, a piece of furniture freely floated above the floor would normally be rejected because of its conflict with extremely well-confirmed physical principles, even in the absence of some specific explanation of the report, say, in terms of deliberate fraud by the medium, or of high suggestibility on the part of the observer. Similarly, the experimental findings reported by the physicist Ehrenhaft, which were claimed to refute the principle that all electric charges are integral multiples of the charge of the electron, did not lead to the overthrow, nor even to a slight modification, of that principle, which is an integral part of a theory with extremely strong and diversified experimental support. Needless to say, such rejection of alleged observation reports by reason of their conflict with well-established theories requires considerable caution; otherwise, a theory, once accepted, could be used to reject all adverse evidence that might subsequently be found—a dogmatic procedure entirely irreconcilable with the objectives and the spirit of scientific inquiry.

Even reports on directly observed phenomena, then, are not irrevocable; they provide no bedrock foundation for the entire system of scientific knowledge. But this by no means precludes the possibility of testing scientific theories by reference to data obtained through direct observation. As we noted, the results obtained by such direct checking cannot be considered as absolutely unquestionable and irrevocable; they are themselves amenable to further tests which may be carried out if there is reason for doubt. But obviously if we are ever to form any beliefs about the world, if we are ever to accept or to reject, even provisionally, some hypothesis or theory, then we must stop the testing process somewhere; we must accept some evidential statements as sufficiently trustworthy not to require further investigation for the time being. And on the basis of such evidence, we can then decide what credence to give to the hypothesis under test, and whether to accept or to reject it.

This aspect of scientific investigation seems to me to have a parallel in the case of sound valuation and rational decision. In order to make a rational choice between several courses of action, we have to consider, first of all, what consequences each of the different alternative choices is likely to have. This affords a basis for certain relative judgments of value that are relevant to our problem. If *this* set of results is to be attained, this course of action ought to be chosen; if *that other* set of results is to be realized, we should choose such and such another course; and so forth. But in order to arrive at a decision, we still have to decide upon the relative values of the alternative sets of consequences attainable to us; and this, as was noted earlier, calls for the acceptance of an unconditional judgment of value, which will then determine our choice. But such acceptance need not be regarded as definitive and irrevocable, as forever binding for all our future decisions: an unconditional judgment of value, once accepted, still remains open to reconsideration and to change. Suppose, for example, that we have to choose, as voters or as members of a city administration, between several alternative social policies, some of which are designed to improve certain material conditions of living, whereas others aim at satisfying cultural needs of various kinds. If we are to arrive at a decision at all, we will have to commit ourselves to assigning a higher value to one or the other of those objectives. But while the judgment thus accepted serves as an unconditional and basic judgment of value for the decision at hand, we are not for that reason committed to it forever—we may well reconsider our standards and reverse our judgment later on; and though this cannot undo the earlier decision, it will lead to different decisions in the future. Thus, if we are to arrive at a decision concerning a moral issue, we have to accept some unconditional judgments of value; but these need not be regarded as ultimate in the absolute sense of being forever binding for all our decisions, any more than the evidence statements relied on in the test of a scientific hypothesis need to be regarded as forever irrevocable. All that is needed in either context are *relative* ultimates, as it were: a set of judgments—moral or descriptive—which are accepted at the time as not in need of further scrutiny. These relative ultimates permit us to keep an open mind in regard to the possibility of making changes in our heretofore unquestioned commitments and beliefs; and surely the experience of the past suggests that if we are to meet the challenge of the present and the future, we will more than ever need undogmatic, critical, and open minds.

20

The Exact Role of Value Judgments in Science

Michael Scriven

0. Introduction

If there is one set of arguments worse than those put forward for 'value-free science', it is those put forward against it. Both sets have one common characteristic, besides a high frequency of invalidity, and that is the failure to make any serious effort at a plausible analysis of the *concept* of 'value judgment', one that will apply to some of the difficult cases, and not just to one paradigm. Although the problem of definition is in this case *extremely* difficult, one can attain quite useful results even from a first step. The analysis proposed here, which goes somewhat beyond that first step, is still some distance from being satisfactory. Nevertheless, we must begin with such an attempt since any other way to start would be laying foundations on sand. And we'll use plenty of prescientific examples, too, to avoid any difficulties with irrelevant technicalities. As we develop the definitions and distinctions, we'll begin putting them to work, so that we will almost complete the argument while seeming to be just straightening out the concepts. For this is really an area where the problem is a conceptual one, rather than an empirical or inferential one.

1. The Nature of Value Judgments

It is presumably a truism that a value judgment is a judgment of value, merit, or worth. From this basic meaning, other uses of the term are often generated by adding some extremely questionable philosophical position to the basic definition. For example, there are many contexts in which 'value judgment' is used as a synonym for 'dubious, unreliable, or biased judgment'. In Webster's Third International Dictionary, the illustrative quotation identifies it with prejudice or intolerance, although the definition given has no such implication. This can hardly make sense unless one assumes the truth of the view that all judgments of merit or worth are in fact biased, unreliable, intolerant. It is a measure of the extent to which the doctrine of value-free science has received support, that social scientists in introductory lectures, as well as in writing, typically make this assumption. One of the aims of this paper will be to demonstrate why the philosophical position just mentioned is unsound: but whether the reader's conclusions are the same or not, it would obviously be inappropriate to *begin* by begging the question, i.e., by accepting the identification of value judgments as a sub-species of unsubstantiated or unreliable judgments.

Again, one sometimes encounters usages which make it clear that people are even prepared to mingle the sense of 'value' in the mathematical descriptive phrase 'the value of a variable' (which refers to a number) with the sense of 'value' which is essentially equivalent to 'merit': we shall in general disregard this 'quantitative' sense of 'value' as an irrelevant ambiguity, though we shall indicate how it can be included in a comprehensive taxonomy of the term 'value'.

A related misconception involves the identification of the results of *any* act of judgment as a value judgment. This is sometimes related to the quantitative sense of 'value', e.g., when what is judged is the magnitude of a variable. But it is sharply distinct from the qualitative sense. There is a great difference between the assertion of a highway patrolman that you are going 'too fast for the conditions', which can reasonably be translated as 'faster than you should or ought to', i.e. (approximately) 'faster than it is right or good or proper to travel in these conditions'; and his judgment that you are traveling above the legal limit, which is 55 mph on this stretch of road. If he has just read a higher figure off a radar device, one probably wouldn't even use the term judgment in the latter case; but highway patrolmen have been known to give someone a ticket after passing them in the opposite direction, because, 'in their judgment', it was obvious that the pilgrim in question was traveling substantially above the speed limit. That's not a *value* judgment in the sense of concern to us here, although it *is* a judgment of the value of a variable, of course. The extreme form of this position takes judgments of *any properties,* e.g., color, as value judgments. This does not seem to be a useful generalization of the term's sense and we will operate with the idea that value judgments are a *sub*-species of judgments. To avoid some of the confusion due to the 'judgment' element, we'll generalize to talk about value *claims* most of the time.

After ruling out these more extreme misconceptions, there still remain four categories of 'judgments' or claims that have a considerably better claim to being called value judgments, although only one of them has an impeccable claim. Let us consider these residual candidates in more detail.

II. Type I: Value-Base Claims

Although I shall call these value-base claims, other terms for them that are in some ways appropriate are 'primary value claim', 'pre-value claim', or 'proto-value claim'. The paradigm here is the assertion 'I value X' or, what may reasonably be taken to be closely equivalent, 'I favor/prefer/endorse/like X'.[1] Such claims need not be in the first person; they may be in the third person singular or plural. They make a claim about what some person (or people) *values* in the sense of value whose principal content is liking (having a favorable attitude towards, etc.) and that does not entail consciousness of liking. Such claims are entirely objective; that is, it is simply a matter of fact whether Jones or the Jones family or Jones's nation likes $X,$ though it is not always a matter of fact that is particularly easy to establish. The only subjectivity of such claims lies in the fact that they refer to subjects. Like belief-statements, liking statements *typically* involve no commitment to the truth or true merit of whatever it is that is believed/valued: but there is a usage in which they do, discussed below. Value-base claims have always been considered a legitimate part of science: anthropology and sociology report on the value bases of social groups with considerable detail and care.

If claims like these were the only paradigm of value judgments, then one could understand how one might be tempted to identify value judgments with matters of taste, and hence perhaps with judgments that are essentially not objective or scientific. However that whole chain of inference is unsound. The assertion of a taste *as* a taste is simply a factual assertion, an assertion about the person or group whose taste is being reported, described, or expressed. Only if what is *properly* a matter of taste is misdescribed as a *universally recognizable* fact *about the object of taste,* has anything improper or unscientific occurred. For example, if somebody says, "The post-impressionists were simply bad painters," when all that can be substantiated is that the speaker (perhaps not alone) doesn't like them, then we have an example of an unjustified value judgment, one that ought to have been replaced by an expression of taste, a value-base claim. Assertions of liking, etc., are primarily assertions about dispositional properties of the liker; they are not primarily assertions about the intrinsic merits of the object liked. In the sequel, I'll call the liker, the 'internal referent' of a liking-assertion; and the thing liked, the 'external referent'. Such claims are really pre-value claims: on a utilitarian account they are the factual data upon which one can construct the system of values, and thus they may be called value-base claims or proto-value claims. It is only when one of these *masquerades* in the language of objective value judgments, supposedly only about the *external* object of reference, that something improper has occurred.

But there is an extremely important complicating feature of the language of value-base claims that goes a long way towards explaining why the common conclusion of relativism about values is often supported by appeal to such examples.

It will be helpful to use an analogy with a crucial ambiguity in another epistemological area. If somebody says, "I think that this hypothesis offers the most promising explanation," he or she is usually not just making an autobiographical remark (analogous to a value-base claim).

As an autobiographical remark, investigation of its truth and falsity require only that one consider the cognitive state of the speaker. But the fact is that someone will often reply to such a remark, by saying "That's not so—there's a *much* better explanation," showing that they understand—I think correctly—the first remark to be a (guarded) endorsement of, indeed a kind of weak *assertion* of, the claim "this hypothesis offers the best explanation." It is too coy to retreat from the attack by saying "I only said I *think* this." For one can say that one *shouldn't* think what is false; that one was *wrong* to have thought such a thing. Thus, such claims are not *just* (auto) biographical; they are also observations/claims about the *external* referent. And not merely in the sense that relational statements are about both terms of the relation; they are what one might call *autonomous* claims about the external referent, albeit guarded ones. ('Autonomous' in the sense that they are put forward to stand on their own, as claims about the nature of the external referent, not as 'mere' relational claims about the attitude of the internal referent to the external one.)

Bringing this to bear on assertions of one's likes; if someone says he or she likes *X,* then of course this does attribute a property to *X,* namely the property that it is liked by this person. That's a kind of back-handed way of attributing properties—markedly different from what happens when someone says "*X* is three meters long." What I mean by saying that a statement of a certain kind is an *autonomous* statement is that it could have been made by anyone without difference in meaning, it has no *essential internal reference.* Now when somebody says "I think this hypothesis offers the best explanation," he or she is not *just* saying that the hypothesis has the property of being-thought-by-him-or-her-to-be-the-best-explanation. I am saying that it is better to construe his or her statement as *also* involving the autonomous claim that this hypothesis *does* provide the best explanation, albeit rather modestly proposed ("It is just *my* view that this is so; I don't claim everyone agrees.") Now this odd two-faced character of assertions that begin with 'I think' or 'I believe' or 'It seems (likely) to me', etc., illuminates some disputes in the value judgment area. If someone says that he or she 'likes the chances' of a horse named Mistral in the fourth race, it would be too narrow a reconstruction to suppose that this is just an autobiographical remark. Generally, it is an expression of a judgment in the usual sense. It is very similar to the belief statements that we have just been talking about in that there is an implied commitment to the autonomous claim about the horse winning or placing. If someone says that he or she likes or prefers or thinks well of the short club convention in contract bridge, there will certainly be contexts in which this is taken as a mere autobiographical comment, as for example when we are simply taking votes on preferences. But there will be other contexts in which it is perfectly appropriate to reply to this by saying "That's a mistake," that is, by taking it as if it were a (weak) assertion about the objective facts of the case. There's a continuum between these interpretations, involving reactions like "That's not very discriminating of you." After all, in many areas, what you like, as well as what you think, is objectively judgeable as sensible or correct.

In one sense, one could say that the reason for the two-facedness of these value claims is that matters of value are not all '*mere* matters of taste', i.e., matters where there is no question of rightness and wrongness. In fact, it is just because there *is* an implicit framework of *objectivity* about values that one can suppose people are right or wrong to have certain values, and hence can respond to their value-base assertions as more than mere autobiographical claims. It is this schizophrenic quality of value claims, as the tidy-minded might call it, that has led to much of the confusion about their status. The relativist, looking at one of these value-base claims, often instinctively realizes that it has the objectivity implications of an autonomous claim, but he also sees that it can be construed as an autobiographical claim, and too hastily concludes that the latter possibility shows the former to be charlatanry. As we shall see, *both* facets of a value-base claim are objective, though they have quite different contents: and they are not incompatible at all. Remember, though, that there are contexts in which a value-base claim is *simply* autobiographical, and it is then completely wrong to complain about its lack of objectivity.

We can see from the preceding discussion that there is an interesting interplay between the notions of subjectivity and objectivity, on the one hand, and internal versus external reference, on the other. The ideal of objectivity is often thought to be an assertion with purely external reference. But we have seen that it is perfectly legitimate to regard a statement with purely *internal* reference as entirely objective, in the sense of testable, reliable, etc. We will next look at another class of value judgments, where there is again a kind of multi-person reference involved, and a kind of autonomy involved, but in a rather different way than the preceding, a way that raises a different kind of doubt about objectivity.

III. Type II: Market-Value Claims

These claims grow naturally from value-base claims, as one extends the group to which the value refers. If one person is said to value something, we know exactly how to check this, construed as an autobiographical claim. If a group of people are said to value something, similarly. But for certain practical purposes we develop a form of value claim which continues to be relative to a social group, but ceases to mention it directly. Thus it resembles a value-base claim in avoiding any commitment to the correctness of the group's judgment — but it has the appearance of an autonomous judgment, an 'objective' judgment. And it *is* autonomous — it does assert a property, about an external referent, in a sense, not just the property of being liked by someone: it ascribes a more abstract quality that emerges from liking.

This kind of claim emerges when we start talking about *the* market value of *things*. The emphasis shifts to the external referent and the internal referent becomes implicit or contextual. Take a claim like 'This is an extremely valuable rug'. This doesn't mean that the speaker values it at all (although he may), nor does it mean that any other identified person values it. It is a reference to the open market price of the rug, i.e., a reference to a *hypothetical population of prospective purchasers*. It is entirely straightforward to test market value claims, though not always easy to make them accurately prior to the test. The claim, though not *explicitly* relativized (like a value-base claim) is only intended to apply in a certain area and for a certain time. We have specialists, called assessors or appraisers, whose task it is to determine the market value of property of various kinds. There is a matter of skilled judgment here, and the judgments made are judgments of worth or value; but they are not the value judgments that most of the fighting has been about, because, to put the matter in one way, the assessor does not endorse the propriety, the rationality, the morality, or, in the relevant sense, the objectivity of the market's judgment of value. He just estimates it. He (or she) claims objectivity for the *estimate,* but not for the judgment (taste) of the public to which the estimate implicitly and internally refers. We have here yet another example of a perfectly objective type of value judgment, once more contrary to the extreme skeptic's condemnation of all value judgments as lacking in objectivity.

Just as a significant part of the task of a certain type of social anthropology and sociology or political science lies in correctly determining the value system of a society, so the assessor performs a similar task in the more specific context of jewelry pricing. A clinical or educational psychologist is often concerned with the values of a small group or individual; and he or she knows how to go about determining them, although it is not an easy task. This is identifying the value-base of a group, and we may think of the market value as a modestly more abstract concept than value-base. Market-value is a 'social fact' in the technical sense of that term. It is typical, but not essential that market value can be expressed in quantitative terms; it is typical but not essential that the language of market value usually refers to objects rather than to qualities. When we talk about the extent to which certain qualities or conditions, such as security or freedom of speech, are valued by a certain group, then we tend to think of this as value-base talk; when we talk about the extent to which Hispano-Suizas are valued, that tends to be market value talk; but at the borderline the classification is not important. Market value always refers to a hypothesized population of potential purchasers but only implicitly; value base refers to an actual population, explicitly, which may consist of only one individual, Market value claims do *not,* whereas the words used to express value-base claims *sometimes* do, imply an autonomous claim of the very important kind we are about to consider. Finally, market-value claims actually incorporate some of that 'distance' that a more precise model of value-base claims would introduce between mere liking and actual *valuing*: for the ascription of market-value has that very modest commitment to stability or generality that is lacking in a mere fancy, a liking of the moment.

IV. Type III: Real-Value Claims

Claims about the *real* value, merit or worth about something—or its 'true' value—often mean something quite different from market value. They refer to 'absolute' value, in one sense. They may indeed be *contrasted* with market value, thereby demonstrating the difference. For example, we may say that the market 'undervalues' used Chrysler Imperial cars because of the prestige appeal of Cadillacs: obviously this only makes sense as a reference to a kind of value that goes beyond describing what people actually value. We may say that antique Bentleys aren't worth that much although they *cost* a lot; thereby thrusting the distinction into an alien context—but no one has difficulty understanding it. These real value claims are what the fight is all about. The great question is what kind of objectivity *they* can have.

What *do* we mean when we say that something isn't worth what we paid for it? Often we would support this claim if it was challenged, by saying things like this:

"Look at what else you could have done with the money—you could have got some things you really need," . . .
"Those prices are just faddist—you could lose half your investment within days." . . .
It just isn't well *made*; look at the poor joinery work, the low grade of teak, etc."

These remarks remind us of the standards of 'true worth'—life and health and satisfaction, money (because of its translatability into other useful goods), constancy, reliability, utility, etc. Far from being ephemeral, subjective, like tastes and the market, *real* value is tangible, durable, useful, multifunctional—which doesn't mean that it's always easily measured, any more than forces and fields. Now these qualities—tangibility, etc.—are not so much values in themselves as guarantors of the reality of whatever the valued properties or objects are to which they are attached. Thus—for many people—good life insurance coverage is very valuable. To assert of such a policy that it really *is* a good one, that is, to make a real value claim about it, requires support in terms of evidence that the company will not go bankrupt, that the cost is not higher (without corresponding increases in benefits) than competitors, etc. For these considerations provide safeguards that increase the probability your investment will bring the most return in *your* currency; in this case, security for yourself and perhaps for others. In these commonsensical examples, it is easy to see just how the process of verification of real value claims goes on and it seems absurd to suppose that no such claims are beyond reasonable doubt.

Is there a general model of real-value claims that will enable us to see how they can be objectively confirmed?

A simple way to conceptualize the logic of confirmation of real value claims is to see them as representing the result of combining two kinds of evidence; the first being value-base evidence, the second performance data. To determine the merit of a particular life-insurance policy, we examine the needs, performances, and economic abilities of the purchaser; and we also examine the alternative policies he or she can purchase, for their performance on these dimensions. Often we can obtain a simple weighted sum of the performance variables (the weights representing the importance of each dimension of performance, inferred from the value-base) as an indicator of merit and the entry with the highest score is the winner. Or one may have to make a synthesizing judgment that cannot be reduced to a simple weighted-sum, as happens to be the case with life insurance where the fine print varies for each of the companies that deserve serious consideration. (See *Consumer Reports,* January, 1974.) Here indeed is a true judgment of value. But there is nothing arbitrary about it, any more than in the judgment of speed. Arguments can be given and assessed for the judgments made, and they can easily have the status of very reliable conclusions. This is what one might call the Consumers Reports (CR) approach.

The degree of generality of such a conclusion of merit will depend on the degree of commonality of the value bases of prospective policy holders, and on the size of the 'winner's' performance advantage. Naturally, conclusions will have to be phrased so as to reflect these constraints; "For young, single, securely employed people, the five-year term policy from X company is the best because . . . ." Even where there are huge disparities in the values base, they may be swamped by huge differences in the performance profiles of the candidates; thus, one may be able to conclude "Geico is the best company for automobile insurance, whatever your needs" (so long, of course, as they don't include the need to have Allstate write your insurance). Hence, very great differences in 'taste' or needs are perfectly compatible with a universally true (across all *people*) real-value claim.

The model so far described is sometimes said to provide inadequate foundation for real value claims on the grounds that it only supports claims of the form "X is what this person will *think* is the best choice" (the rational decision-making model). But the CR model does *not* make predictive claims, since it is not committed to the further assumption predictions would require, the assumption that people are in fact rational. It is a normative model; it tells us what people *should* buy, prefer, etc. And it really can tell us what they *should* buy, not "what someone should buy *if* they have such and such preferences." For we can, as social scientists, determine whether they do in fact have those preferences, hence we can make a categorical and not just a conditional assertion about the 'Best Buy' for them, a typical real-value claim.

The only really serious threat to the objectivity of value judgments generated by the CR model is sometimes put as follows: "The model generates true value judgments only if the values that people have in their value bases are themselves *correct*. But *those* values cannot be shown to be correct by the CR model, since that would involve an infinite regression (to support them, we'd have to appeal to yet other values held by these people, which could in turn be called into question). And clearly the values people actually have are *not* always correct; certainly not morally, but also not even prudentially; i.e., egocentrically. So the CR model only does produce conditional evaluations and the conditions are very questionable. Moreover, there is no *other* way to certify the rightness of the value-base values—at least no way that has any claim to scientific objectivity. So the CR approach simply fails to do what you claim it can do.''

If we had to choose the one argument that has had the most effect on the most intelligent of those self-styled 'empiricists', 'behaviorists', or neo-positivists who have continued to believe in the ideal of value-free science, this argument is it. The argument is completely wrong, but not because of a formal fallacy. It is wrong in the way most common with philosophical arguments—it misrepresents a point of philosophical interest as one of practical significance (cf. the arguments about certainty, the external world, etc.). In this case, the argument is of sufficient importance to be criticized in two ways; first, by *reductio ad absurdum,* and then by detailed analysis.

First, then, it will be argued that the same argument would destroy the descriptive part of science. Given that the argument is put forward by those who view the descriptive part of science as the paradigm of well-founded knowledge, this constitutes a reductio of the form of argument.

Such an application of the argument would go as follows. In order to support any descriptive conclusion in science that is not directly observable, we often attempt to construct an argument for it using premises that *are* observable, and we believe this can be done in all such cases. But it is well known that even observations can be mistaken, especially since we are well aware that what we call observation-statements are quite often impregnated with theoretical language and implicit assumptions. Thus circularity or infinite regression are involved. Since these built-in assumptions are not themselves reliable, and since there is no other way to establish descriptive assertions besides observation and inference from observations, it follows that there is no room in science for descriptive statements: for descriptive statements are either observably true or not, and we have just demonstrated weakness in both kinds.

The refutation of *this* argument rests on the fact that one can *increase* the probability of statements, to the point where they are beyond reasonable doubt, by *accumulating* observations and inferences that bear upon them. Only if one was engaged in a search for certainty that transcends the *logical* possibility of error could one complain about the reliability of well-supported scientific claims: and the search for *that* kind of super-certainty is irrelevant to the quest for scientific knowledge which has never laid claim to infallibility. Hence the argument does not show that descriptive claims have no place in science, only that they share with all scientific claims the possibility of error.

Applying the analogy to the attack on real-value claims, let's take a simple case. Suppose that we are evaluating methods of resuscitation and we conclude that the mouth-to-mouth method is the best one. This conclusion would be based on synthesizing the data on the relative performance in restoring the vital signs of alternative methods such as the Schaefer method; and the value-base data about the relative strength of people's desire to live versus their antipathy to catching a cold from the germs in the would-be deliverer's lungs, etc.

The skeptic's attack involves reminding us that these values may be in error. It is possible, he might argue, that it is better for the nearly-drowned victim to die than to live since it is possible that the victim will live only to contract an agonizing and ultimately fatal disease, etc. Hence one cannot *assume* that *survival* is good, as the victim and rescuer both do. And without that assumption, one can't conclude that mouth-to-mouth is a *better* treatment here.

But one *can* make that assumption, indeed one should; *exactly* as—in the descriptive area—one should assume that the man whom you believe you have just seen shoot a policeman is dangerous. For assumptions should be made when the evidence strongly supports them. It will never do more than that, and it is a complete mistake to suppose it could and hence to suppose that assumptions should only be made when they are infallible.

The nature of all scientific inference and legal inference is to proceed via what are sometimes called *prima facie* cases, i.e. arguments that establish a *presumption* of something being the case, arguments (or observations) that make it 'reasonably probable'.

And that is all we do in the values area. 'Smoking is bad for (harmful to) your health' is a value judgment and it's fallible—but only like any other scientific claim, including the 'pure performance' claims about carcinogens on which this claim is partly based. The other part of the basis is no more fallible; it's simply a series of value-base claims about the physical states people prefer to be in, from which the concept of health is constructed by combination with medical knowledge.

Skepticism about real-value claims is not more appropriate than skepticism about any other claims. When the coroner determines the cause of death, there always *might* be something else that's overlooked. But part of good medical (scientific) practice is exactly to search for the presence of other possible causes. When the physician recommends a certain treatment, there may of course be a side-effect not yet identified, which will prove serious for this patient. What appears to be for the best may deceive us. Science can only reduce uncertainty, not eliminate it.

Now *one* subspecies of real-value claims is the set of moral judgments, the field of ethics. Another is aesthetics. Another is wine connoisseurship. The credentials of those fields require separate investigation, but the objectivity of real-value claims in general is in no way dependent on them. There is a moral version of the question whether any particular person's life or health should be saved, but that in no way shows that there isn't a non-moral values issue involved, what one might call the purely medical issue; and the question of the 'best treatment' that we have just discussed is of that kind.

Naturally enough, there are borderline cases in medicine where the moral issues become deeply intertwined with the medical ones. But in general it is easy (and even there it is possible) to separate off, not the scientific from the values issues, but the medical values issues from the moral ones. Eventually both must be recombined prior to justified action, but the evidence and expertize for each is different and it's usually helpful to work on them separately.

Let's look back at the line of argument in this sub-section for a moment, to see if we've committed any obvious logical crimes. The most threatening candidate is the possibility that we've extracted 'ought' from 'is', i.e., value claims from factual ones, which Hume argued is always fallacious. And it's true that we've argued real-value claims are synthesized out of value-base and performance claims. That is, they have no other evidential base, and need none. But this doesn't violate Hume, though it's often thought to, since our synthesizing process may include a value premise as long as it's true by definition. For example, it might include the premise that whatever is valued is valuable, which is false, but useful to illustrate the point. If that premise were analytic (i.e., definitionally true) then we could legitimately perform the proposed synthesis, and produce conclusions about what is valuable from premises about value-bases.

The *actual* synthesizing principle to which we are committed does not admit of complete condensation into a brief formula—but it can be *approximated* by 'Being valued establishes a prima facie case for being valuable'. That is, unless countervailing considerations can be produced and substantiated, the fact that Smith wants to live provides a sound basis for the prima facie (real-value) conclusion that he should give up smoking.

The 'raw values' of a person or group are of course often inconsistent or inappropriate in view of the facts. The resources of logic and science are often severely strained to determine what *is* the best law to pass concerning, e.g., heroin pushers, even if we already know that 95% of the population favors the death penalty for this. But our difficulties in coming to the value conclusion in such a case do not for a moment show that we are appealing to some other basis for our values besides the value-bases of the people involved. We are simply taking into account *all* of the value-base of the people involved, and the relevant evidence from penology, physiology, etc. A doctor can justify his or her recommendation to the patient to give up a favorite recreation, thereby rejecting an element in the patient's value-base, simply by appeal to other elements in that same value-base—plus medical knowledge. The appeal may be to a single other element which is held more strongly or it may be several others of the same or less importance; or it may simply be to the facts, "Do you really know what it means to become addicted to barbiturates/to have a mongoloid child/to undergo irreversible sterilization? If you did know, you wouldn't continue to do what you've been doing—you wouldn't think the immediate gratification was worth it."

There are, of course, vicious forms of the argument that Doctor Knows Best, and part of the charm of the value-free ideal for some people is that it appears to leave the choices more squarely in the hands of the individual. But our concern here is not with the problem of minimizing the abuses of the correct position; it is only in determining that position. And just as people have long recognized the possibility that a great spiritual leader may lead them to see what is called 'the error of their ways', i.e., of their values, so it is entirely appropriate for a scientist, in certain circumstances, to identify errors of attitude, practice, and judgment in the values domain.

V. Type IV: Valued-Performance Claims

So far, we have discussed claims which can be classified by direct inspection of content. But there is a class of value claims which can only be identified by consideration of context as well as content. These might be called 'quasi-real-value claims', because they are treated—in the particular context—with all the respect (or irritation, etc.) afforded a real-value claim. But they are intrinsically 'merely descriptive', i.e., they have just the same characteristics as a simple performance claim, which is only one of the two kinds of assertion normally synthesized into a real-value claim (the other being a value-base claim). The special feature of *these* performance claims is that the performance they refer to is one that—from the context—we can see is obviously held to be valuable. Hence in ascribing this property (performance) to an entity, the ascriber is clearly honoring it, embuing it with merit. A good example is the ascription of 'exceptional acceleration' to a sportscar, which is usually done in a context where this is assumed by all to be desirable. That is, they treat assertions about it as if they were real-value claims. Exactly the same occurs in technical areas when terms such as 'phase-locked', 'digital', 'stratified-change', 'transistorized', or 'Bugatti' become endowed with the aura of quality, usually through a close association, despite the fact that their cognitive content remains (essentially) the same. Nagel is so persuaded by this contextual aura that he actually classifies claims using such terms as 'appraising' value judgments (his term, which is otherwise equivalent to real-value judgments), thereby showing a sensitivity to context that is rare amongst the neo-positivists who were and are notable for their content-orientation.

But it is surely worth distinguishing, at least pro tem, between assertions which *necessarily* involve the ascription of merit and those which do so only in and only because of some contexts, and do not impute merit in others. Putting it another way, if there is any distinction at all between descriptive and evaluative claims, we have to mark it here; for almost *any* property can be a mark of value to someone in some circumstances, and if *that* is enough justification to classify the ascription of the property as a value judgment, *simpliciter,* then of course science consists largely of value judgments, at least of assertions that are value judgments in some contexts. That view of the matter trivializes the whole dispute about value-free science to a degree it does not deserve. It has chiefly been an argument about whether *real*-value claims are scientifically legitimate, not about whether the usual scientific properties can ever be valued by anyone. Obviously, 'weighing 236 pounds' can be valued (positively or negatively in different cases) and it is a scientific predicate, which scarcely proves science cannot be value free.

If one says that such claims are value judgments *only in those contexts* where the ascribed properties are treated as valuable by the author of the claim, then it will still be the case that science is infused with value judgments since even 'carcinogenic' will sometimes be a term of (contextual) appro-bation, e.g., when searching for the cause of occupational skin cancer.

The truth is that the use of 'purely descriptive' terms as if they were valua-tional tells us something about the value-base of the user, and thus they have as much claim to be called 'quasi-value-base' as 'quasi-real-value'. I suggest we give them the neutral title 'valued-performance claims'.

This concludes our classification of value claims, of claims which involve the term value or its cognates in any way. For completeness, we could call claims involving the 'quantitative' use of value ("The impedance had a value that ranged from 2.6–8 ohms") Type 0 claims. We have still excluded artificial uses of 'value judgment' as equivalent to (a) unreliable or debatable judgment or matter of taste, (b) any *judgment* (as opposed to simple observation or measurement or calculation). It will be clear that Types I and II *report* value judgments, while Types III and IV *make* them: that Types I and IV are very close to the 'descriptive' level, while Types II and III are more theoretical in nature. In fact, there is a continuum of increasing abstraction running from I to III which corresponds roughly to the continuum from phenomenological to group-illusion to real-world claims in the basic epistemological dimension. And so on. But Type III is what it's all about, and the other types are set up to clarify its nature by contrast.

VI. Problem Cases

There is no suggestion that the preceding categories are sharply distinct. It may be helpful, however, to see some more ways in which they are related (to each other and to their opposites) and do merge into one another. At one extreme, we have the 'merely descriptive' or pure performance claim; it can become a *kind* of value claim in certain contexts, as we have just seen: a valued perfor-mance claim. A particular kind of descriptive claim also happens to be a description of the values of a person or group, in which case we have called it a value-base or — another variety — a market-value claim. At the other 'extreme', we have the paradigm value judgment, the real-value claim, directly ascribing merit or worth. Between the two we have crypto-real-value claims galore, which look like mere descriptions, but turn out to have terms of merit buried in their meaning. Thus, 'intelligence' can plausibly be said to require reference to being *good* at problem-solving, and/or conceptualizing, etc. Given the credentials of a well-administered and well-validated IQ test, one can regard the determination of intelligence to be both a 'mere matter of measurement' and a real-value judgment. There are other cases where evaluation has been standardized by introducing scales or checklists as, e.g., with diamonds and certain educational products. To a die-hard facts/values 'segregationist', this poses a threat, usually handled by saying that the claim of merit for something to which these special measures apply is "not really a value judgment: it is equivalent to saying that the thing has the following properties." The general version of this I shall call the 'translatability thesis'; the view that the real meaning of certain (putative) value judgments can be given in terms of an assemblage of purely factual judgments. But the translatability thesis is a two-edged sword. If the equivalence does hold, then one could just as well conclude that some value judgments are entirely factual. Indeed, unless one wishes to *define* value judgments as unreliable, this is just what one must conclude. And to define them in that way is to trivialize (misrepresent) the dispute, and to make a perfectly useful term entirely redundant and misleading (since judgments of value are no longer *ipso facto* value judgments).

It is much more sensible to say that real-value claims are sometimes judgmental, sometimes not; sometimes reliable, sometimes not; sometimes pretty well determined by the results of measurement and sometimes not.

Indeed, to put the matter bluntly, it simply won't do to argue that 'facts' and 'descriptions' are classes that exclude real-value claims. If someone is asked to describe a telescope which he is offering for sale, the following would be entirely appropriate; "It's a Schmidt–Cassegrain 8″ reflector, in good-as-new condition, with a set of very good eyepieces covering 6 mm–40 mm." Descriptions often do and sometimes should include value judgments. Facts are often facts about merit. There is scarcely anything in the world more factual than Einstein's merit as a physicist.

The fact/value distinction does exist, but only in a fixed context. In a context, we do distinguish between the data, the known, the accepted; and the conclusion, the step beyond, the synthesis. Where that conclusion is a value-claim, it's colloquial to say *it* is the value judgment and the data are the fact. *But,* among those facts—if one looks closely—one will often find other value judgments. They deserve to be called facts in this context, because they are not in question for this issue. Yet they are plainly real-value claims themselves. Thus in arguing about the relative merits of a refractor against a reflector for planetary studies, one will often appeal to the superior color rendering of the reflector, so important in planetary work.

An interesting borderline case is provided by the predicate 'true'. In some cases—as in characterizing the answers to exam questions—it seems fair to treat this as equivalent to 'correct' or 'right'. And surely right/wrong is a pair of real-value predicates. The connections between good/bad, ought/should, and right/wrong are very close, outside the moral domain as well as inside. The identification of superior *worth or merit* implies the identification of the *right and wrong* choices or actions, which are the ones a person *ought to* make or *should* do. If one *should* do *X*, if *X* is the *right* thing to do, then it surely follows definitionally that it's *better* to do *X* than something else, that doing *X* has greater *merit* than the alternatives, etc. If the implication is not quite absolute, then at least it's prima facie sound, and on definitional grounds.

If these close connections are as suggested, then assertions about the truth of a claim must be regarded as a species of value judgment, even in our rather carefully delimited sense of the term. And that, of course, puts the values cat amongst the scientific pigeons with a vengeance, since assertions about truth can scarcely be excluded from science.

The concept of 'significance', in its use in, e.g., political science and in statistics, is often hard to classify—sometimes its ascription is a mere performance claim, sometimes a real-value claim, sometimes even a moral real-value claim. It is often quite useful to tease out the components in these predicates, not just to tie the right label on them, but to understand better the nature of the claim itself and thus what evidence will be required in order to establish it. It is interesting to consider such predicates as 'careful', 'thorough', 'sufficient', 'comprehensive', 'explanation', 'probable' (= well-supported), 'plausible', 'important', 'relevant', 'successful', 'confirmed', 'valid' (in the technical sense), 'harmful', etc.

VII. Value Judgments in Science

The groundwork we have laid can quickly provide an answer to the main question about the occurrence and role of value judgments in science. We can sum it up by saying that in the sense of the term 'value judgment' that is used in the strongest non-trivial arguments for a value-free science, the fact is that science is *essentially* evaluative, would not be science if it could not make and thoroughly support a whole range of value judgments.

Value judgments were classified in the last section in terms of their logical characteristics. They can also be classified in terms of subject-matter, or general type of backing—one might say, domain. Thus there are practical value judgments ("You should never cross the leads when jump-starting an alternator-equipped automobile."), moral value judgments, and etiquette value judgments ("You should never smoke in the ring at sheep-dog trials."). The value language is not, properly speaking, ambiguous just because it can be used to express value judgments from such different domains with such varied force. It is simply context-dependent. The terms 'large' and 'small' are not ambiguous—even though they may be somewhat imprecise—just because a very large mouse is smaller than a very small elephant. Value-language, like size-relative language, should be thought of—understood—in terms of its *function* rather than some specific content. Its function is the same in all contexts, its specific content varies. (Cf. 'it', 'him', 'now'.)

Now one subject-matter category of value judgment is the methodological value judgment, that is the assessment of the merit or worth of methodological entities, such as experimental designs, theories, observations, explanations, estimates, (curve-) fits; and we will extend this concept to cover the hardware and software of method, so it will include the assessment of instruments, computers, and programs. A closely related category of value judgments, which I separate off only to avoid shudders from humanists, consists of judgments of the scientific merit of scientific performances, including the writing of articles, the presentation of papers, the doing of an experiment, etc. Of course, there are methodological value judgments in areas other than science, for example the dance and the law; and there are extra-scientific performances. It is a natural step from performance to performer, and the scientific sub-species include the Great Scientist Game, the Nobel selection and the activities of other award and appointment committees.

I submit that learning to be a scientist involves learning to distinguish between good theories and bad theories, between good experimental designs and bad ones. Someone who could not make such discriminations could not distinguish good science from bad science, science from nonscience; and therefore could not be a scientist. Science itself is an evaluative term, in one of its principal uses: but even if it were not, the practice of science is an evaluative activity. Nor are the scientific evaluations only methodological: performance evaluations and performer evaluations, evaluations of projects, proposals and personnel by scientific standards are an essential part of the scientific activity.

Faced with these charges, it is a temptation for the value-free enthusiast to react by saying that that position never involved denying that *this* kind of activity went on. It was rather that *moral* value judgments, or *ultimate* value judgments, or some other special kind of value judgment were being disbarred.

Charitable we should be, but it is well to remember that *the reasons* usually given for the value-free position belie this defense. The reasons given were usually something about the impossibility of getting value judgments from facts, which was combined with the assumption that science's natural basic concern was with facts. We have now argued not only that in the relevant sense one *can* get value judgments from facts, but also that some facts are valuational, *and* that science is necessarily involved in making value judgments as part of its *basic* concerns. So that position looks a little tatterdemalion. Alternative versions speak of science as concerned with *descriptions,* which cannot yield *recommendations*; or of science as concerned with *means* and not with *ends*—and perhaps this one deserves direct comment.

First, science *is* concerned with ends; value-base claims describé them and have always been conceded to be scientifically legitimate. The claim must be refined to read: "Science is concerned only with the *assessment* of means, not that of ends. Ends can only be judged by reference to further ends; and to do this is to judge them as means. Ends which can't be thus judged cannot be judged scientifically. It is the evaluation of these ultimate ends which must be excluded from science."

The best reply to this is to roll with the punch and simply add; there are no ultimate ends, hence science can assess all ends as well as all means. Hence this distinction imposes no limits on science. There *are* no ends which cannot be related to, assessed in terms of their consequences for other ends. Of course, justice and art are often, in fact, by a particular person pursued as "ends in themselves. But science or — in the more general case — reason is not restricted by the limitations of one person. Justice can be assessed in terms of its contributions to social harmony, individual safety, and so on." (And the reverse is true.) So all ends are assessable.

Some people have a very strong sense that systems of knowledge must have a basically axiomatic or hierarchical structure. There *must,* they feel, be some *basic* ends or axioms from which all the rest follow. The power of this model has driven philosophers to phenomenology and logical empiricism, and ultimately to skepticism, in the epistemological area — and to relativism in axiology (value theory). The defects of the model are well known. If the axioms are definitional, they can't lead to knowledge about the world; if empirical, they need further justification ad infinitum. It is clear that the hierarchical reconstruction of knowledge, though a useful artifice for some purposes, is fundamentally misleading. One must adopt a 'bootstraps' model, or — as I prefer to call it — a network model of mutually interconnecting and reinforcing nodes of knowledge, linked by both analytic and empirical connections, both to the ground (reality) and to other nodes. This extremely vague picture is nevertheless sometimes enough to break the hold of the hierarchy model, which — even when one can see that it won't work, for the reasons given a moment ago — still has a bewitching force. But justification in science and in law, is not a linear process; indeed, it is not in mathematics either, except within an *artificial* system.

So science has no need to assume ultimate ends to justify value judgments, since they *can* be justified in other ways. Justification is always a context-dependent process, only making sense insofar as it can connect the acceptable with the debatable, and we are never concerned in science to provide justifications whose premises are infallible, only those whose premises are acceptable to the rational person, i.e., immune to the sources of error that we are trying to avoid by means of a justification. So we justify claims about the carcinogenic properties of cigarettes by appealing to data that is more reliable; and similarly we can justify claims about the harmfulness of smoking. In neither case do we need to find premises that are beyond all possibility of error for eternity.

Hence, *even if* science were unable to justify claims about ultimate ends, it could justify value judgments in just the way it justifies non-value judgments.

But in fact science can also provide some ultimate premises from which value judgments can be inferred: and these are *definitionally true value judgments.* For example, it can reasonably be argued that a watch is a time-keeping device by definition and hence that being an *accurate* timekeeper will be a *merit* in a watch. (Similarly with regard to portability, legibility, etc.) We need only combine this value-premise with some performance data to generate value judgments. The whole idea that value judgments must be built—ultimately—on an *arbitrary* assumption becomes absurd with regard to such cases. Nor are they unusual. Methodological value judgments are *typically* built on definitional or quasidefinitional premises. The 'ultimate values' lying behind assertions about the merit of a particular hypothesis are claims such as "Good hypotheses explain or predict or summarize more phenomena, or do it more simply, than bad ones." Even if we have to elaborate this under the pressure of counterexamples, it's a pretty close approximation to a definitional truth.

Quite apart from its use as a basis for value judgments about specific hypotheses—for which purposes, of course, we have to combine it with 'performance' data about the latter—the simple fact is that propositions like the one just given are value judgments themselves and definitionally true. Hence they constitute an exceptionally powerful counter-example to the skeptical view about the objectivity of value judgments.

Both the specific and the definitional false judgments we are now discussing thus provide an independent line of attack on the doctrine of value-free science. And a simpler line than the one we elaborated as the CR model, where the 'ultimate value' was an empirical value-base claim. We covered that case first because it shows us much more about the real logic of evaluation than the present rather 'lucky' cases where the chain of justification strikes gold in the form of a definitional premise.

Another justification procedure besides the theoretical regress, the CR model, and the definitional one is the functional one. If one can identify the function of a device or social institution, then one can argue that the better it is performed, the better the device (etc.) is—other things being equal. The functional analysis will usually not be a matter of definition itself—though in the case of the watch it is—but it may be possible to establish it beyond reasonable doubt. It is easier with artifacts, harder with social institutions, as long experience in anthropology has shown; but it is a proper task of science to do functional analysis.

We can conclude this section with a few other brief arguments about the role of value judgments in science. The examples we have been discussing so far apply just as well to the purest of pure science as to applied science. Even the mathematician has to make methodological and performance value judgments. But one should of course add that the applied scientist must often make many more. When the cancer researcher has to evaluate proposed treatments, he or she gets into what should be a scientific procedure of identifying relevant criteria from value-base and general performance considerations. Then these criteria are combined with the specific performance data on the treatments, to yield value judgments. What is evaluated here is not itself a component of science; and the value-base data here do not refer to the values of scientists qua scientists, as they do when we are evaluating, e.g., the relative merit of bubble against cloud chambers for a particular kind of investigation. But—'external' to science though the object under evaluation may be—the *process* is (should be) entirely scientific.

It is of particular interest that in some few cases of applied science—medicine (especially psychiatry) and education are perhaps the most obvious examples—the evaluative criteria apparently *must* include some moral considerations. It doesn't make much sense to talk of the patient's overall improvement after psychotherapy, no matter how much better the patient feels, if his behavior towards others has become totally ruthless, even if he can get away with it; or to talk of a successful outcome of a remedial reading program which has incidentally indoctrinated the students with violent racist attitudes. These cases appear to be quite different from the commonplace situation in which the application of science, e.g., to pest control or breeder reactor design, will (probably always) raise moral problems. The apparent difference is that in these latter cases the criteria for scientific success, the definition of the problem in scientific terms, does not necessarily involve moral issues. The scientific problem, it appears, can be separated from the social problem of when to employ the scientific solution.

The distinction will not really hold up, but the correct way to treat the matter is not the traditional one. We could define a 'purely scientific' problem in the psychotherapy or education case, as in the reactor or agricultural case. But to do so is immensely inefficient, for it leads to the development of solutions—at great expense—that either cannot be used, or require expensive modification to be used. Applied science *involves* the skill of problem-specification. It takes a good applied scientist to identify and conceptualize all the relevant considerations that make up the problem. And since the effects of a pesticide on farm workers, crop-dusting pilots, river and bay aquatic life, and city water drinkers are serious, the *problem* is to create or discover one that minimizes these (and other) undesirable side-effects while maximizing the destruction of pests affecting the crops. In determining how heavily to weight the effects to be minimized, the scientist is entering the moral domain and *there is no one who can do that for him.* The value-free ideal presented a picture of the applied scientist as receiving the value judgments from a shadowy figure in the wings, adding these to the scientific parameters to get the specification of the problem, and then bringing in his or her scientific skills to create the solution. Now what of course happened in practice was that the scientist ignored any external values unless someone was yelling about them very loudly, specified what looked like the 'practical' problem ('kill pests') in scientific terms and solved that, if possible. I am not recommending getting the scientist into the moral issues on the grounds that they otherwise get under-emphasized, though that might independently be a good reason, but because the previous procedure was incompetent. The problem to be solved had been incorrectly specified, and the error did not lie with the 'layman', whoever that might be, but with the scientist, who knows very well from many sad experiences that one cannot usually rely on 'outsiders'' specifications of problems, whether the outsider is lay or a specialist in another area; one must understand the problems and analyze them oneself as part of the scientific project. True, there can sometimes be minions who can be given micro-problems to solve without being expected to check on the formulation, but we are talking about huge university research departments (in the agricultural case, for example), mostly working independently of each other. More of *them* should have been criticizing the over-limited formulation of the problems being proposed by industry even if such formulations could not be questioned by the chemists actually on the industries' payrolls.

It is, then, an (applied) scientific necessity to look into the moral dimension of problems because that dimension is part of the problem: it will be so, even if morality is just an arbitrary system of conventions.

The behavior modification enthusiasts in education could probably solve the 'scientific problem' of classroom discipline, by wiring students to an electro-shock network controlled from the teacher's desk, but this is simply not the correct formulation of the problem—and for moral reasons.

Bomb design and chemical-biological warfare provide extreme examples of the same point; but it should not be confused with the moral but not—in the same sense—scientific problem of whether such work is ever justified. Deciding *that* is an obligation on any person, scientist or not, engaged in such a project. Deciding how to conceptualize and weight destructive power of a bomb, a morally significant variable, is—on the other hand—part of the scientific problem.

An extreme case of great interest affects political science, although the 'empiricist' political scientists kept wishing it wouldn't. This is the problem of justification and criticism of the forms of government, such as democracy. It would take a book to document the conceptual confusion surrounding this issue over the past four or five decades, most of it stemming from the clash of the two facts: (a) justification involves values and hence on the prevailing views could have no place in political science, (b) if political science in a political system called a democracy wasn't prepared to discuss the question of the merits of political systems, including democracies, it could in no way deserve its title. After all, the noblest task of physics has always been the critical scrutiny of the prevailing system of physics, of psychiatry the prevailing concept of mental health, etc. The touching idea that such matters could be left to political philosophy didn't seem quite satisfactory; after all, had Einstein or Schrodinger left the questions of the foundations of their subjects to philosophers of science? Should they have? Obviously not. The justification of basic positions in a science is a task for *both* the scientist and the philosopher.

Thus I am here suggesting a considerable broadening of the scope of science, compared to the neo-positivist conception. It could be illustrated with a dozen different examples. It is nothing short of tragic to read van der Graaf's farewell to welfare economics, a subject he abandoned after making significant contributions because he saw that the basic problems could not be handled without leaving the realm of science for the realm of value theory, which he took to be forbidden territory. The literature of theoretical psychiatry is rent by the cleft between people like Szasz who see the moral dimension as an essential part of the problem, and those who see it as scientifically irrelevant.

VIII. Ethics as a Science

The most powerful way to prove that value judgments have a place in science is to prove that ethics is a science. I believe it is, potentially at least, though not for the usual reasons. Given the preceding arguments, it is clear that one can construct a type of ethics by addressing the applied social science problem of determining the optimal set of social rules and attitudes for a society facing given economic, psychological, and environmental constraints, where 'optimal' is defined pre-morally, i.e., only with regard to the value-bases of the elements in the society. Since the system resulting will probably cover about the same domain of behavior, etc., as what is traditionally called ethics, and will involve about as many of the traditional moral precepts as any two traditional ethical systems share, and since it will not involve anything notably different, it is entitled to be called ethics. And since it can be scientifically justified, it's entitled to be viewed as the only defensible system of ethics, alternative justifications for ethics having been long since exposed as untenable. (Detailed support for this long string of controversial assertions is suggested in *Primary Philosophy,* McGraw-Hill, 1966).

Put the matter another way. It is part of science, especially sociology and anthropology, to do functional analysis of social institutions. Very well; let one be done of the ethical system of a given society. It is part of science to identify the value-base of social beings; let this be done for the same society. It will now be *entailed* that some of the ethical rules of that society cannot be justified with regard to their needs/wants/values; or that other parts can; or both. If, for example, their rule against killing is functional, then it's justified and it *is* wrong to kill in that society. (Note that any society interacting with others cannot have the functionality of its rules determined only by considering its own value-base, since disregard of the other value-base may well lead to disastrous results.)

Of course, there are many troublesome details that could turn into disasters for the above account, though in fact I believe they do not (and so argue in the reference given). But I would stress the following warning. The crudest common-sense consideration makes it obvious that there is an excellent justification for a society to have a system of law. The simplest science reveals that much of a given system of ethics can be treated as an internalized extension of law. Hence, unless a clear proof of impossibility can be given, it's obvious that a good slice of ethics can be social-scientifically justified and (probably) another slice rejected, which also can be justified. Incidentally, the justification for, e.g., accepting or banning adultery and homosexuality, certainly requires scientific investigation of their consequences. Conversely, that scientific investigation plus one into the value-base of the society, provides a *prima facie* reason for (or against) the existing moral rule, i.e., a scientific justification for (a part of) ethics. I repeat that this isn't a proof, it's a sketch. My criticism is that scientists have for decades been trained in a way that made them incapable of seeing the possibility of such an argument, *even though* they were not trained to see what was wrong with it. That is, they thought the idea of ethics as a science was absurd, but had no good reasons for the position. It is no wonder that, without cracking a smile, without any sense of absurdity, without ever looking at the possibility of self-refutation, they taught their students: "It is scientifically improper to make value judgments." They might as well have said, "Thou shalt not ever say: 'Thou shalt not ever say'."

One should not close this section without stressing that philosophers were not much better off. In the fifties, the 'in' view was emotivism or non-cognitivism, the doctrine that value judgments were mere expressions of feeling, lacking any propositional content at all, and hence incapable of verification or falsification. Even today, tawdry 'refutations' of pragmatism or utilitarianism are bandied about without serious examination, and most philosophers still think the so-called naturalistic fallacy of Moore and the classic argument of Hume make the facts/value or ought/is distinction secure. But John Searle's crusade has shaken this stance lately (in *Speech Acts,* Oxford, 1971) and to some extent Rawls' *Theory of Justice* (Harvard, 1972) and the 'good reasons' school in ethics have provided an alternative approach.

IX. The Bad Reasons

In the end one has to ask how it came about that scientists should have accepted a view of their own subject which is so patently unsound. There are, I think, many contributory factors. One major factor is the very poor training in self-scrutiny that characterizes most scientists' background, something that becomes embarrassingly obvious when they move beyond their own specialty without noticing that useful precepts of scientific practice in quantum physics are not 'the scientific method' and do not apply at all in certain other areas of science or non-science. (I have in mind here the particle physicist's tendency to assume the universal merit of the frequency theory of probability, or the macro-physicist's insistence upon repeatability as a criterion which makes para-psychology unscientific.) The lack of concern with the history of e.g., psychology, in which many issues fought over later were comprehensively disposed of (the 'puzzle' of inversion of the retinal image, for example) is another index of lack of self-scrutiny by the scientists.

Allied with this innocence was the powerful pressure of the desire that science be value-free, a desire held for much the same reasons that Weber had when he first proposed the thesis. The values area is messy, controversial, and you can get on the wrong side and hurt yourself politically. Why rock the boat? The society has treated science well, or tolerated its extravagances. Why start criticizing *it*?

Of course, to concede on values in science is not to concede on politics in science; but perhaps scientists instinctively felt it was better to defend the forward position and fall back if necessary. But when the enemy gets the momentum of attack going, he doesn't wait for you to regroup or even prepare the long-neglected yet defensible position. And indeed you find that you have forgotten exactly where it is.

On the other side, the arguments were at least as bad. If the doctrine of value-free science is now in considerable disrepute amongst the intelligentsia, it is mainly for an irrelevant reason. That is, what has brought it down is the recognition of the huge social costs and commitments of science. "Science is not value-free," the radical is fond of saying: "Its values are those of the establishment." True enough, true of scientists, and of science as a social phenomenon. But never denied by the value-free supporter, who was only arguing that the *content* of science is value-free, not that its *effect* has no social significance. Nor was the claim ever made that scientists could somehow choose to be scientists (or physicists, etc.) without having and thereby exhibiting their values; the claim was that *after* that choice had been made, while they were in the laboratory, they were free of the need (or possibility) of making value judgments *in the name of science.*

Many crimes have been committed in the name of science, but none so serious as this withholding of that name.

Berkeley, Calif.

NOTE

1. The differences are of some interest, although too limited to concern us now. For example, one may talk of valuing someone's friendship, where one would not talk of liking it, and one may like a view without valuing it. Valuing involves an element of enshrining, of respect. But our first task here is brush cutting, not pruning.

STUDY QUESTIONS FOR PART FIVE

1. B. F. Skinner (in *Beyond Freedom and Dignity*) claims that the survival of our culture is an ultimate value, and that all other values, e.g., freedom, are only instrumental to this intrinsic value. (He often argues that we can discover, "experimentally," what is valuable.) What do you think of this claim? What are some of its implications?

2. Many authors (including Skinner) think that values do not embody knowledge claims—so that ethics is not the search for moral knowledge—but rather are concerned with what Dewey calls "experiments in living." It is often held that on this view science and ethics become very closely connected, especially for those (such as Dewey and Skinner) who think of science itself as an instrument for problem-solving, and not as a search for truth or knowledge in some abstract sense. Can you think of any kind of values, and any kinds of "experiments in living," which might illustrate and lend support to this view?

3. What does the (alleged) distinction between "pure" and "applied" science have to do with the question of whether science is value-free?

4. In what sense might it be correct to say that knowledge and truth are, or are manifestations of, moral ideals? *What* moral ideals?

5. What are the connections between the idea of value-neutral science and ethical relativism? Give some examples. How do these views support the idea that value judgments are merely scientifically explainable features of a culture or an individual's "behavioral repertoire"?

6. A number of recent authors suggest that ethical criteria should be used to decide whether a proposed hypothesis or experiment should be tested or pursued, the assumption being that some hypotheses should not be, if it turns out they are either intrinsically or potentially immoral. Is there such a thing as forbidden knowledge or truth? For instance, G. B. Shaw said: "No man is allowed to put his mother in the stove because he desires to know how long an adult woman will survive the temperature of 500°, no matter how important or interesting that particular addition to the store of human knowledge may be" (*The Doctor's Dilemma*). Do you agree or disagree with this view?

7. In light of the question above, what are we to say to critics of, say, Darwin's theory, or Arthur Jensen's view that intelligence is genetically inherited, who argue that these views must be false because they are, say, incompatible with accepted religious beliefs, or are racist, or anti-egalitarian, and so forth?

8. Read A. Gerwirth's article, "Positive Ethics and Normative Science." Briefly summarize his argument and explain how this combination of ideas is an outgrowth of the concept of objectivity discussed in the introduction to this section.

9. Rescher's main theme is that "evaluative, and more specifically *ethical,* problems crop up at numerous points within the framework of scientific research." Summarize at least three ethical problems that occur within scientific research, explaining why they are *ethical* problems. Can you think of any ethical problems of scientific research which Rescher has missed? Does the existence of

ethical problems within science cast any doubt on the belief that science is a rational enterprise which yields "objective" results? Why or why not?

10. The "fundamental problem" that Hempel addresses in his essay concerns the extent to which "valuational questions" can be answered by means of the "objective methods of empirical science." In particular, he asks whether these methods can serve to "establish objective criteria of right and wrong and thus to provide moral norms for the proper conduct of our individual and social affairs." Briefly discuss the relevance of the following to Hempel's discussion of these questions: (a) instrumental judgments of value; (b) categorical judgments of value; (c) rational choice; (d) "absolutely" ultimate ends and values. Do you think that Hempel is an advocate of the "objectivist" theory of science and values discussed in the introduction to this part? Explain and defend your answer.

11. At the end of his article, Rudner says, "What is being proposed here is that objectivity for science lies . . . in becoming precise about what value judgments are being and might have been made in a given inquiry—and even . . . what value decisions ought to be made; in short that a science of ethics is a necessary requirement if science's progress toward objectivity is to be continuous." Briefly explain what Rudner means by this remark, and summarize his argument for it.

12. Scriven tries to argue that science is "essentially" evaluative, in that there could not be any science at all unless it makes what he calls "real value judgments" (i.e., what he calls "Type III" judgments of value) all the time. Briefly summarize and evaluate Scriven's defense of this claim, taking into account the examples he gives to support it. (You might want to bring in examples of your own, to either support or challenge Scriven's claim.) What, according to Scriven, are some of the "bad reasons" advanced pro and con the view that science is value-free? *Are* these "bad reasons"?

SELECTED BIBLIOGRAPHY

[1] Bronowski, J. *Science and Human Values.* New York: Harper & Row, 1965, rev. ed. A seminal work on the interconnections and parallelisms between science and values.

[2] Caws, P. *Science and the Theory of Value.* New York: Random House, 1967. An interesting attempt at providing a systematic "naturalistic" account of value judgments, in the tradition of Dewey's *Quest for Certainty.*

[3] Diesing, P. *Reason in Society.* Urbana: Univ. of Illinois Press, 1962. An overview of the role of the cost-benefit model of rationality in economics, sociology, law, political science, management science.

[4] Edel, A. *Ethical Judgment.* New York: The Free Press, 1955. A useful, but in some ways outdated, discussion of the relationship between results in the various special sciences and the issue of ethical relativism.

[5] Gerwirth, A. "'Positive' Ethics and 'Normative' Science." *Philosophical Review* 69, no. 3 (1960): 311–330. A provocative critique of those who distinguish between good and bad science, but who fail to distinguish between good and bad ethical views and theories.

[6] Polanyi, M. *Personal Knowledge.* New York: Harper & Row, 1958. A stimulating account of knowledge which argues (among other things) that truth is a moral ideal and that tacit knowledge is more pervasive and fundamental than many accounts recognize.

[7] Riley, G., ed. *Values, Objectivity and the Social Sciences.* Reading, Mass.: Addison Wesley, 1974. An excellent collection of recent and some classical discussions of the issue of value-neutrality in the social and policy sciences.

[8] Skinner, B. F. *Beyond Freedom and Dignity.* New York: Knopf, 1971. A forceful and provocative attempt to combine a utilitarian account of morality with a basically operationalist view of science in defense of a revaluation of our culture's values.

Part Six

Science and Culture

Part 6: Science and Culture

Introduction

I. Science, Purpose and Morality

Since the general topic of science and culture takes in far too many issues to be even mentioned in an introduction, it will be necessary to limit this discussion to one or two central issues, so as to make the readings intelligible. The most obvious place to begin is with some of the issues that were mentioned, but not discussed, in the introduction to Part 5.

The first issue concerns the notion of purpose or order in the universe. As was mentioned earlier, for the Greeks it is impossible to eliminate any reference to purpose, ordering, patterning, beauty or the good in trying to understand the cosmos and our place in it, since what is natural is good. To be sure, these are not ("subjective") *human* purposes or goods, although knowing the universe is needed to know what is good for humans, and what is a human purpose or "function." But the point remains: The mechanical picture of objectivism begins by eliminating purpose from science, because the world is seen as devoid of it. Later developments, e.g., in the social sciences, try to eliminate or "reduce" human purposes and values, or else treat these as merely subjective labels which we tack on to an objective world. Hence, critics of science and its effect on our culture (see Roszak) take modern science to portray the world, and human life, as devoid of purpose or meaning. Such views often appeal to the platonic vision as an alternative to this picture, which, they claim, gives rise to the idea that life is meaningless, and reflects the extent to which modern life is dehumanizing, alienating, and immoral or amoral.

The second issue concerns the relationship between science and morality, especially the relationship between the uses (and abuses) of science and our moral beliefs and principles. As we have said, some authors hold the view that knowledge, truth, and inquiry—and thus science itself—either express or are motivated by moral ideals, such as honesty, fairness, and truthfulness. This dimension of science is addressed by Rescher in Part 5, and will not be pursued further. It is to the moral issues concerning the uses of science and technology that we shall now turn.

A convenient point of departure into this important and complex topic is furnished by our earlier discussion of the "objectivist" accounts of value. This account, which is basically nihilistic, takes two forms: (A) Value judgments express arbitrary personal acts of the will or commitments, which are irrational and subjective, or social commitments with the same status; or (B) Value judgments are a species of scientific statement which explain or describe personal and social behavior (e.g., in terms of some variant of deterministic laws in history, psychology, sociology, economics, and so forth). The only real (nonexplanatory) function of "objective," "rational" value judgments is to predict or calculate the best means of achieving a given end or purpose. Such ends are, on either (A) or (B), to be taken as given, and certainly as beyond rational dispute. In either case, morality leads to a kind of relativism, or else to the idea that (at least) *ultimate* values cannot be challenged or even discussed in any rational manner.

Nihilism about values thus dovetails with the idea that the universe, and thus life itself, is meaningless or devoid of purpose. At least this is the way these two ideas are connected for critics of modern "scientism" and "objectivism." ("Scientism," which is being used here as a rough equivalent to "objectivism," is often defined as the view that science and its methods provide universal standards for rationality and objectivity, as well as universally valid beliefs about the world.) The "bottom line," so to speak, of this outlook, is that "'tis not contrary to reason to prefer the destruction of the world to the scratching of my little finger," which is the eighteenth-century philosopher David Hume's claim about the relation between the ultimate "facts" about human nature and the role of reason in morality.

What does this claim really mean? One thing it seems to lead to is the idea that if our ultimate values and purposes cannot be rationally discussed or evaluated, but must be taken as either conceptual axioms or conventions or as biological, psychological, historical, or personal "ultimates," "reason" only plays an "instrumental role" in life. This amounts to the idea that value judgments only tell us *how* to maximize our goals; they cannot help us to decide which goals to pursue—and which goals not to pursue. As a result, the criteria of morality and those of efficiency or expediency become merged. Coupled with the scientific attempt to reduce or eliminate human purposes to some kind of "objective" variables—e.g., economic or psychological—we are faced with a picture of values and purposes in which the ultimate values and purposes of a person or a group must be accepted as brute facts or ultimate postulates. Thus, many scientists and philosophers interpret the ultimates as facts about human psychology (all human actions are motivated by the desire for pleasure or the avoidance of pain), or else as facts about the value of a culture's survival. Given these assumptions, value judgments become the search for "objective" evidence about what will produce a maximum of pleasure (or a net balance of pleasure over pain) for a given individual or a society. Or ethics becomes the search for principles which tell us how to predict what the rational means are for promoting the survival of a culture. In short, "practical success" or efficiency, measured by standards that purport to be scientifically justified, becomes the hallmark of the adequacy and, indeed, the very meaning of values. Utilitarianism, cost-benefit models of decision-making, and so on, operate upon the assumptions embedded in this view. The only alternative to it (i.e., to (B) above) is supposed to be (A): the idea that value judgments are totally irrational acts of subjective preference, which can no more be rationally assessed than can one's preferences for chocolate ice cream or red convertibles.

II. The Cultural Impact of "Scientism" and "Objectivism"

Two vivid expressions of the connections between these issues about values and the larger themes raised by the label "science and culture" find expression in the writings of Orwell (*1984*) and the psychoanalyst Erich Fromm. Fromm says this:

> [Technological civilization] is programmed by two principles that direct the efforts and thoughts of everyone working in it: The first principle is the maxim that something *ought* to be done because it is technically possible to do it. If it is possible to build nuclear weapons they must be built even if they might destroy us all. . . . This principle means the negation of all values which the humanist tradition has developed. This tradition says that something should be done because it is needed for man, for his growth, joy, reason, because it is beautiful, good, true. Once the principle is accepted that something ought to be done because it is technically possible to do it, all other values are dethroned, and technological development becomes the foundation of ethics.

The second principle is that of maximal efficiency and output. (See [9], p. 15.) According to this view, the understanding of values and the cost-benefit model of rational decision-making are but manifestations of a turning away from the humanist tradition—roughly speaking, the tradition embodied in the Platonic ideals outlined in the introduction to Part 5, and carried through into the religious, philosophical, and common-sense tradition of Western culture. The scientific world picture, as outlined in the introduction to Part 5, leads to a situation in which the only "objective" variables are those which can be treated scientifically, which means only those variables which can be controlled and manipulated. Coupled with the understanding of value judgments that is attendant upon this idea of objectivity, Western culture comes to be dominated by the "technological imperative" (of Fromm's first maxim).

According to critics, science was originally rooted in certain practical, moral, and purposive contexts. Seen in its larger context, scientific knowledge was valued instrumentally, as a means for realizing the practical and moral ideals of civilization. Later, in "modern" (Western) history, science lost its roots in such concerns and began to develop values and techniques which proceeded in a cultural vacuum. Once this was done, this account continues, the progress of science came to be measured by standards that actually undermined the social utility of knowledge, because science and its values came to be seen as ends in themselves, rather than as means to human goals such as happiness and well-being.

To be more specific, two assumptions influenced this change: the assumption that knowledge is power (pragmatism) and the assumption that reality can be defined in terms of what can be measured and controlled by human beings (this assumption is shared by objectivism *and* its critics). When science was still a practical instrument for social and moral progress, these two assumptions could be used as standards for the use of knowledge. The more we learned, the more power we had to bring about the ends which science was designed to fulfill. However, suppose we imagine that science becomes isolated from the social and moral concerns which were to guide its developments. Then these two assumptions are no longer merely instrumentally valuable as means to ends; but they become ends in themselves, i.e., knowledge, power, and the ability to control nature become ends in themselves. Once the social and policy sciences develop, the extension of these values amounts to the idea that it is intrinsically valuable to be able to control human behavior. Coupled with the assumption that all there is to things is what can be measured and controlled, the uses of the social sciences and technology come to be guided by the maxims Fromm discusses. The result is that science and technology come to control our lives, instead of functioning as means to human ends. This is alienation in a nutshell.

For instance, on the assumption that scientific knowledge should be seen in a larger social framework, the ability to build more sophisticated weapons would have to be justified in terms of whether such activities further what Fromm calls "humanistic values." But on the assumption that the ability to produce technological improvements is an end in itself, i.e., worthwhile for its own sake, the very ability to build such weapons becomes a justification for doing so. Taken together with certain other assumptions, e.g., that science and technology are either morally neutral or a substitute for morality, or that values and ends cannot be rationally evaluated, or that the expedient and the right are the same, we come full circle to Fromm's initial remarks.

Another avenue to the same point is furnished by the following remarks of Orwell, which relate to the way in which science and technology have become ideologies for a totalitarian society in *1984*. According to Orwell:

Science and technology were developing at a prodigious speed, and it seemed natural to assume that they would go on developing. This failed to happen . . . partly because scientific and technological progress depended on the empirical habit of thought, which could not survive in a strictly regimented society. . . . The search for new weapons continues unceasingly, and is one of the very few remaining activities in which the inventive or speculative type of mind can find any outlet. In Oceania at the present day, Science, in the old sense, has almost ceased to exist. In Newspeak there is no word for "science." The empirical method of thought, on which all the scientific achievements of the past were founded, is opposed to the most fundamental principles of Ingsoc. And even technological progress only happens when its products can in some way be used for the diminution of human liberty . . . in matters of vital importance— meaning, in effect, war and police espionage—the empirical approach is still encouraged, or at least tolerated. The two aims of the party are to conquer the whole surface of the earth and to extinguish once and for all the possibility of independent thought. There are, therefore, two great problems which the party is concerned to solve. One is how to discover, against his will, what another human being is thinking, and the other is how to kill several hundred million people in a few seconds without giving warning beforehand. In so far as scientific research still continues, this is its subject matter. The scientist of today is either a mixture of psychologist and inquisitor, studying with extraordinary minuteness the meaning of facial expressions, etc. and testing the truth-producing effects of drugs, shock therapy, hypnosis and physical torture; or he is a chemist, physicist or biologist concerned only with such branches of his special subject which are relevant to the taking of life. . . .

Here, in quite a striking passage, is the logical outcome of Fromm's maxims. What has gone wrong? To belabor the obvious, let me mention some of the factors that perverted science and technology. For seeing that they have been perverted will, we think, sustain the view that there is nothing instrinsically or inevitably dehumanizing, manipulative, or alienating in them.

1. The empirical habit of thought is stifled; i.e., free scientific inquiry, diversity of opinions, lack of authoritarianism, and the need for independent evidence are lost. Science is an ideological tool which must serve as a means to preordained goals.

2. The loss of free thinking—which is considered both useless and dangerous—turns science into a purely practical concern. But there are no standards to tell whether "what works" is good or bad. Rather, the standards for utility are the survival of the society and the perpetuation of its goal, no matter how morally objectionable.

3. Scientific research is not, although it might be claimed to be, morally neutral; and the search for truth, together with democratic procedures for

finding it, are absent in science and the rest of society. (This is a classic illustration of how the open society and free scientific inquiry stand or fall together.)

4. What is real is a function of what is manipulable; indeed, one of the doctrines in *1984* is that reality is what is in the mind of the party; i.e., subjective idealism, coupled with the lack of any independent evidence, or independent thought, conspire to keep people in ignorance, and thus in slavery. (Idealism is actually the product of objectivism's view that reality is defined by what can be manipulated and controlled.)

5. Morality and truth are defined in terms of what some authoritarian fiat defines as "moral" or "true." That is, an extreme form of relativism, coupled with an equally radical variant of idealism, lead to the absence of both knowledge and morality. Whatever is expedient for the stated aims of the party become true and moral.

6. Science and technology are no longer paths to utopia, nor even important in the practical affairs of life for everyone, but are the property of dictators and their lackeys.

To summarize: Science and technology, according to these interpretations, have come to be dominated by ideals of rationality, values, and objectivity that are rooted in the conceptions of knowledge as power and of reality as a system of manipulable variables. When dislodged from the humanist tradition, or when attached to a social and political system guided by an ideology of domination, the technological maxim takes on a life of its own; i.e., it becomes a self-sustaining and self-certifying criterion of rational choice and action — an intrinsic value is placed upon control, technique, and efficiency, or at any rate technological or instrumental rationality comes to dominate all thinking, in which case we effectively get the same result.

Other factors operating in our culture play a role in these developments: e.g., (a) the loss of "purpose" or "meaning," (b) the nihilistic account of morality embedded in both the objectivist and subjectivist variants on nihilism (i.e., (A) and (B) above), (c) the social division of labor and responsibility [(c) often leads to the use of evasions to avoid responsibility and ignore the misuses of science and technology and the emphasis on expediency and efficiency that are so pervasive in our society]. The prevalence of these factors provide additional support for Fromm's remark that the "technological maxim" underlies our conscious and unconscious behavior to an extent we often are unwilling or unable to recognize.

III. The "Counterculture" Alternative to "Scientism"

As an alternative to "scientism" and "objectivism," critics (such as Roszak and Fromm) often appeal to the "humanistic" tradition, embodied in the platonic vision articulated in the introduction to Part 5. Underlying this ideal is a view of the cosmos and human life in which order, purpose, and "the good" are ineliminable for purposes of both understanding and action. The attempt to understand the cosmos, and the idea that knowledge is power, are completely separated. Even if we could dominate and control, either as part of, or in addition to, the goal of understanding the world, we should not do so. At best, we should let things be; at a minimum, we should live in such a way as not to manipulate and dominate nature or other persons or cultures in an effort to achieve knowledge and power, even if our motives are noble. (Often they are far from noble! A corollary of this ideal, of course, is an end to the ideal of human domination over nature, or the assumption that other forms of life have value only if they serve human purposes.)

The religious themes underpinning this view are fairly obvious, despite the fact that many of its proponents are critical of the biblical idea that man is created in God's image and therefore is entitled to rule the earth. Interestingly enough, this theological idea is the guiding principle behind the Baconian conception of science, which forms the basis of modern utopianism—up to and including Skinner's *Beyond Freedom and Dignity* and recent efforts toward genetic engineering of the course of evolution. (Such "utopian" efforts are a combination (roughly speaking) of what has been referred to here indifferently as "scientism" and "objectivism" and some form of a utilitarian or cost-benefit model of value judgments and rational choice.)

It is thus no accident that critics of recent manifestations of this combination of religious utopianism and technological rationality use phrases such as "playing God" and "the divine right to pursue research" in their attacks. However, critics of science, such as Roszak, owe us a more detailed account of the alternatives they propose and of why their view is more in keeping with the humanist tradition than those of, say, Skinner or genetic engineers, given their (evident) common roots.

One dimension to this debate is the question whether knowledge, or at least modern science, is intrinsically evil. Roszak and other advocates of a "countercultural" alternative to "scientism" seem to give an affirmative response to this question, perhaps echoing the Old Testament story about the fall from Eden, precipitated, of course, when Eve eats from the tree of knowledge. Rousseau (the eighteenth-century philosopher), in his *Discourse on the Arts and Sciences,* develops this idea further. (Rousseau is often labeled a "Romanticist," as are Roszak and others.) The sciences are grounded in human vanity, pride, idle curiosity, and egotism; they are the products of a corrupt and morally decadent civilization, in which the simpler, nobler virtues—family, citizenship, religion, love and the simpler life—are stifled, and in which "natural man" is distorted beyond recognition. While this theme is echoed in anti-utopian literature, e.g., in Huxley's *Brave New World,* whose (anti?) hero is known as the (noble?) Savage, it is also a main theme of Skinner's *Walden Two,* a utopian commune designed by behavioral engineers. This is another piece of evidence for the ambiguous legacy of our religious and philosophical tradition of humanism.

Another attitude toward these developments is the idea that there is nothing intrinsically evil, dehumanizing, or alienating about knowledge, science, or technology. This is the view of Orwell and Fromm, as well as respondents to Roszak (see the L. Marx reading). It is true that science and technology have been misused. But there is nothing unavoidable about this, so that if we can come (through political and moral means) to control the uses of technology and science, and if scientists and the people develop critical attitudes concerning its uses, all may not be lost. (This route is not without its own problems, e.g.: the politcal problem of democratizing the decision-making process in spite of a growing tendency toward a technocracy governed by cost-benefit criteria and Orwellian techniques of manipulation and control; the need for an educated public, together with scientists committed to accepting responsibility for their work, which means refusing to look the other way or make up excuses to avoid responsibility for what they are doing.)

Still others seek to steer a mid-course between these two views. On this account, knowledge is not pure contemplation; nor is it defined along crudely pragmatist lines. Rather, knowing, making, and doing are inseparable components by which we must comport ourselves in the world, while at the same time letting things be, i.e., adapting ourselves to our natural environment as best we can. Perhaps this approach, if we have time enough to develop it and act upon it, may prove the best hope for the future. In any event, one place to begin thinking about these issues is from a broad historical perspective on the humanist, religious, and scientific traditions that have molded our culture and which still, in ways too often unnoticed, shape its continued development.

IV. The Readings

Roszak challenges the ideals of objectivity, rationality, and knowledge—he calls it "argumentative knowledge"—embedded in modern science, on the grounds that it denudes the universe, and thus human life, of meaning or purpose. We must, he holds, supplement these scientific ideals with those of the humanist tradition (as exemplified, e.g., in the Platonic ideals mentioned in the introduction to Part 5, as well as the "humanism" of Fromm), which emphasize the qualitative and purposive dimensions of experience, and which articulate an augmentative view of knowledge.

Leo Marx, in his reaction to Roszak's claim that the ideals of knowledge, objectivity, and rationality in modern science render the universe meaningless, and are immoral for other reasons as well (e.g., because of the connections between modern science, capitalism, and industrialization), outlines the historical background of humanism and romanticism, and thus puts Roszak's view within a broad historical context. Finally, Baier takes up the question whether science, and especially scientific explanations, do render the universe less intelligible, or meaningful, than religious explanations. He further discusses the idea that human life becomes meaningless, and morality impossible, within a scientific outlook. He tries to show that morality, and indeed the value of life, take on even more significance within a scientific world view, if only because life in "this world" is not denegrated, as it is on the religious view of things.

The reader may note that the issues raised in this section take us full circle, back to the questions about the relations between science and nonscience, to issues about explanations and purpose, and to questions about values and morality. At the same time, they place many of these issues into a unified historical perspective, so as to enrich our thinking about them. This is at least a first step in any attempt to come to grips with the role of science in contemporary society.

R.H.

21

The Monster and the Titan:
Science, Knowledge, and Gnosis

Theodore Roszak

The title or the book was *Frankenstein*. The subtitle was *The Modern Prometheus*.

An inspired moment when Mary Shelley decided that a maker of monsters could nonetheless be a Titan of discovery—one whose research might, in our time, win him the laurels of Nobel. She claimed the story broke upon her in a "waking dream." It may well have been by benefit of some privileged awareness that one so young fused into a single dramatic image the warring qualities that made Victor Frankenstein both mad doctor and demigod. A girl of only nineteen, but by virtue of that one, rare insight, she joined the ranks of history's great myth makers. What else but a myth could tell the truth so shrewdly, capturing definitively the full moral tension of this strange intellectual passion we call science? And how darkly prophetic that science, the fairest child of the Enlightenment, should find the classic statement of its myth in a Gothic tale of charnel houses and graveyards, nightmares and bloody murder.

Asked to nominate a worthy successor to Victor Frankenstein's macabre brainchild, what should we choose from our contemporary inventory of terrors? The bomb? The cyborg? The genetically synthesized android? The behavioral brain washer? The despot computer? Modern science provides us with a surfeit of monsters, does it not?

I realize there are many scientists—perhaps the majority of them—who believe that these and a thousand other perversions of their genius have been laid unjustly at their doorstep. These monsters, they would insist, are the bastards of technology: sins of applied, not pure science. Perhaps it comforts their conscience somewhat to invoke this much muddled division of labor, though I must confess that the line which segregates research from development within the industrial process these days looks to me like one of gossamer fineness, hardly like a moral *condon sanitaire.*

I realize, too, that there are some—those who champion a "science for the people"—who believe that mad doctors are an aberration of science that can be wholly charged to the account of military desperados and corporate profiteers. Their enemies are also mine; I write in full recognition of how the wrong-headed power elites of the world corrupt the promise of science. But I fear there are more unholy curiosities at work in their colleagues' laboratories than capitalism, its war lords, and hucksters can be made the culprits for. Certainly they must share my troubled concern to see the worst excesses of behavioral psychology and reductionist materialism become unquestionable orthodoxies in the socialist societies.

I will grant to both these views some measure of validity (less to the first, much more to the second). But here and now I have no wish to pursue the issues they raise, because I have another monster in mind that troubles me as much as all the others—one who is nobody's child but the scientist's own and whose taming is no political task. I mean an invisible demon who works by subtle poison, not upon the flesh and bone, but upon the spirit. I refer to the monster of meaninglessness. The psychic malaise. The existential void where modern man searches in vain for his soul.

Of course there are few scientists who will readily accept this unlovely charge upon their paternity. The creature I name wears the face of despair; its lineaments are those of spiritual desperation; in its bleak features scientists will see none of their own exhilaration and bouyant morale. They forget with what high hopes and dizzy fascination Victor Frankenstein pursued his research. He too undertook the adventure of discovery with feverish delight, intending to invent a new and superior race of beings, creatures of majesty and angelic beauty. It was only when his work was done and he stepped back to view it as a whole that its true—and terrifying—character appeared.

The pride of science has always been its great-hearted humanism. What place, one may wonder, is there in the humanist's philosophy for despair? But there is more than one species of humanism, though the fact is too often brushed over. In the modern West, we have, during the past three centuries, run a dark, downhill course from an early morning humanism to a midnight humanism; from a humanism of celebration to a humanism of resignation. The humanism of celebration—the humanism of Pico and Michelangelo, of Bacon and Newton—stems from an experience of man's congruency with the divine. But for the humanism of resignation, there is no experience of the divine, only the experience of man's infinite aloneness. And from that is born a desperate and anxious humanism, one that clings to the human as if it were a raft adrift in an uncharted sea. In that condition of forsakenness, we are not humanists by choice, but by default—humanists because there is nothing else we have the conviction to be, humanists because the only alternative is the nihilist abyss.

If I say it is science that has led us from the one humanism to the other, that it is science which has made our universe an unbounded theater of the absurd . . . does that sound like an accusation? Perhaps. But I intend no condemnation, because I believe that, at every step, the intentions of the scientists have been wholly honest and honorable. They have pursued the truth and followed bravely where it took them, even when its destination became the inhuman void. In any case, I say no more than thoughtful scientists have themselves recognized to be true—in some cases with no little pride. Thus, Jacques Monod:

> By a single stroke [science] claimed to sweep away the tradition of a hundred thousand years, which had become one with human nature itself. It wrote an end to the ancient animist covenant between man and nature, leaving nothing in place of that precious bond but an anxious quest in a frozen universe of solitude.[1]

Or, as Steven Weinberg puts it elsewhere in this volume [*Daedalus,* vol. 103 (1974)]:

> The laws of nature are as impersonal and free of human values as the rules of arithmetic. We didn't want it to come out this way, but it did. . . . The whole system of the visible stars stands revealed as only a small part of the spiral arm of one of a huge number of galaxies, extending away from us in all directions. Nowhere do we see human value or human meaning.[2]

Our universe. The only universe science can comprehend and endorse. "A universe," Julian Huxley has called it, "of appalling vastness, appalling age, appalling meaninglessness." But not for that reason, an uninteresting universe. On the contrary, it is immensely, inexhaustibly *interesting.* There is no reason, after all, why what is wholly alien to us should not be wholly absorbing. Nor is there any reason why, in such a universe, we should not make up meanings for ourselves—whatever meanings we please and as many as we can imagine. Is this not the favorite preoccupation of modern culture, the intellectual challenge that adds the spice of variety to our lifestyle? We may even decide to regard science itself as the most meaningful way of all to pass the time. All we need remember—if we are to remain scientifically accountable— is that none of these meanings resides in nature. They simply express a subjective peculiarity of our species. They are arbitrary constructions having no point of reference "Out There.' Which is to say: the universe we inhabit— insofar as we let it be the universe science tells us we inhabit—is an *inhuman* universe. We share some minute portion of its dead matter, but it shares no portion of our living mind. It is (again to quote Jacques Monod) an "unfeeling immensity, out of which [man] emerged only by chance. . . ." and where "like a gypsy, he lives on the boundary of an alien world, a world that is deaf to his music, just as indifferent to his hopes as it is to his suffering or his crimes.''

Perhaps not every reader agrees with me that meaninglessness is a monster. If not, then our sensibilities are of a radically different order and we may have to part company from this point forward, for this is not the place to try closing the gap between us. But I believe more than a few scientists have looked out at times upon the "unfeeling immensity" of their universe with some unease. Note Weinberg's phrase, "We didn't want it to come out this way. . . ."

Perhaps not every reader regards the degradation of meaning in nature as a *moral* issue. But I do. Because meaninglessness breeds despair, and despair, I think, is a secret destroyer of the human spirit, as real and as deadly a menace to our cultural sanity as the misused power of the atoms is to our physical survival. By my lights at least, to kill old gods is as terrible a transgression of conscience as to concoct new babies in a test tube.

But even if scientists should agree that their discipline buys its progress at a dear price in existential meaning, what are they to do? Steven Weinberg faces the question squarely in his essay and offers an answer which would, I suspect, be endorsed by many of his colleagues. He tells us that "other modes of knowledge" (the example he gives is aesthetic perception) might be accommodated alongside science in a position of coexistence, but they cannot be given a place *within* science as part of a radical shift of sensibilities.

> . . . science cannot change in this way without destroying itself, because however much human values are involved in the scientific process or are affected by the results of scientific research, there is an essential element in science that is cold, objective, and nonhuman. . . . Having committed ourselves to the scientific standard of truth, we have thus been forced, not by our own choosing, away from the rhapsodic sensibility . . . In the end, the choice is a moral, or even a religious, one. Having once committed ourselves to look at nature on its own terms, it is something like a point of honor not to flinch at what we see.

"The universe," Weinberg insists, "is what it is." And science, as the definitive natural philosophy, can have no choice but to tell it like it is, and "not to flinch."

One cannot help admiring the candor of such an answer—and grieving a little for the pathos of its resignation. But it is, in any case, a Promethean answer, one that reminds us that the free pursuit of knowledge *is,* after all, a supreme value, a need of the mind as urgent as the body's need for food. However much one may upbraid science for having disenchanted our lives, sooner or later one must come to grips with the animating spirit of the discipline, the myth that touches it with an epic grandeur. Call up the monster, and the scientist calls up the Titan. Press the claims of spiritual need, and the scientist presses the claims of mind as, in their own right, a sovereign good.

Any critique of science that challenges the paramount good of knowledge risks becoming a crucifixion of the intellect. If Prometheus is to stop producing monsters, it must not be at the sacrifice of his Titanic virtues. The search for knowledge must be a free adventure; yet is must not choose, in its freedom, to do us harm in body, mind, or spirit. One no sooner states the matter in this way than it seems like an impossible dilemma. We are asking that the mind in search of knowledge should be left wholly free and yet be morally disciplined at the same time. Is this possible?

I believe it is, but only if we recognize that there are *styles* of knowledge as well as *bodies* of knowledge. Besides *what* we know, there is *how* we know it—how wisely, how gracefully, how life-enhancingly. The life of the mind is a constant dialogue between knowing and being, each shaping the other. This is what makes it possible to raise a question which, at first sight, is apt to appear odd in the extreme. *Can we be sure that what science gives us is indeed knowledge?*

Plato, Don Juan, and Gnosis

For most Western intellectuals that might seem a preposterous question, since for the better part of three centuries now science has served as the measure of knowledge in our society. But to raise it is only to recall the Platonic tradition, within which our science would have been regarded as an intellectual transaction distinctly beneath the level of knowledge. There is no telling for sure how highly Plato might have rated the spectacular theoretical work of the modern world's best scientific brains, but I suspect he would have respected it as "information"—a coherent, factually related account of the physical structure and function of things: a clever scheme for "saving the appearances," as Plato liked to characterize the astronomy of his day. Here we have a demanding and creditable labor of the intellect; but on Plato's well-known four-step ladder of the mind, science would be placed somewhere between the second and third levels of the hierarchy—above mere uninformed "opinion," but distinctly below "knowledge."

Easy enough to dismiss Plato as backward or plain perverse for refusing to rate science any higher on the scale. But how much more interesting to let the mind follow where his gesture takes it when he invites us to look beyond experiment, theory, and mathematical formulation to a higher object of knowledge which he calls "the essential nature of the Good . . . from which everything that is good and right derives its value for us."

Significantly, when Plato tried to put this object of knowledge into words, his habit, like that of many another mystic, was to enlist the services of myth and allegory, or to warn how much must be left unsaid. "There is no writing of mine about these matters," he tells us in the Seventh Epistle in a passage that might be a description of the Zen Buddhist Satori, "nor will there ever be one. For this knowledge is not something that can be put into words like other sciences; but after long-continued intercourse between teacher and pupil, in joint pursuit of the subject, suddenly, like light flashing forth when a fire is kindled, it is born in the soil and straightaway nourishes itself." No doubt, at first glance, such an elusive conception of knowledge is bound to seem objectionable to many scientists. But in light of all that Michael Polanyi has written about the "personal knowledge" and the "tacit dimension" involved in science, Plato's remarks should not seem wholly alien. Plato is reminding us of those subtleties that can only be conveyed between person and person at some nonverbal level; to force such insights into words or into a formal pedagogy would be to destroy them. If we are to learn them at all, there is no way around intimate association with a guru who can alone make sure that each realization is sensibly adapted to the time, and the place, and the person. So too in science, as in every craft and art. Is not much that is essential to the study left to be learned from one's master by way of nuance and hint, personal taste and emotional texture? And does this not include the most important matters of all: the spirit of the enterprise, the choice of a problem deemed worth studying, the instinctive sense of what is and what is not a reputable scientific approach to any subject, the decision as to when a hypothesis has been sufficiently demonstrated to merit publication? How much of all this is taught by the glint in the eye or the inflection in the voice, by subtle ridicule or the merest gesture of approval? Even the exact sciences could not do without their elements of taste and intuitive judgment, talents which students learn by doing or from the living example before them.

Plato is, of course, pushing the uses of reticence much further. He contends that, if only the tacit dimension of instruction between guru and student is exploited to its fullest, we can find our way to a knowledge, properly so-called, which grasps the nature and *the value* of things as a whole, and so raises us to a level at which intellect and conscience become one and inseparable in the act of knowing. "Without that knowledge," he insists, "to know everything else, however well, would be of no value to us, just as it is of no use to possess anything without getting the good of it."

Again, I suspect that Plato is not so far removed in his pursuit from a familiar scientific experience, one which comes in the wake of any significant discovery. It is the sense that, over and above what the particular discovery in question has shown to be factually so, this activity of the mind has proved itself *good*; it has, as a human project, elevated us to a level of supremely satisfying existence. One has not only found out something correct (perhaps that is the least of it, in the long run) but one has *been* something worth being. It is an experience many people have known, at least fleetingly, in their work as artists, craftsmen, teachers, athletes, doctors, etc. We might call it "an experience of excellence," and let it go at that. But what Plato wished to do was to isolate that experience as an object of knowledge, and to treat it, not as the by-product of some other, lesser activity, but as a goal in its own right. He wished to know the Good in itself which we only seem to brush against now and again in passing as we move from one occasional task to another. Nothing in modern science would have appalled Plato more than the way in which a professional scientific paper seeks, in the name of objectivity, to depersonalize itself to the point of leaving out all reference to that "experience of excellence"—that fleeting glimpse of the higher Good. For, I believe Plato would have objected, if no such experience was there, then the work was not worth doing; and if it was, then why leave it out, since it must surely be the whole meaning and value of science? Once you omit *that,* you have nothing left except . . . information.

If I invoke Plato here, it is not because I wish to endorse his theory of knowledge, but only to use him as a convenient point of departure. I recognize the logical blemishes that have dogged his epistemology through the centuries—and regard many of them as unanswerable within the framework Plato erected for his work. He is, however, the most renowned philosophical spokesman for a style of knowledge which is far older than formal philosophy; in his work we confront a visionary tradition which runs through nearly every culture, civilized and primitive. The prime value of Plato—so it has always seemed to me—lies not so much in the intellectual territory he occupied and surveyed, as in his stubborn determination to keep open a passage through which the mind might cross over from philosophy to ecstasy; from intellect to illumination. His dialogues stand on the border of a transrational sensibility whose charm seems a constant feature of human culture—a sensibility perhaps as old as the mind itself, and yet as contemporary as the latest bestseller list. Recall what the Yaqui Indian shaman Don Juan calls himself in Carlos Castaneda's recent popular reports: "a man of *knowledge.*" And, for all the differences of personal style and lore that part the two men, the old sorcerer means knowledge in exactly the way Plato meant it, as an ecstatic insight into the purpose and place of human existence in the universe, a glimpse of the eternal.

What both Plato the philosopher and Don Juan the sorcerer seek as knowledge is precisely that *meaningfulness* of things which science has been unable to find as an "objective" feature of nature. To follow where such a conception of knowledge takes us is not to denigrate the value or fascination of information. It is to be neither antiscientific nor antirational. It leads us not to an either/or choice, but to a recognition of priorities within an integrated philosophical context. Information can be exciting to collect; it can be urgently useful: a tool for our survival. But it is not the same as the knowledge we take with us into the crises of life. Where ethical decision, death, suffering, failure confront us, or in those moments when the awesome vastness of nature presses in upon us, making us seem frail and transient, what the mind cries out for is the meaning of things, the purpose they teach, the enduring significance they give our existence. And that, I take it, is Plato's knowledge of the Good.

To call this *another kind* of knowledge may seem a convenient compromise or a generous concession. But I submit that either as compromise or concession, this policy of Cartesian apartheid is treacherous. At best, it asks for the sort of schizophrenic coexistence that divides the personality cruelly between fact and feeling. At worst, it is the first step toward denying the "other knowledge" any status as knowledge at all—toward considering it a sort of irrational spasm devoid of any claim to truth or reality, perhaps an infantile weakness of the ego that is only forgiveable because it is so universally human. At that point we are not far from treating the need for meaning as a purely subjective question for which there is no objective answer—as an unfortunate behavioral trait which we leave psychologists or brain physiologists to stake out for investigation. Once it ceases to be the basis for knowledge, it may finish as an occasion for therapy.

My purpose here is to call back to mind the traditional style of knowledge for which the nature of things was as much a reservoir of meanings as of facts—a style of knowledge which science is now aggressively replacing in every society on earth. Let us call this knowledge "gnosis," borrowing the word not to designate a second and separate kind of knowledge, but an *older* and *larger* kind of knowledge from which our style of knowledge derives by way of a sudden and startling transformation of the sensibilities over the past three centuries. My contention is that this process of derivation has been spiritually impoverishing and psychically distorting. It has resulted in a narrowing of our full human potentialities and has left us—especially in science—with a diminished Titanism that falsely borrows upon the myth it champions. When the modern Prometheus reaches for knowledge, it is not the torch of gnosis he brings back or even searches for, but the many candles of information. Yet not a million of those candles will equal the light of that torch, for these are fires of a different order.

Augmentative Knowledge

I will not try to characterize gnosis here as an "alternative cognitive system" in any programmatic way—as if to offer a new methodology or curriculum. Rather, I want to speak of gnosis as a different sense of what knowledge is than science provides. When we search for knowledge, it is a certain texture of intelligibility we first and most decisively seek, a *feeling* in the mind that tells us, "Yes, here is what we are looking for. This has meaning and significance to it." Though it may work well below the level of deliberate awareness, this touchstone of the mind is what makes the persuasive difference in our thinking. Indeed, science itself arose in just this way, when men of Galileo's generation came to feel, with an uncanny spontaneity, that to know was to measure, that all else was subjective and unreal, a realm of "secondary qualities."

In the broadest sense, gnosis is *augmentative* knowledge, in contrast to the *reductive* knowledge characteristic of the sciences. It is a hospitality of the mind that allows the object of study to expand itself and become as much as it might become, with no attempt to restrict or delimit. Gnosis invites every object to swell with personal implications, to become special, wondrous, perhaps a turning point in one's life, "a moment of truth." Paul Tillich has called gnosis "knowledge by participation . . . as intimate as the relation between husband and wife." Gnosis, he tells us, "is not the knowledge resulting from analytic and synthetic research. It is the knowledge of union and salvation, existential knowledge in contrast to scientific knowledge."

It is the guiding principle of gnosis that only augmentative knowledge is adequate to its object. As long as we, at our most open and sensitive, feel there is something left over or left out of any account we give of any object, we have fallen short of gnosis. Gnosis is that nagging whisper at the edge of the mind which tells us, whenever we seek completeness of understanding or pretend to premature comprehension, "not yet . . . not quite." It is our immediate awareness, often at a level deeper than intellect, that we seem not to have done justice to the object—not because there remains quantitatively more of the object to be investigated, but because its essential *quality* still eludes us.

I speak here of the experience many people have known when faced with some brutally reductionist explanation of human conduct. We feel the explanation "reduces" precisely because it leaves out so much of what we spontaneously know about humanness from inside our own experience. We look at the behaviorist's model and we know—as immediately as our eye would know that a circle is not a square—that this is not *us*. It may not even be an important part or piece of us, but only a degraded figment. Even if such knowledge "worked"—in the sense that it allowed others to manipulate our conduct as precisely as an engineer can manipulate mechanical and electrical forms of energy—would we not still protest that *knowing how* to dangle us like a puppet on a string is not *knowing* us at all? Might we not insist that such "knowledge" works in the very opposite direction—that it is an ignorant, insulting violation of our nature? As Abraham Maslow once observed of his own experience in behavioral psychology: "When I can predict what a person will do under certain circumstances, this person tends to resent it. . . . He tends to feel dominated, controlled, outwitted."[3] Between "knowing" and "knowing how" there can be a fearful discord—like Bach being played on skillets and soup kettles: more mockery than music.

That discord shows up readily enough when we ourselves are the specimens under study. In that case, the standard of adequacy is provided by the object of investigation. We can speak for ourselves and fend off the assault upon our dignity. But what about the nonhuman objects of the world? Does it make any sense to say that our scientific knowledge of them may be *qualitatively* inadequate?

To answer that question, let us begin with a familiar comparison: that between art and science. The coincidence of the two fields has been observed many times, especially in so far as they share a common fascination for form and structure in nature. Yet, while there is an overlap, it is, from the scientist's point of view, an overlap of interest only, not of intellectual competence. Both art and science find an aesthetic aspect in nature (though of course many scientists have done significant research without pausing over that aspect). But for the scientist, the aesthetic appearance is a *surface*; knowledge stands behind that surface in some underlying mechanism or activity requiring analysis. What the artist sees is not regarded by science as knowledge of what is *in* the object as one of its constituent properties. Instead, what preoccupies the artist is called "beauty" (though often it would better be understood as awe, conceivably mixed with fear, anxiety, dread). Beauty is, for science, a sort of subjective supplement to knowledge, a decoration the mind supplies before or after the act of cognition, and which can or even ought to be omitted from professional publication. Aesthetic fascination may attract us to the object; it may later help flavor popularized accounts of research. From the scientist's viewpoint, however, only further study (dissection, deep analysis, comparison, experiment, measurement) allows us to find out something about the object, something demonstrable, predictive, useful. Compared to such hard fact, the artist's perception is merely dumb wonder, which, apparently, artists have not the intellectual rigor to go beyond. Jacob Bronowski has, for example, referred to the artist's response to nature as "a strangled, unformed and unfounded experience." But, he goes on, "science is a base for [that experience] which constantly renews the experience and gives it a coherent meaning."[4]

If this were not the supposition, we might imagine an entire specialization in science devoted to studying the nature poets and painters: biologists sprinkling their research with quotations from Wordsworth or Goethe . . . neophyte botanists taking required courses in landscape painting . . . astronomers drawing hypotheses from Van Gogh's "Starry Night" . . . theoretical physicists pondering the bizarre conceptions of time and space one finds in the serial tone row, cubism, constructivism, or Joyce's *Finnegans Wake*. Of course, nothing forbids scientists from wandering into these exotic realms, but what curriculum *requires* that they do so?

From the viewpoint of gnosis, however, what artists find in nature is decidedly knowledge of the object, indeed knowledge of a uniquely valuable kind. It is not repeatable or quantitative, nor is it open to experimentation or utilitarian application. It is usually not logically articulable; that is why special languages of sound, color, line, texture, metaphor and symbol have been invented to carry the message, in much the same way that mathematics has been developed as the special language of objective consciousness. But that message is as much knowledge as when, in addition to knowing your chemical composition, I discern that you are noble or base, lovable or vicious. So artists discover the communicative mood and quality that attach to form, color, sound, image. They teach us those qualities, and these become an inseparable part of our total response to the world.

Of course, these qualities can be screened out if our interest is directed to something less than the whole, but this does not mean the sensuous and aesthetic qualities are not really there as a constituent property of the world—a property that is being artfully *displayed* to us. Would it not, in fact, be truer to our experience to conceive of the world about us as a *theater,* rather than as a mechanism or a randomized aggregation of events? It is surely striking how often science quite naturally presents its discoveries as if it were unfolding a spectacle before us, thus borrowing heavily on sensibilities that have been educated by the dramatists and story-tellers. All cosmology is talked about in this way, and even a good deal of high energy physics and molecular biology. Everything we have lately discovered about the evolution of stars is, quite spontaneously, cast in the mode of biography: birth, youth, maturity, senility, death, and at last the mysterious transformation into an afterlife called "the black hole." Or, take the classic example of aesthetic perception in science. Can there by any doubt that much of the cogency of Darwin's theory of natural selection stemmed from the pure drama of the idea? Natural selection was presented as a billion-year-old epic of struggle, tragic disasters, lucky escapes, triumph, ingenious survival. Behind the sensibility to which Darwin's theory appealed lay three generations of Romantic art which had pioneered the perception of strife, dynamism, and unfolding process in nature. Behind Darwin stand Byron's Manfred, Goethe's Faust, Constable's cloud-swept landscapes, Beethoven's tempestuous quartets and sonatas. All this became an integral part of the Darwinian insight. I doubt there is anyone who does not still bring to the study of evolution this Romantic taste for effortful growth, conflict, and self-realization. The qualities are not only in the idea, but also in the phenomenon. It is not that these dramatic qualities have been "read" into nature by us, but rather that *nature* has read them into *us* and now summons them forth by the spectacle of evolution we find displayed around us.

We should by now be well aware of the price we pay for regarding aesthetic quality as arbitrary and purely subjective rather than as a real property of the object. Such a view opens the way to that brutishness which feels licensed to devastate the environment on the grounds that beauty is only "a matter of taste." And since one person's taste is as good as another's, who is to say—as a matter of *fact*—that the hard cash of a strip mine counts for less than the grandeur of an untouched mountain? Is such barbarism to be "blamed" on science? Obviously not in any direct way. But it is deeply rooted in a scientized reality principle that treats quantities as objective knowledge and qualities as a matter of subjective preference.

The Spectrum of Gnosis

Now to push the point a little further. If art overlaps science at one wing, it overlaps visionary religion at the other. If artists have found the cool beauty of orderly structure in nature, they have also found there the burning presence of the sacred. For some artists, as for the Deist scientists of Newton's day, God's imprint has appeared in the rhythmic cycles and stately regularities of nature. For other artists—Trahern, Blake, Keats, Hopkins—the divine grandeur of the world appears all at once, in an ecstatic flash, a jolt, a "high." Here we find the artist becoming seer and prophet. For such sensibilities, a burning bush, a storm-battered mountaintop can be, by the sheer awesomeness of the event, an immediate encounter with the divine.

To know God from the order of things is a deduction, a shaky one perhaps in the eyes of skeptical logicians, but at least remotely scientific in character.

To know God from the power of the moment is an epiphany, a knowledge that takes us a long way from scientific respectability. Yet here is where gnosis mounts to its heights, becoming knowledge willingly obedient to the discipline of the sacred. It does not close itself to the epiphanies life offers by regarding them as "merely subjective." Rather, it allows, it *invites* experience to expand and become all that it can. After all, if Galileo was right to call those men fools who refused to view the moon through a telescope, what shall we say of those who refuse Blake's invitation to see eternity in a grain of sand? Gnosis seeks to integrate these moments of ecstatic wonder; it regards them as an advance upon reality, and by far the most exciting advance the mind has undertaken. For here is the reality that gives transcendent meaning to our lives.

Perhaps the best way to summarize what I have said so far is to conceive of the mind as a spectrum of possibilities, all of which properly blend into one another—unless we insist on erecting barriers across the natural flow of our experience. At one end, we have the hard, bright lights of science; here we find information. In the center we have the sensuous hues of art; here we find the aesthetic shape of the world. At the far end, we have the dark, shadowy tones of religious experience, shading off into wave lengths beyond all perception; here we find meaning. Science is properly part of this spectrum. *But gnosis is the whole spectrum.*

If, in the past, gnosis has been more heavily weighted on the side of meaning than information, it should not be difficult to understand why. Our ancestors saw fit to put first things first. Before they felt the need to know how fire burns or how seeds germinate, they needed to know the place and purpose of their own strange existence in the universe. And this they found generously offered to them in the nature of things. Yet, I know of no visionary tradition that has ever refused to agree that natural objects possess a structure and function worthy of study. Certainly none of these traditions has been as adamantly closed to the technical level of knowledge as our science has been closed to gnosis. Plato may have wanted the mind to rise to a level of ecstatic illumination, but he never said there was no such thing as information or that its pursuit was a sign of madness or intellectual incompetence. Similarly, the alchemists may have sought their spiritual regeneration in natural phenomena, but they never refused to examine the way nature works. Undeniably, where gnosis becomes our standard of knowledge, science and technology proceed at a much slower rate than the wild pace we accept (or suffer with) as normal. This is not to say, however, that gnosis is without its practical aspect, but rather that its sense of practicality embraces spirit as well as body, the need for psychic as much as for physical sustenance.

The most familiar examples we have of culture dominated by gnosis are in the world's primitive and pagan societies. Many of these societies have been capable of investing agrarian and hunting technologies every bit as ingenious as the machine technics of modern times. But, in stark contrast to the culture of urban-industrialism, their technology blended at every step with poetic insight and the worship of the elements. The tools and routines of daily life normally participated in the religious sensibility of the society, functioning as symbols of life's higher significance. From the viewpoint of the modern West, such a culture may look like a hodge-podge of wholly unrelated factors. In reality, it is an ideal expression of gnosis, for it expresses a unitary vision bringing together art, religion, science, and technics. Our habit, in dealing with such cultures, is to interpret their technics as lucky accidents and their aesthetic-religious context as an encumbrance. But by at least one critical standard, these "underdeveloped" cultures have proved more technically successful than our own may. They have *endured,* in some cases a hundred times longer than urban industrialism may yet endure. Surely that is some measure of how well a culture understands its place in nature.

Most of the world's mystic and occult traditions have been worked up from the gnosis of primitive and pagan cultures. At bottom, these traditions are sophisticated, speculative adaptations of the old folk religions, which preserve in some form their antique wisdom and modes of experience. Behind the Cabbala and Hermeticism, we can still see the shadowy forms of ritual magic and fertility rites, symbols of a sacred continuum binding man to nature and prescribing value. In all these mystic traditions, to know the real is to know the good, the beautiful, and the sacred at the same time.

This is not to say that all who followed these traditions achieved gnosis. The human mind goes wrong in many ways. It can go mad with ecstasies as well as with logic. Discriminating among the levels and directives of transrational experience is a project in its own right—one I do not even touch upon here, for the discussion would be far too premature at this point. There are disciplines of the visionary mind as well as of the rational intellect, as anyone will know who has done more than scratch the surface of the great mystic traditions. All I stress here is the difference between a taste for gnosis and a taste for knowledge whose visionary overtones have been systematically stilled as a supposed "distortion" of reality.

The Visionary Origins of Science

Our science, having cut itself adrift from gnosis, contents itself to move along the behavioral surface of the real—measuring, comparing, systematizing, but never penetrating to the visionary possibilities of experience. Its very standard of knowledge is a rejection of gnosis, any trace of whose presence is regarded as a subjective taint. Yet, ironically, the scientific revolution of the sixteenth and seventeenth centuries was in large part launched by men whose thought was significantly colored by lingering elements of gnosis in our culture, most of which survived in various subterranean occult streams. Copernicus very nearly resorted to pagan sun worship as a means of supporting his heliocentric theory, the sheer aesthetic beauty of which seems to have been as persuasive for him as its mathematical precision. Kepler's astronomy emerges from a search for the Pythagorean music of the spheres. Newton was a life-long alchemist and student of Jacob Boehme. The scholarship on early science finds more and more hidden continuities between the scientific revolution and the occult currents of the Renaissance. Frances Yates has gone so far as to suggest that science only flourished in those societies where there had been a strong, free influx of Hermetic and Cabbalistic studies.[5] From this origin came the number magic and nature mysticism which were to be assimilated into science as we know it. These historical links have yet to be fully traced, but certainly the key paradigm of "law"—that mysterious sense of natural right order without which early science could never have gotten off the ground—carried with it in the thinking of early physicists unmistakable moral and theological reverberations. It was the concept of universal law that made the study of nature as a celebration of the grandeur of God compatible with the Christian doctrine.

What this confabulation with occult tradition suggests is that many lively minds of the seventeenth century, including some founding fathers of modern science, looked forward to seeing the New Philosophy become a true gnosis, possibly to replace the rigid, decaying dogmatism of Christianity. The trouble was that their exciting new approach to nature progressively screened out the very dimension of consciousness in which gnosis can alone take root: visionary insight. In seeking to externalize gnosis by raising it to a wholly articulate and mathematical level of expression, the New Philosophers left behind the mystic and meditative disciplines which might have taught them that introspective silence and transcendent symbolism are necessary media of gnosis. It was as if someone had invented an ingenious musical instrument with which he hoped to replace the full orchestra, with the result that thereafter all orchestral music had to be scaled down to the capacities of his instrument. And once that had been done, he and his audience began to lose their ear for the harmonies and overtones that only the orchestra can achieve. Quantification is just such an instrument of severely reduced resonance.

There is a haunting and troubling strangeness about this interval in our history. One might almost believe that perverse forces which baffle the understanding were at work beneath the surface of events turning science into something that did not square with the personalities of its creators. What was it, for example, that inspired Descartes to regard mathematics as the new key to nature? An "angel of truth" who appeared to him in a series of numinous dreams on three successive nights. But in his writing, he never once mentions the epistemological status of dreams or visionary experience. Instead, he turns his back on all that is not strict logic, opting for a philosophy of knowledge wholly subordinated to geometrical precision. Yet that philosophy purchases its apparent simplicity by an appalling brutalization of the very existential subtleties and psychic complexities that are the living substance of Descartes' own autobiography. Newton, a man of stormy psychological depths, spent a major portion of his life in theological and alchemical speculation; but all this he carefully edited from his natural philosophy and his public life. He even allowed himself to be talked out of attending the meetings of occult societies in London, lest he damage his reputation as a scientist. Arthur Koestler is not wide of the mark in calling the early scientists "sleepwalkers," men who unwittingly led our society into a universe whose eventual godlessness they might well have rejected vehemently.

This much of the problem stands out prominently enough: the mystic disciplines, on which gnosis depends, have never been as highly refined and widely practiced in the West as in the oriental cultures. In large part, they have suffered neglect because they cut across the doctrinal grain of conventional Christianity with its insistent emphasis on historicity and dogmatic theology. (I often think that few positivists realize how great a debt they owe to the peculiarly one-dimensional religious psychology of mainstream Christianity; its literalism and verbal rigidities paved the way for the secular skepticism of the religion's deadliest critics.) Still, in the Hermetic, Cabbalistic, Neo-Platonic, and alchemical schools of the Renaissance, at least a promising foundation existed for the building of a true gnosis. In these currents of thought we find an appreciation of myth, symbol, meditative stillness, and rhapsodic intellect that might, with maturity, have matched the finest flights of Tantric or Taoist mysticism.

But if these elements were mixed with early science in many exotic combinations, they were soon enough filtered out as violations of that strict objectivity which is the distinguishing feature of the Western scientific sensibility. It was Galileo's quantitative austerity and Descartes' dualism that carried the day with science, casting out of nature everything that was not matter in motion mathematically expressed. Here was the crucial point at which scientific knowledge ceased to be gnosis. Value, quality, soul, spirit, animist communion were all ruthlessly cut away from scientific thought like so much excess fat. What remained was the world-machine—sleek, dead, and alien. However much physics has, in our time, modified the mechanistic imagery of its classical period, the impersonality of the Newtonian world view continues to dominate the scientist's vision of nature. The models and metaphors of science may alter, but the sensibility of the discipline remains what it was. Since the quantum revolution, modern physics has ceased to be mechanistic, but it has scarcely become in any sense "mystical." The telling fact is that both in style and content it serves today as an ideal foundation for molecular biology and behavioral psychology, sciences which have of late become as mechanistic as the crudest reductionism of the seventeenth century. Almost universally these days, biologists regard the cell as a "chemical factory" run by "information-transfer" technology. And, at the same time, the arch-behaviorist B. F. Skinner suggests that since physics only began to make progress when it "stopped personifying things," psychology is not apt to gain a firm scientific footing until it likewise purges itself of "careless references to purpose" and ceases "to trace behavior to states of mind, feelings, traits of character, human nature, and so on"[6]—meaning, one gathers, that the way forward for psychology is to stop personifying people . . . and to begin mechanizing them.

The Suppression of Gnosis

Why has science taken this course toward even more aggressive depersonalization? Perhaps the myth of Dr. Frankenstein suggests an answer—a tragic answer. Where did the doctor's great project go wrong? Not in his intentions, which were beneficent, but in the dangerous haste and egotistic myopia with which he pursued his goal. It is both a beautiful and a terrible aspect of our humanity, this capacity to be carried away by an idea. For all the best reasons, Victor Frankenstein wished to create a new and improved human type. What he knew was the secret of his creature's physical assemblage; he knew how to manipulate the material parts of nature to achieve an astonishing result. What he did not know was the secret of personality in nature. Yet he raced ahead, eager to play God, without knowing God's most divine mystery. So he created something that was soulless. And when that monstrous thing appealed to him for the one gift that might redeem it from monstrosity, Frankenstein discovered to his horror that, for all his genius, it was not within him to provide that gift. Nothing in his science comprehended it. The gift was love. The doctor knew everything there was to know about his creature—except how to love it as a person.

To find the cultural meaning of modern science, for *"Frankenstein's monster,"* read "nature-at-large" as we in the modern West experience it.

In the early days of the scientific revolution, Robert Boyle, convinced of the "excellency" of the new "mechanical hypothesis," insisted that nature, if it was to be mastered, must be treated like an "engine" or an "admirably contrived automaton." His argument prophetically relegated to the dustbin every lingering effort to personify nature, even by remote metaphor.

> The veneration, wherewith men are imbued for what they call nature, has been a discouraging impediment to the empire of man over the inferior creatures of God. For many have not only looked upon it as an impossible thing to compass, but as something impious to attempt, the removing of those boundaries which nature seems to have put and settled among her productions; and whilst they look upon her as such a venerable thing, some make a kind of scruple of conscience to endeavor so to emulate her works as to excel them.[7]

Here was a deliberate effort—and by a devout Christian believer—to cut science off from every trace of Hermetic or alchemical influence, from every connection with animist sympathy and visionary tradition. Boyle—like Bacon, Descartes, Galileo, and Hobbes—realized that herein lay the promise of material power. From that point on, it became permissible for the scientist to admire the mechanical intricacy of nature, but not to love it as a living presence endowed with soul and reflecting a higher order of reality. A machine can be studied zealously, but it cannot be loved. By virtue of that change of sensibilities—which may of course have transpired at a subliminal level of consciousness—the New Philosophy could lay claim to power (at least short-term manipulative power) but it had lost the *anima mundi,* which, as an object of love, belongs only to gnosis.

Still, from time to time, something of the spirit of gnosis intrudes itself into scientific thought, if only as a passing reflection upon some aspect of design in nature which hints that there is indeed *something* more to be known than conventional research can reveal. Science is not without such moments. But they appear only as autobiographical minutiae along the margins of "knowledge," modest confessions of faith, personal eccentricities, a bit of subprofessional self-indulgence on the part of established great names. These ethical, aesthetic, and visionary aspects have long since become human interest sideshows of science, the sort of anecdotal material that never makes it into the textbooks or the standard curriculum, except perhaps as a whimsical footnote.

And yet, have scientists never noticed how the lay public hangs upon these professions of wonder and ultimate belief, seemingly drawn to them with even more fascination than to the great discoveries? If people want more from science than fact and theory, it is because there lingers on in all of us the need for gnosis. We want to know the meaning of our existence, and we want that meaning to ennoble our lives in a way that makes an enduring difference in the universe. We want that meaning not out of childish weakness of mind, but because we sense in the depths of us that it is *there,* a truth that belongs to us and completes our condition. And we know that others have found it, and that it has seized them with an intoxication we envy.

It is precisely at this point—where we turn to our scientists for a clue to our destiny—that they have indeed a Promethean role to perform, as has every artist, sage, and seer. If people license the scientist's unrestricted pursuit of knowledge as a good in its own right, it is because they hope to see the scientists yet discharge that role; they hope to find gnosis in the scientist's knowledge. To the extent that scientists refuse that role, to the extent that their conception of what science is prevents them from seeking to join knowledge to wisdom, they are confessing that science is not gnosis, but something far less. And to that extent they forfeit—deservedly—the trust and allegiance of their society.

Dr. Faustus, Dr. Frankenstein, Dr. Moreau, Dr. Jekyll, Dr. Cyclops, Dr. Caligari, Dr. Strangelove. The scientist who does not face up to the warning in this persistent folklore of mad doctors is himself the worst enemy of science. In these images of our popular culture resides a legitimate public fear of the scientist's stripped-down, depersonalized conception of knowledge—a fear that our scientists, well-intentioned and decent men and women all, will go on being titans who create monsters.

What is a monster? The child of knowledge without gnosis, of power without spiritual intelligence.

The reason one despairs of discussing "alternative cognitive systems" with scientists is that scientists inevitably want an alternative system to do exactly what science already does—to produce predictive, manipulative information about the structure and function of nature—only perhaps to do so more prolifically and more rapidly. What they fail to understand is that no amount of information on earth would have taught Victor Frankenstein how to redeem his flawed creation from monstrosity.

But there is, in the Hermetic tradition we have left far behind us, a myth which teaches how nature may, by meditation, prayer, and sacrifice, be magically transmuted into the living presence of the divine. That was the object of the alchemist's Great Work, a labor of the spirit undertaken in love whose purpose was the mutual perfection of the macrocosm, which is the universe, and the microcosm, which is the human soul.

> And what if all of animated nature
> Be but organic Harps diversely fram'd
> That tremble into thought, as o'er them sweeps
> Plastic and vast, one intellectual breeze
> At once the Soul of each and God of all?
>
> Samuel Taylor Coleridge

NOTES

1. Jacques Monod, *Chance and Necessity* (New York: Knopf, 1971), p. 172.
2. Steven Weinberg, "Reflections of a Working Scientist." *Daedalus,* vol. 103 (1974), p. 43.
3. Abraham Maslow, *The Psychology of Science* (New York: Harper and Row, 1966), p. 42.
4. J. Bronowsky, *Science and Human Values* (New York: Harper Torchbooks, 1965), p. 95.
5. Frances Yates, *Rosicrucian Enlightenment* (London: Routledge, 1972). See also P. M. Rattansi, "The Social Interpretation of Science in the Seventeenth Century." *Science and Society 1600-1900* (Cambridge: Cambridge University Press, 1972).
6. B. F. Skinner, *Beyond Freedom and Dignity* (New York: Knopf, 1971), pp. 5-7.
7. Robert Boyle, "A Free Inquiry into the Received Notion of Nature," *Works* (1744), Ch. IV, p. 363.

22

Reflections on the Neo-Romantic Critique of Science

Leo Marx

I have insisted that there is something radically and systematically wrong with our culture, a flaw that lies deeper than any class or race analysis probes and which frustrates our best efforts to achieve wholeness. I am convinced it is our ingrained commitment to the scientific picture of nature that hangs us up.

The scientific style of mind has become the one form of experience our society is willing to dignify as knowledge. It is our reality principle, and as such the governing mystique of urban industrial culture.

Theodore Roszak[1]

I

Serious, widespread criticism of science is a relatively recent development in the United States. Until World War II the national faith in the identity of scientific and social progress remained largely unshaken. Most Americans, even after the Great Depression of the 1930s, continued to regard the life-enhancing value of scientific knowledge as self-evident. But since Hiroshima, public anxiety about the consequences of scientific discovery has risen steadily. The nuclear arms race; the polluting and carcinogenic effects of new petrochemicals and other products of science-based industry; the actual and possible uses of electronic devices as instruments of social control; the prominent part played by certain science-based technologies of a particularly revolting kind in the prosecution of the American war in Southeast Asia; the potentialities for genetic engineering created by advances in molecular biology—these are only the more conspicuous causes for the rising public alarm about the results of scientific research. It is now evident that the American belief in the inherently beneficial character of science no longer can be taken for granted.[2] Judging by current discussion of the subject, however, one might infer that the legitimacy of science—by which I do not mean its lawfulness in any narrow sense, but rather its compatibility with accepted standards and purposes—is now being called into question for the first time.

But in Western culture the legitimacy of modern science has been in question since its emergence in the seventeenth century. To be sure, certain major themes in the legacy from the earlier critique of science have since lost their credibility and all but disappeared. The learned clergy, for example, no longer attacks science as a threat to the churches, or as a deflection from the primacy of theological knowledge. Today no responsible clergyman would think of opposing a scientific research project on the ground that the worship of God precludes the study of nature. But it is necessary to add that serious criticism of science based on religious values, though not expressly identified as such, retains an immense appeal. Some of the more effective of the currently popular arguments against science prove, on inspection, to be secularized versions of an essentially religious or teleological, conception of knowledge.

Another ancient theme in the critique of science which has virtually disappeared is in effect a defense of the older, aristocratic, humanism. On this view, widely held in the age of Pope and Swift, the proper study of mankind is man, not nature. At stake then was the moral instruction of a small, privileged, ruling class for whom scientific education was deemed inappropriate—which is to say, vulgar, unedifying, merely useful. Like the antagonism toward science grounded in theological preconceptions, this avowedly patrician argument is no longer invoked, for obvious reasons, by critics of science.

But the same cannot be said about the general ideas embodied in imaginative literature beginning in the late eighteenth century. On the contrary, most of the themes which figure prominently in the current criticism of science were anticipated by the writers of the romantic era. They called into question the legitimacy of science both as a mode of cognition and as a social institution.[3] To question the legitimacy of science as a mode of cognition means, I assume, to ask whether the conception of reality implicit in the scientific method is adequate to our experience. Is it reliable, coherent, sufficient? Insofar as it is not sufficient, does it mesh with what we know by means of other modes of knowledge? To question the legitimacy of science as an institution is to ask whether the methods (and products) of scientific inquiry are compatible with the expressed and tacit goals of society. Can the technological consequences of scientific discovery be assimilated, for example, to a more just, healthful, and peaceful social order?

Both kinds of question, to repeat, were implicit in the literary response to scientific rationalism and the innovations, technological and social, associated with it. At the core of the romantic reaction, in the well-known formulation of Alfred North Whitehead, was "a protest on behalf of the organic view of nature, and also a protest against the exclusion of value from the essence of matter of fact."[4] Implicit in each of these "protests" is a negative, or potentially negative, answer to the questions about the legitimacy of science raised above. The protest on behalf of the organic view of nature is directed against the presumed epistemological insufficiency of science. Scientific method is thus held to be inadequate to the (unified) nature of nature, which is assumed to be a whole distinct from the sum of its parts, and hence not apprehensible by means of the piecemeal, or analytic, procedures which dominate (normal) scientific inquiry.

The second "protest" identified by Whitehead (against "the exclusion of value from the essence of matter of fact") is applicable to both the cognitive and institutional senses of "science." It means that as a method of knowledge science lends insufficient expression to the distinctively human attributes of reality, those which are properties of mind rather than merely of natural objects. But "exclusion of value" also may be taken as a reference to the negative social and political results of scientific neutrality. It anticipates the now familiar charge that scientists do not assume adequate responsibility for the social consequences of their work. The substantive moral neutrality of natural science as a method of inquiry is not a warrant, in this view, for the morally uncommitted posture of scientists outside their laboratories or classrooms. My point, in any case, is that much of today's criticism of science, including the antagonistic viewpoint widely disseminated by spokesmen for the dissident movement, or counterculture, of the 1960s, may be traced to the double-barreled romantic reaction of European intellectuals which began in the late eighteenth century.

II

The mainstream of the European critique of science entered American literary thought under the auspices of Ralph Waldo Emerson. The English writers who chiefly influenced his thinking on the subject—Wordsworth, Coleridge, Carlyle—had in turn been influenced by the several versions of post-Kantian idealism then being imported into England from Germany. If we accept for the moment the standard, over-simplified handbook view of the spectrum of English literary attitudes toward science as extending from the hostility of Blake at one extreme to the admiration of Shelley at the other, then the writers who were congenial to Emerson must be accounted middle-of-the-roaders. In view of today's assumptions about the antagonism between the "two cultures," in fact, these English moderates would seem, like Goethe, to have been remarkably hospitable toward the claims and prospects of science. Although in one way or another they all recognized the limitations of scientific rationalism, they expressed no serious doubts about the inherent validity of the scientific method. Nor were they frightened by the prospect of the revolution in the conditions of life soon to result from the application of the new science to the fulfillment of economic needs. Their optimism was most directly expressed in repeated assertions about the essential compatibility between scientific and other modes of perception, especially aesthetic or literary, like this well-known statement of Wordsworth's in the *Lyrical Ballads* preface of 1800:

> If the labors of men of science should ever create any material revolution, direct or indirect, in our condition, and in the impressions which we habitually receive, the poet will sleep then no more than at present; he will be ready to follow the steps of the man of science, not only in those general indirect effects, but he will be at his side, carrying sensation into the midst of the objects of the science itself. The remotest discoveries of the chemist, the botanist, or mineralogist will be as proper objects of the poet's art as any upon which it can be employed. . . . If the time should ever come when what is now called science, thus familiarized to men, shall be ready to put on, as it were, a form of flesh and blood, the poet will lend his divine spirit to aid the transfiguration, and will welcome the being thus produced, as a dear and genuine inmate of the household of man.[5]

Beyond such optimism about the future collaboration between science and poetry ("poetry" usually taken to represent aesthetic and moral discourse generally) was the assumption that the two modes of perception stand in a potentially complementary relation to each other. Thus at least the majority of scientists were presumed to be operating at the level of the "Understanding." This is the empirical mode of apprehending external reality, based upon sense perception, as set forth in John Locke's "sensational" theory of knowledge. This theory had proven, by Wordsworth's time, to be ideally suited to negotiations, theoretical and practical, with the world of material objects. As Emerson put it, in what came to be recognized as the philosophic manifesto of American transcendentalism, *Nature* (1836), the Understanding is the capacity of mind which "adds, divides, combines, and measures," whereas the Reason, a mythopoeic, analogizing, intuitive mode of perception, "transfers all these lessons [of the empirical Understanding] into its own world of thought, by perceiving the analogy that marries Matter and Mind."[6]

Assertions on the plane of the Understanding are in effect data-bound, and require only literal language, whereas assertions on the plane of Reason transcend the "natural facts," require figurative language, and thus contribute to epochal rearrangements of thought and feeling. Emerson's favorite illustration of this point, following the analogous distinction between the Fancy and the Imagination, is the difference between the practice of a merely fanciful poet and that of a truly imaginative genius, a Virgil, Dante, or Milton, whose work effects a symbolic reconstruction of reality. But Emerson clearly meant the distinction to apply, by analogy, to the work of scientists as well. It is the difference between routine science, which merely elaborates, confirms, and refines an established theoretical structure, and the revolutionary synthesis of a Galileo or a Newton. So far as Emerson's theory of knowledge can be taken as tacit criticism of science, it is directed against inquiry confined to the plane of the empirical Understanding. Besides, he is only questioning the sufficiency, not the reliability, of such knowledge. But the distinction between the two modes opens the way for those critics who would charge science with encouraging a dangerous imbalance on the side of the instrumental, empirical Understanding.

Thomas Carlyle was one of the first writers to invoke a similar distinction between two modes of knowledge as a way of calling into question the legitimacy of modern science as a social institution. In "Signs of the Times" (1829), he locates the governing spirit of the "Age of Machinery" in the empirical philosophy of John Locke. Locke's "whole doctrine," he asserts, "is mechanical, in its aim and origin, its method and its results. It is not a philosophy of the mind; it is a mere discussion concerning the origins of our consciousness, or ideas . . . a genetic history of what we see *in* the mind." By "mechanical" Carlyle refers to Locke's emphasis upon the accumulation and sorting of external data (accomplished by the Understanding), and a minimizing of the active, synthetic, and transformational power of mind (accomplished by Reason). Because he conceives of the contents of the mind as contingent upon sense experience, upon facts flowing in from the outside, Locke tends to reduce thought to a reflex of the environment. This way of knowing is extremely useful for manipulating physical reality, but it leads to a quietist abdication, or fatalism, with respect to the controlling purposes of man's newly acquired power over nature. "The science of the age . . . is physical, chemical, physiological; in all shapes mechanical," hence the image of a machine best characterizes the dawning era of instrumental reason.

But the "mechanical philosophy," as Carlyle describes it, need not be destructive. On the contrary, it could be advantageous to mankind, and in fact he admits that in certain respects the age is advancing. The trouble is, however, that the advance is grossly unbalanced: while the physical sciences are thriving, the moral and metaphysical sciences are falling into decay. Carlyle's complaint in effect belongs to a later stage in the response to the advance of science which Lynn White has discerned as early as the fourteenth century. As scientists tended "to narrow their research methods to the mathematical, and their topics to the physical," according to White, there was a decline in conviction of the cogency of "trivial" argument, that is an argument in the essentially rhetorical language of the "trivium" as against argument in the essentially mensurative language of the "quadrivium."[7] In Carlyle's view, in any case, scientific rationality is spreading far beyond its proper sphere, and the result is that the culture is permeated by "mechanical" or technological thinking. The age of machinery overvalues those aspects of life which are congenial to the "quadrivial" mode of thought, to use the terminology of the earlier age, which is to say they are quantifiable, calculable, manipulatable. By the same token it downgrades the sphere of the moral, aesthetic, affective and imaginative—all that springs from the inner resources of the psyche: "the primary, unmodified forces and energies of man, the mysterious springs of Love, and Fear, and Wonder, of Enthusiasm, Poetry, Religion, all which have a truly vital and *infinite* character. . ." This neglected province, the antithesis of the mechanical, Carlyle calls "dynamical." His entire criticism of science rests upon the conviction that we need to develop both of these "great departments of knowledge," and indeed "only in the right coordination of the two, and the vigorous forwarding of *both,* does our true line of action lie." Carlyle continues:

> Undue cultivation of the inward or Dynamical province leads to idle, visionary impracticable courses . . . to Superstition and Fanaticism, with their long train of baleful and well-known evils. Undue cultivation of the outward, again, though less immediately prejudicial, and even for the time productive of many palpable benefits, must, in the long-run, by destroying Moral Force, which is the parent of all other Force, prove not less certainly, and perhaps still more hopelessly, pernicious. This, we take it, is the grand characteristic of our age.[8]

This "grand characteristic" takes the form both of a cognitive and of an institutional imbalance. Within science it manifests itself in a neglect of what once was called Natural Philosophy, and a preference for piecemeal analysis: breaking complex problems down into small, simple, particularized elements, thereby anticipating the tendency of scientific inquiry in our own time in which Gerald Holton recognizes "an asymmetry between analysis and synthesis."[9] Implicit here, too, is a homology between the analytic mode of scientific inquiry and the new principles of economic and social organization. The secular, fragmenting, particularizing tendency within science has its counterpart in the management of the market economy and the new political bureaucracies. All in all, therefore, the mechanical philosophy is producing "a mighty change in our whole manner of existence."

> By our skill in Mechanism, it has come to pass, that in the management of external things we excel all other ages; while in whatever respects the pure moral nature, in true dignity of soul and character, we are perhaps inferior to most civilized ages.[8]

III

Among the misconceptions fostered by C. P. Snow's "two cultures" thesis is the notion that in the twentieth century the humanities have been the province of unqualified hostility to science. My impression is that a comprehensive survey of literary thought, at least, would reveal a spectrum not unlike that which emerged in the age of Emerson and Carlyle. At one extreme, exemplified by the early writings of I. A. Richards, we find what amounts to the emulation of scientific "objectivity" or positivism. In his influential *Principles of Literary Criticism* (1924), and in *Science and Poetry* (1926), Richards embraced a virtual dichotomy between two uses of language, one scientific, the other "emotive." In science, which he conceives as a largely autonomous activity ("the impulses developed in it are modified only by one another, with a view to the greatest possible completeness and systemization, and for the facilitation of further references"), statements are made only for the sake of the reference, true or false, which they allow. This austerely denotative use of language corresponds to that which Emerson and other post-Kantian idealists associated with the Understanding. And like them, Richards opposes it to a connotative or suprareferential use of language for the sake of the effects, both in feeling and in attitude, it occasions. This he calls the "emotive" use of language, a term eloquent in its apparent acceptance of an invidious distinction between the disinterested, "objective" character of scientific statements and the personal bias or "subjectivity" of all other kinds of statements.[10]

It is true of course that Richards later changed his mind and repudiated this early, positivistic phase in his thinking. But it had a long afterlife, particularly within the formalistic "new criticism" which played a leading role in Anglo-American literary thought between, roughly, 1930 and 1960. Whatever their express ideas about the natural sciences (they sometimes were markedly antagonistic), the proponents of this analytic critical method often tended to emulate the posture of the dispassionate, impersonal, scientific observer. Their primary concern was with the "how" as against the "what" or "why" of literature, and in their effort to arrive at precise, neutral, verifiable knowledge, they tended to treat the literary text as comparable, in its susceptibility to precise analysis and in its virtual autonomy, to the isolatable data studied by physicists. Certain academic exponents of the "new criticism" carried the doctrine to extremes never envisaged by theorists like I. A. Richards. They taught students to confine assertions about literary texts to "the words on the page" and to heed the Blakean motto (taken out of context): "To generalize is to be an idiot." (Humanists often seem to associate scientific rigor with an extreme nominalism and avoidance of generalization.) The tacit aim of this kind of literary study, moreover, was chiefly to enhance the methodological power of specialists in literary study. Instead of being thought of as a capacity of general culture, available to all educated people, the ability to read imaginative literature was recast by the more extreme practitioners of this new formalism into an arcane skill, like the abilty to do physics, accessible only to a tiny minority of expertly trained initiates.

The point about scientism within the humanities—a misplaced application of the assumptions and methods of the natural sciences—is that it can be misleading to gauge the attitudes of humanists toward science by their express opinions alone. It is important to examine what they do (their tacit aims and methods and principles of organization) as well as what they say about science. For they often manage to combine an overt hostility toward the activities of professional scientists, toward science as an institution, with an uncritical and sometimes unconscious emulation of scientific assumptions and procedures. In the realm of literary criticism and scholarship, in any event, the scientistic impulse has remained powerful. After the "new criticism" had lost its vitality, in the 1960s, the yearning of humanists for exact "objective" knowledge, which is to say for a "scientific" critical method, reappeared in such new and ambitious forms as semiotics and a variety of methodological adaptations of "structuralist" principles derived from the latest developments in linguistics.

But the scientistic bent of humanists within the academy was a relatively inconspicuous feature of the cultural history of the recent past. Far more prominent was the new wave of antiscientific thinking that arose in the same period. Following Hiroshima, to repeat, a whole series of problematic science-based innovations had aroused public anxiety about the consequences of scientific discovery. Then the civil crises of the Vietnam era alienated a large segment of the best-educated American youth from any mental work, but especially scientific and technological work, performed in the service of the government or other basic institutions. By the late 1960s, therefore, a large audience was prepared to accept the neo-romantic critique of science at the core of the dissident counterculture.

IV

The viewpoint of Theodore Roszak belongs at the other end of the spectrum of humanist scholars' attitudes toward science from that represented by I. A. Richards and the scientistic literary critics. Taken together, Roszak's two influential books, *The Making of a Counter Culture: Reflections on the Technocratic Society* (1969), and *Where the Wasteland Ends: Politics and Transcendence in Postindustrial Society* (1972),[11] comprise the most systematic effort to formulate a reasoned, coherent ideology expressive of the diffuse antagonism toward science, technology, and scientific rationalism within the dissident "movement" or counterculture which arose during the 1960s.

At first sight Roszak's epistemology would seem to be diametrically opposed to that of those humanists who aspire to exact, "objective" knowledge like that of natural scientists. Yet the striking fact is that a literary theorist like Richards, who in his early writing had endorsed the superior truth value of scientific statements, and Roszak, who regards them as dangerously inadequate, share certain basic assumptions. They both take for granted the antithetical character of objective and subjective, and therefore of scientific and moral (or aesthetic), modes of thought. They both assume that scientific statements are, or come close to being, or provide a compelling illusion of being, "objective." Or, to put it even more subtly, if total objectivity is not finally attainable, Roszak asserts, the fact remains that scientists can still feel and behave as if it were. If, he says, "an epistemology of total objectivity is unattainable, a *psychology* of objectivity is not. There is a way to *feel* and *behave* objectively, even if one cannot *know* objectively."[12]

The apparent objectivity of scientific knowledge is crucial, it would seem, to both the emulation and the antagonism it elicits from humanists. Whereas Richards (in his positivist phase) implied that objectivity conferred a superior authority upon statements made by scientists, Roszak believes just the opposite. He concedes that in the eyes of the gullible public scientific knowledge has immense authority, but it is a misplaced authority. To insist upon the quantifiability and verifiability of knowledge is, on this view, to insure its shallowness and triviality. The experimentally verifiable results of scientific inquiry comprise a body of information of undeniable utility for the mastery and manipulation of the bio-physical world. But when such information is deferred to as the exemplar of true knowledge the results can be disastrous. For the "scientific picture of nature" it provides effectively screens out all qualities of mind and nature, all modes of perception and of being, except those with instrumental value.

> Objective knowing gives a new assembly line system of knowledge, one which relieves us of the necessity to integrate what we study into a moral or metaphysical context which will contribute existential value. We need no longer waste valuable research time and energy seeking for wisdom or depth, since these are qualities of the person. We are free to become specialists.[13]

The scientific style of mind, devoted as it is to "objective knowing," is the radical flaw in the culture of urban industrialism. Roszak calls this style, after William Blake, "the single vision": a one-dimensional, technologically useful, but humanly impoverished world view. Although it happens to be the one form of experience now dignifiable as knowledge, it should not in his view be called that. It would be more accurate, Roszak contends, to call what science reveals to us about nature "information," and to reserve the term "knowledge" for those holistic, often ecstatic syntheses of fact and value—of nature, spirit, and self—which are properly called "gnosis."[14] His epistemology, therefore, must be distinguished from that held by the moderate romantics (Wordsworth, Coleridge, Emerson) who envisaged an accommodation between the two modes of perception, empirical and transcendental, which could effect a "marriage," in Emerson's figure, between matter and mind. Again, Roszak's version of the neo-romantic critique of science is like the positivism of a Richards in seeming to rule out the potential complementarity between the two kinds of knowing.

Although Roszak draws upon the romantic poets, especially Blake, and indeed upon the entire legacy of visionary and prophetic literature going back to the Old Testament, he presses the case against practical reason to a new extreme. Many other writers have insisted upon the superiority of intuitive, nonrational ways of knowing; many have pointed out the severe limitations of scientific rationalism; but few before now have singled out the scientific world view as *the* root cause of what is most alarming about modern societies. According to Roszak, however, it is the critical variable in an essentially destructive, perhaps suicidal, pattern of collective behavior.

> Undeniably, those who defend rationality speak for a valuable human quality. But they often seem not to realize that Reason as they honor it is the god-word of a specific and highly impassioned

ideology handed down to us from our ancestors of the Enlightenment as part of a total cultural and political program. Tied to that
ideology is an aggressive dedication to the urban-industrialization of
the world and to the scientist's universe as the only sane reality. And
tied to the global urban industrialism is an unavoidable technocratic
elitism.[13]

The notion that rationality, or the quasireligious belief in Reason, is the
motive force behind urban industrialism exemplifies Roszak's idealistic theory
of history. Unlike most contemporary historians, he imputes to ideas an
almost exclusive efficacy in social change. He therefore portrays the contemporary world as the scene of an all-encompassing Manichean struggle between
opposed views of reality, each marked by an ideal type of knowing: scientific
rationalism and gnosis. One is reductive, partial, analytic; the other augmentative, holistic, synthetic. The social forms accompanying each are, for his
purposes, largely irrelevant. So far from being important determinants of
human behavior, indeed, social structures and processes are for Roszak
relatively inconsequential reflections of the dominant mental style. What
chiefly accounts for differences in ways of life, accordingly, are differences in
the ruling conceptions of reality. Roszak's theory of history might be called a
form of metaphysical or, to be more specific, epistemological determinism.

Assuming that theories of knowledge are the prime movers of history,
Roszak deals with them apart from the social groups which embraced them or
the functions they served in actual historical situations. This enables him to
discuss the ecstatic, visionary mode of cognition (gnosis) "found in the
world's primitive and pagan societies" without reference to its uses as an
instrument of minority rule or, in many instances, of tyranny. The political
role of the shaman is, so far as he is concerned, largely irrelevant; what matters
is the shaman's "unitary vision bringing together art, religion, science, and
technics."[15] Similarly, he discusses the emergence of the purposive-rational
way of knowing in the Enlightenment without any reference to the larger vision
of scientific progress as a corollary of political, social, and psychological
liberation. My impression is that Roszak's apolitical sense of history as a
battlefield of free-floating ideas is characteristic of the view held by many
adherents of today's counterculture.

The inadequacy of this simple, single-factor mode of historical explanation
is nowhere more apparent than in Roszak's attempt to account for the destructive uses to which our society puts scientific knowledge. Whereas Whitehead's
description of the romantic protest against science had allowed for the distinction between science as a mode of cognition and science as it functions in a
particular social setting, Roszak's protest does not. On the contrary, his
fundamental charge against a science grounded in instrumental reason is that
the evil uses to which it is put follow from its epistemological inadequacy. But
it is not clear whether he considers those evils a necessary or merely a possible
consequence. Although his generalizations imply that scientific knowledge

leads inevitably to flagrant abuses of mankind and of nature, his specific examples are ambiguous.

> We should by now be well aware of the price we pay for regarding aesthetic quality as arbitrary and purely subjective rather than as a real property of the object. Such a view opens the way to that brutishness which feels licensed to devastate the environment on the grounds that beauty is only "a matter of taste." And since one person's taste is as good as another's, who is to say—as a matter of *fact*—that the hard cash of a strip mine counts for less than the grandeur of an untouched mountain?[16]

What does it mean to say that rationality "opens the way" for strip mining? Roszak's point is that the sharp instrumental focus of modern science ignores the aesthetic attributes of the object. To protect the environment, to give mountains and trees adequate "standing" in our culture, we need to restore a sense of the absolute value inherent in natural objects comparable to the divinity imputed to them by primitive (animistic) modes of thought. But Roszak does not give us much help in imagining a mode of knowledge capable of coordinating a geological understanding of a mountain with an apprehension of its allegedly inherent beauty. We have reason to believe, for one thing, that the beauty is not inherent. As Marjorie Hope Nicolson demonstrated years ago, most English travelers before the late sixteenth century regarded the Alps as ugly excrescences on the faces of nature.[17] Had they been practitioners of gnosis they presumably would have advocated strip mining on the slopes of Mont Blanc. In one sense, admittedly, the example is absurd. Roszak's point is that gnosis, by definition, entails a world view incompatible with either modern geology or strip mining. Yet the absurdity does serve to illustrate the all-or-nothing character of the choice we are being invited to make. No accommodation between science as we know it and Roszak's conception of an adequate epistemology is conceivable.

To say that rationality "opens the way" for strip mining is in any case far from saying that the resulting devastation is attributable to science. Let us suppose that advances in geology are among the factors that have contributed directly to the feasibility of strip mining. It is still necessary to consider the relative influence of geological knowledge and economic profitability (Roszak's "hard cash") as motive forces here. Granted that technical competence (equated with "rationality" hence "science" in this lexicon) makes strip mining possible, the fact remains that a business corporation conceives and organizes the operation and a juridical system legitimizes it. In what sense, therefore, is science accountable here?

Roszak's answer embodies the crux of the countercultural critique of science. It is worth noting, incidentally, that he did not address the question in the original draft of this passage. Following the conference in which it was criticized, however, he added this telling afterthought:

> Is such barbarism [i.e., strip mining] to be 'blamed' on science?
> Obviously not in any direct way. But it is deeply rooted in a scien-
> tized reality principle that treats quantities as objective knowledge
> and qualities as a matter of subjective preference.[18]

In other words, the technique of the mining engineer and the economic calculations of the corporate management, like the scientific information of the
geologist, may be thought of as products of the same "scientized reality principle." An old-fashioned historian's distinction between the enabling power of
scientific knowledge and the motives generated and sanctioned by socioeconomic institutions is not meaningful to Roszak. According to his idealistic
interpretation of history, *all* of these activities are traceable to the one root cause:
a rationalistic world view. Since the domination of that ideology is a result of the
advances of scientific knowledge, Roszak is in fact putting the ultimate blame
for the destructiveness of modern society upon science.

To sum up, then, the strip miner's brutish devastation of the landscape
typifies this conception of the way our flawed metaphysic issues in social evil.
The epistemological flaw, again, is the reductionism characteristic of science:
the screening out of those qualities of mind and nature, in this case aesthetic
qualities, not useful to the purpose at hand. If one accepts the major premise
of Roszak's metaphysical determinism, his apocalyptic conclusion follows
logically enough. The scientific view of man's relations to nature is conducive
to a kind of institutionalized moral nihilism. Hence the destructiveness of
urban industrial society is irremediable, and our only hope is to replace it and
the conception of reality from which it derives.

V

Today's criticism of science has a long history in Western thought. To be sure,
a series of shocking events following World War II aroused widespread public
anxiety about the latest advances in research. In one sense, therefore, the
counterculture's attack upon rationality may be interpreted as an extreme
expression of a current mood. But it is necessary to recognize that this recent
development also is a new phase of the "romantic reaction" that began some
two centuries ago. Then, as now, the reliability of scientific knowledge within
its own proper sphere was not in question. Many thinkers noted, however, that
the scientific view of reality imputes excessive importance to the small part of
life susceptible to experimental and logical methods of analysis. Even a writer
like Emerson, who retained much of his Enlightenment faith in scientific progress, expressed a characteristic post-Kantian skepticism about the sufficiency
of practical reason. He did not doubt the absolute validity of "natural facts,"
and since he regarded "Nature" as "the present expositor of the divine
mind,"[19] he believed that knowledge grounded in empirical facticity could in
theory yield the kind of certainty and authority hitherto claimed for religious
truth. But in order to satisfy the full range of human needs, it would be necessary

to "marry" the neutral data to value-laden concepts arrived at by the other (intuitive, mythopoeic, holistic) way of knowing.

By thus insisting upon a coordination of the two modes of cognition, even the more optimistic critics of empiricism helped to prepare the way for today's neo-romantic attack upon science. For the anticipated marriage of fact and value, matter and mind, did not occur. So far from effecting a closer, more meaningful and harmonious relationship between man and nature, science ir the context of Western industrialism is perceived by its latter-day critics as having divested nature of its ultimate, or teleological, significance. At bottom, then, this critique of science is rooted in an essentially religious, suprasensual conception of man's relation with nature. We must recover our capacity for gnosis, according to Roszak, because we desperately require access to the value, meaning, and purpose presumed to reside in "things as a whole."[20] The monster created by science is meaninglessness. Instead of providing the unconditioned meaning sought by mankind, science in the nineteenth century came to be identified with new, more acute forms of alienation from nature.

The dislocation attendant upon the rapidly accelerating rate of industrialization and urbanization was destined, in the long run, to undermine confidence in science. Whether science "caused" these changes, or whether specific technological innovations did in fact derive from scientific research, is largely beside the point. Scientists, inventors, and entrepreneurs, as they functioned within an expanding capitalist society, would appear to critics of that society as kindred embodiments of the same predominantly analytic, secular, matter-of-fact mentality. The emergence of entire industries manifestly based upon recently acquired scientific knowledge, the electrical and chemical industries in particular, subtly eroded the ideal of science as the disinterested pursuit of truth. Idolators of "progress" boasted about the complicity of scientists in changing the face of nature. By the end of the nineteenth century the partnership of science, engineering, and capitalism was acknowledged by its defenders and critics alike. The old distinction between pure and applied science had lost much of its force long before the neo-romantic critique of science had been formulated.

These developments also were to lend more and more credence, as time went on, to Carlyle's choice of the machine as the cardinal metaphor for the emergent industrial system. Scientific knowledge, according to that figure, is the intellectual fuel upon which an expanding machinelike society runs. In the iconography of antiindustrialism, therefore, machines represent the most conspicuous products, both physical and cultural, of modern science. They simultaneously represent several aspects of science-based technology: (1) a new kind of apparatus and technique; (2) the rational organization of work and of economic activity generally; (3) the principles underlying the first two senses of "technology," which is to say the analytic mode of thought and, by extension, the social order typified by a perfection of means and a diminished control over ends.

Beginning with Carlyle's generation, novelists and poets invoked the imagery of mechanization to convey a sense of dwindling human agency, or what may be called, in retrospect, the incipient totalitarianism of industrial society. By the 1960s adherents of the dissident movement throughout the West were invoking the terms "technology" and "the system" and "the machine" more or less interchangeably. All referred to the controlling network of large-scale institutions (government, business corporations, universities) in whose services most scientists do their work. The "machine," in other words, is coterminous with organized society. We are reminded of the inception of the Berkeley uprising in 1964 when one of the leaders, after describing "the operation of the machine" as intolerably odious, called upon his fellow students at the University of California to throw their bodies upon it, if necessary, to make it stop.[21] What is striking about the episode is the extent to which the audience, and the rebellious youth movement in general, seems to have accepted the meaning tacitly imparted to "The Machine."

The received iconographical convention by which "the machine" of science-based high technology is equated with organized society bears witness to the deep-rooted, historical basis for the antiscientific strain in contemporary culture. As Whitehead noted years ago, the literary reaction to the scientific revolution of the eighteenth century expressed the "deep intuitions of mankind penetrating into what is universal in concrete fact." He insisted, moreover, upon the philosophic cogency of the great body of imaginative literature which testifies to the discord between those intuitions—aesthetic, moral, metaphysical—and the "mechanism of science."[22] The discord reflects an increasingly obvious discrepancy between what science provides in the way of certain, verifiable knowledge, and what mankind would have in the way of a meaningful existence. In view of the history of the half-century since Whitehead made these observations, and of the ambiguous part that science has played in that history, it is not surprising that the discord has grown sharper. Nor is there any reason to doubt that it will continue to do so. It would be a serious mistake, accordingly, for those concerned about the future of science to underestimate the appeal, or the force, of the neo-romantic critique of science as a mode of knowing built upon an inadequate metaphysical foundation.

NOTES

1. "Some Thought on the Other Side of This Life," *The New York Times* (April 12, 1973): 45.

2. In 1974, a survey commissioned by the National Science Foundation showed that 39 percent of the American people expressing any opinion do not agree with the proposition that "Overall, science and technology do more good than harm." See Loren Graham, "Concerns about Science and Attempts to Regulate Inquiry," *Daedalus* (Spring 1978).

3. Although the writers I have in mind seldom formulated the distinction between the two referents of the word "science"—science as a mode of perception or inquiry, a way of knowing the world, and science as organized activity, a way of behaving in society—it is often implicit

in their thought. By now, in any case, its importance cannot be overstated; my impression is that much of the confusion surrounding discussions of the "limitations of science" derives from the tendency to conflate these two meanings of "science."

4. *Science and the Modern World* (New York: Macmillan, 1947), p. 138.

5. M. H. Abrams (ed.), *Norton Anthology of English Literature* Vol. 2 (New York: Norton, 1962), p. 89.

6. *Complete Works,* Vol. 1 of 12 vols. (Boston: Houghton, Mifflin, 1884), p. 42.

7. "Science and the Sense of Self: The Medieval Background of a Modern Confrontation," *Daedalus* (Spring 1978).

8. "Signs of the Times," *Critical and Miscellaneous Essays* (Chicago: Bedford, Clarke, n.d.), p. 21.

9. "Analysis and Synthesis as Methodological Themata," *Methodology and Science* 10 (March, 1977): 3–33.

10. See especially chapter 34, "The Two Uses of Language," *Principles of Literary Criticism* (New York: Harcourt Brace, 1950), pp. 261–271.

11. *The Making of a Counter Culture: Reflections on the Technocratic Society* (New York: Doubleday, 1969); *Where the Wasteland Ends: Politics and Transcendence in Postindustrial Society* (Garden City, N.Y.: Doubleday, 1972).

12. *Ibid,* p. 167.

13. *Ibid,* p. 171.

14. "The Monster and the Titan: Science, Knowledge, and Gnosis," *Daedalus* (Summer 1974): 17–32.

15. *Ibid,* p. 27.

16. *Ibid,* pp. 25–26.

17. *Mountain Gloom and Mountain Glory* (Ithaca, N.Y.: Cornell University Press, 1959).

18. "The Monster and the Titan," *op. cit.,* p. 26. This passage did not appear in the first mimeographed draft of the paper which was discussed by the present writer at a conference sponsored by *Daedalus.*

19. *Complete Works, op. cit.,* p. 68.

20. "The Monster and the Titan," *op. cit.,* p. 21.

21. Seymour Martin Lipset and Sheldon Wolin (eds.), *The Berkeley Student Revolt* (New York: Doubleday Anchor, 1965), p. 163. For the background of this iconographical tradition, in American and English literature respectively, see Leo Marx, *The Machine in the Garden: Technology and the Pastoral Ideal in America* (New York: Oxford, 1964), and Herbert Sussman, *Victorians and the Machine* (Cambridge, Mass.: Harvard University Press, 1968).

22. *Science and the Modern World, op. cit.,* pp. 126–127.

23

The Meaning of Life

Kurt Baier

Tolstoy, in his autobiographical work, "A Confession," reports how, when he was fifty and at the height of his literary success, he came to be obsessed by the fear that life was meaningless.

"At first I experienced moments of perplexity and arrest of life, as though I did not know what to do or how to live; and I felt lost and became dejected. But this passed, and I went on living as before. Then these moments of perplexity began to recur oftener and oftener, and always in the same form. They were always expressed by the questions: What is it for? What does it lead to? At first it seemed to me that these were aimless and irrelevant questions. I thought that it was all well known, and that if I should ever wish to deal with the solution it would not cost me much effort; just at present I had no time for it, but when I wanted to, I should be able to find the answer. The questions however began to repeat themselves frequently, and to demand replies more and more insistently, and like drops of ink always falling on one place they ran together into one black blot."[1]

A Christian living in the Middle Ages would not have felt any serious doubts about Tolstoy's questions. To him it would have seemed quite certain that life had a meaning and quite clear what it was. The medieval Christian world picture assigned to man a highly significant, indeed the central part in the grand scheme of things. The universe was made for the express purpose of providing a stage on which to enact a drama starring Man in the title role.

To be exact, the world was created by God in the year 4004 B.C. Man was the last and the crown of this creation, made in the likeness of God, placed in the Garden of Eden on earth, the fixed centre of the universe, round which revolved the nine heavens of the sun, the moon, the planets and the fixed stars, producing as they revolved in their orbits the heavenly harmony of the spheres.

And this gigantic universe was created for the enjoyment of man, who was originally put in control of it. Pain and death were unknown in paradise. But this state of bliss was not to last. Adam and Eve ate of the forbidden tree of knowledge, and life on this earth turned into a death-march through a vale of tears. Then, with the birth of Jesus, new hope came into the world. After He had died on the cross, it became at least possible to wash away with the purifying water of baptism some of the effects of Original Sin and to achieve salvation. That is to say, on condition of obedience to the law of God, man could now enter heaven and regain the state of everlasting, deathless bliss, from which he had been excluded because of the sin of Adam and Eve.

To the medieval Christian the meaning of human life was therefore perfectly clear. The stretch on earth is only a short interlude, a temporary incarceration of the soul in the prison of the body, a brief trial and test, fated to end in death, the release from pain and suffering. What really matters, is the life after the death of the body. One's existence acquires meaning not by gaining what this life can offer but by saving one's immortal soul from death and eternal torture, by gaining eternal life and everlasting bliss.

The scientific world picture which has found ever more general acceptance from the beginning of the modern era onwards is in profound conflict with all this. At first, the Christian conception of the world was discovered to be erroneous in various important details. The Copernican theory showed up the earth as merely one of several planets revolving round the sun, and the sun itself was later seen to be merely one of many fixed stars each of which is itself the nucleus of a solar system similar to our own. Man, instead of occupying the centre of creation, proved to be merely the inhabitant of a celestial body no different from millions of others. Furthermore, geological investigations revealed that the universe was not created a few thousand years ago, but was probably millions of years old.

Disagreements over details of the world picture, however, are only superficial aspects of a much deeper conflict. The appropriateness of the whole Christian outlook is at issue. For Christianity, the world must be regarded as the "creation" of a kind of Superman, a person possessing all the human excellences to an infinite degree and none of the human weaknesses, Who has made man in His image, a feeble, mortal, foolish copy of Himself. In creating the universe, God acts as a sort of playwright-cum-legislator-cum-judge-cum-executioner. In the capacity of playwright, He creates the historical world process, including man. He erects the stage and writes, in outline, the plot. He creates the *dramatis personae* and watches over them with the eye partly of a father, partly of the law. While on stage, the actors are free to extemporise, but if they infringe the divine commandments, they are later dealt with by their creator in His capacity of judge and executioner.

Within such a framework, the Christian attitudes towards the world are natural and sound: it is natural and sound to think that all is arranged for the best even if appearances belie it; to resign oneself cheerfully to one's lot; to be filled with awe and veneration in regard to anything and everything that happens; to want to fall on one's knees and worship and praise the Lord. These are wholly fitting attitudes within the framework of the world view just outlined. And this world view must have seemed wholly sound and acceptable because it offered the best explanation which was then available of all the observed phenomena of nature.

As the natural sciences developed, however, more and more things in the universe came to be explained without the assumption of a supernatural creator. Science, moreover, could explain them better, that is, more accurately and more reliably. The Christian hypothesis of a supernatural maker, whatever other needs it was capable of satisfying, was at any rate no longer indispensable for the purpose of explaining the existence or occurrence of anything. In fact, scientific explanations do not seem to leave any room for this hypothesis. The scientific approach demands that we look for a natural explanation of anything and everything. The scientific way of looking at and explaining things has yielded an immensely greater measure of understanding of, and control over, the universe than any other way. And when one looks at the world in this scientific way, there seems to be no room for a personal relationship between human beings and a supernatural perfect being ruling and guiding men. Hence many scientists and educated men have come to feel that the Christian attitudes towards the world and human existence are inappropriate. They have become convinced that the universe and human existence in it are without a purpose and therefore devoid of meaning.[2]

1. The Explanation of the Universe

Such beliefs are disheartening and unplausible. It is natural to keep looking for the error that must have crept into our arguments. And if an error has crept in, then it is most likely to have crept in with science. For before the rise of science, people did not entertain such melancholy beliefs, while the scientific world picture seems literally to force them on us.

There is one argument which seems to offer the desired way out. It runs somewhat as follows. Science and religion are not really in conflict. They are, on the contrary, mutually complementary, each doing an entirely different job. Science gives provisional, if precise, explanations of small parts of the universe, religion gives final and over-all, if comparatively vague, explanations of the universe as a whole. The objectionable conclusion, that human existence is devoid of meaning, follows only if we use scientific explanations where they do not apply, namely, where total explanations of the whole universe are concerned.[3]

After all, the argument continues, the scientific world picture is the inevitable outcome of rigid adherence to scientific method and explanation, but scientific, that is, causal explanations for their very nature are incapable of producing real illumination. They can at best tell us *how* things are or have come about, but never *why*. They are incapable of making the universe intelligible, comprehensible, meaningful to us. They represent the universe as meaningless, not because it *is* meaningless, but because scientific explanations are not designed to yield answers to investigations into the why and wherefore, into the meaning, purpose, or point of things. Scientific explanations (this argument continues) began, harmlessly enough, as partial and provisional explanations of the movement of material bodies, in particular the planets, within the general framework of the medieval world picture. Newton thought of the universe as a clock made, originally wound up, and occasionally set right by God. His laws of motion only revealed the ways in which the heavenly machinery worked. Explaining the movement of the planets by these laws was analogous to explaining the machinery of a watch. Such explanations showed *how* the thing worked, but not *what it was for or why* it existed. Just as the explanation of how a watch works can help our understanding of the watch only if, in addition, we assume that there is a watchmaker who has designed it for a purpose, made it, and wound it up, so the Newtonian explanation of the solar system helps our understanding of it only on the similar assumption that there is some divine artificer who has designed and made this heavenly clockwork for some purpose, has wound it up, and perhaps even occasionally sets it right, when it is out or order.

Socrates, in the Phaedo complained that only explanations of a thing showing the good or purpose for which it existed could offer a *real* explanation of it. He rejected the kind of explanation we now call "causal" as no more than mentioning "that without which a cause could not be a cause," that is, as merely a necessary condition, but not the *real* cause, the real explanation.[4] In other words, Socrates held that *all* things can be explained in two different ways: either by mentioning merely a necessary condition, or by giving the *real* cause. The former is not an elucidation of the explicandum, not really a help in understanding it, in grasping its "why" and "wherefore."

This Socratic view, however, is wrong. It is not the case that there are two kinds of explanation for everything, one partial, preliminary, and not really clarifying, the other full, final, and illuminating. The truth is that these two kinds of explanation are equally explanatory, equally illuminating, and equally full and final, but that they are appropriate for different kinds of explicanda.

When in an uninhabited forest we find what looks like houses, paved streets, temples, cooking utensils, and the like, it is no great risk to say that these things are the ruins of a deserted city, that is to say, of something man-made. In such a case, the appropriate explanation is teleological, that is, in terms of the purposes of the builders of that city. On the other hand, when a comet approaches the earth, it is similarly a safe bet that, unlike the city in the forest, it was not manufactured by intelligent creatures and that, therefore, a teleological explanation would be out of place, whereas a causal one is suitable.

It is easy to see that in some cases causal, and in others teleological explanations are appropriate. A small satellite circling the earth may or may not have been made by man. We may never know which is the true explanation, but either hypothesis is equally explanatory. It would be wrong to say that only a teleological explanation can *really* explain it. Either explanation would yield complete clarity although, of course, only one can be true. Teleological explanation is only one of several that are possible.

It may indeed be strictly correct to say that the question "*Why* is there a satellite circling the earth?" can only be answered by a teleological explanation. It may be true that "Why?"-questions can really be used properly only in order to elicit *someone's reasons for* doing something. If this is so, it would explain our dissatisfaction with causal answers to "Why?"-questions. But even if it is so, it does not show that "Why is the satellite there?" *must be answered by a teleological explanation.* It shows only that either it must be so answered or it must not be asked. The question "Why have you stopped beating your wife?" can be answered only by a teleological explanation, but if you have never beaten her, it is an improper question. Similarly, if the satellite is not man-made, "Why is there a satellite?" is improper since it implies an origin it did not have. Natural science can indeed only tell us *how* things in nature have come about and not *why,* but this is so not because something else can tell us the *why* and *wherefore,* but because there is none.

There is, however, another point which has not yet been answered. The objection just stated was that causal explanations did not even set out to answer the crucial question. We ask the question "Why?" but science returns an answer to the question "How?" It might now be conceded that this is no ground for a complaint, but perhaps it will instead be said that causal explanations do not give complete or full answers even to that latter question. In causal explanations, it will be objected, the existence of one thing is explained by reference to its cause, but this involves asking for the cause of that cause, and so on, ad infinitum. There is no resting place which is not as much in need of explanation as what has already been explained. Nothing at all is ever fully and completely explained by this sort of explanation.

Leibniz has made this point very persuasively. "Let us suppose a book of the elements of geometry to have been eternal, one copy always having been taken down from an earlier one; it is evident that, even though a reason can be given for the present book out of a past one, nevertheless, out of any number of books, taken in order, going backwards, we shall never come upon *a full* reason; though we might well always wonder why there should have been such books from all time — why there were books at all, and why they were written in this manner. What is true of books is true also of the different states of the world; for what follows is in some way copied from what precedes. . . . And so, however far you go back to earlier states, you will never find in those states *a full reason* why there should be any world rather than none, and why it should be such as it is."[5]

However, a moment's reflection will show that if any type of explanation is merely preliminary and provisional, it is teleological explanation, since it presupposes a background which itself stands in need of explanation. If I account for the existence of the man-made satellite by saying that it was made by some scientists for a certain purpose, then such an explanation can clarify the existence of the satellite only if I assume that there existed materials out of which the satellite was made, and scientists who made it for some purpose. It therefore does not matter what type of explanation we give, whether causal or teleological: either type, any type of explanation, will imply the existence of something by reference to which the explicandum can be explained. And this in turn must be accounted for in the same way, and so on for ever.

But is not God a necessary being? Do we not escape the infinite regress as soon as we reach God? It is often maintained that, unlike ordinary intelligent beings, God is eternal and necessary hence His existence, unlike theirs, is not in need of explanation. For what is it that creates the vicious regress just mentioned? It is that, if we accept the principle of sufficient reason (that there must be an explanation for the existence of anything and everything the existence of which is not logically necessary, but merely contingent[6]), the existence of all the things referred to in any explanation requires itself to be explained. If, however, God is a logically necessary being, then His existence requires no explanation. Hence the vicious regress comes to an end with God.

Now, it need not be denied that God is a necessary being in some sense of that expression. In one of these senses, I, for instance, am a necessary being: it is impossible that I should not exist, because it is self-refuting to say "I do not exist." The same is true of the English language and of the universe. It is self-refuting to say "There is no such thing as the English language" because this sentence is in the English language, or "There is no such thing as the universe" because whatever there is, *is* the universe. It is impossible that these things should not in fact exist since it is impossible that we should be mistaken in thinking that they exist. For what possible occurrence could even throw doubt on our being right on these matters, let alone show that we are wrong? I, the English language, and the universe, are necessary beings, simply in the sense in which all is necessarily true which has been *proved* to be true. The occurrence of utterances such as "I exist," "The English language exists" and "the universe exists" is in itself sufficient proof of their truth. These remarks are therefore necessarily true, hence the things asserted to exist are necessary things.

But this sort of necessity will not satisfy the principle of sufficient reason, because it is only hypothetical or consequential necessity.[7] *Given that* someone says "I exist," then it is logically impossible that *he* should not exist. Given the evidence we have, the English language and the universe most certainly do exist. But there is no necessity about the evidence. On the principle of sufficient reason, we must explain the existence of the evidence, for its existence is not logically necessary.

In other words, the only sense of "necessary being" capable of terminating the vicious regress is "logically necessary being," but it is no longer seriously in dispute that the notion of a logically necessary being is self-contradictory.[8] Whatever can be conceived of as existing can equally be conceived of as not existing.

However, even if per impossibile, there were such a thing as a logically necessary being, we could still not make out a case for the superiority of teleological over causal explanation. The existence of the universe cannot be explained in accordance with the familiar model of manufacture by a crafts-man. For that model presupposes the existence of materials out of which the product is fashioned. God, on the other hand, must create the materials as well. Moreover, although we have a simple model of "creation out of nothing," for composers create tunes out of nothing, yet there is a great differ-ence between creating *something to be sung,* and making the sounds which are a singing of it, or producing the piano on which to play it. Let us, however, waive all these objections and admit, for argument's sake, that creation out of nothing is conceivable. Surely, even so, no one can claim that it is the kind of explanation which yields the clearest and fullest understanding. Surely, to round off scientific explanations of the origin of the universe with creation out of nothing, does not add anything to our *understanding*. There may be merit of some sort in this way of speaking, but whatever it is, it is not greater clarity or explanatory power.[9]

What then, does all this amount to? Merely to the claim that scientific explanations are no worse than any other. All that has been shown is that all explanations suffer from the same defect: all involve a vicious infinite regress. In other words, no type of human explanation can help us to unravel the ultimate, unanswerable mystery. Christian ways of looking at things may not be able to render the world any more lucid than science can, but at least they do not pretend that there are no impenetrable mysteries. On the contrary, they point out untiringly that the claims of science to be able to elucidate everything are hollow. They remind us that science is not merely limited to the exploration of a tiny corner of the universe but that, however far our probing instruments may eventually reach, we can never even approach the answers to the last ques-tions: "Why is there a world at all rather than nothing?" and "Why is the world such as it is and not different?" Here our finite human intellect bumps against its own boundary walls.

Is it true that scientific explanations involve an infinite vicious regress? Are scientific explanations really only provisional and incomplete? The crucial point will be this. Do *all* contingent truths call for explanation? Is the principle of sufficient reason sound? Can scientific explanations never come to a definite end? It will be seen that with a clear grasp of the nature and purpose of explanation we can answer these questions.[10]

Explaining something to someone is making him understand it. This involves bringing together in his mind two things, a model which is accepted as already simple and clear, and that which is to be explained, the explicandum, which is not so. Understanding the explicandum is seeing that it belongs to a range of things which could legitimately have been expected by anyone familiar with the model and with certain facts.

There are, however, two fundamentally different positions which a person may occupy relative to some explicandum. He may not be familiar with any model capable of leading him to expect the phenomenon to be explained. Most of us, for instance, are in that position in relation to the phenomena occurring in a good séance. With regard to other things people will differ. Someone who can play chess, already understands chess, already has such a model. Someone who has never seen a game of chess has not. He sees the moves on the board but he cannot understand, cannot follow, cannot make sense of what is happening. Explaining the game to him is giving him an explanation, is making him understand. He can understand or follow chess moves only if he can see them as conforming to a model of a chess game. In order to acquire such a model, he will, of course, need to know the constitutive rules of chess, that is, the permissible moves. But that is not all. He must know that a normal game of chess is a competition (not all games are) between two people, each trying to win, and he must know what it is to win at chess: to manoeuvre the opponent's king into a position of check-mate. Finally, he must acquire some knowledge of what is and what is not conducive to winning: the tactical rules or canons of the game.

A person who has been given such an explanation and who has mastered it—which may take quite a long time—has now reached understanding, in the sense of the ability to follow each move. A person cannot in that sense understand merely one single move of chess and no other. If he does not understand any other moves, we must say that he has not yet mastered the explanation, that he does not really understand the single move either. If he has mastered the explanation, then he understands all those moves which he can see as being in accordance with the model of the game inculcated in him during the explanation.

However, even though a person who has mastered such an explanation will understand many, perhaps most, moves of any game of chess he cares to watch, he will not necessarily understand them all, as some moves of a player may not be in accordance with his model of the game. White, let us say, at his fifteenth move, exposes his queen to capture by Black's knight. Though in accordance with the constitutive rules of the game, this move is nevertheless perplexing and calls for explanation, because it is not conducive to the achievement by White of what must be assumed to be his aim: to win the game. The queen is a much more valuable piece than the knight against which he is offering to exchange.

An onlooker who has mastered chess may fail to understand this move, be perplexed by it, and wish for an explanation. Of course he may fail to be perplexed, for if he is a very inexperienced player he may not *see* the disadvantageousness of the move. But there is such a need whether anyone sees it or not. The move *calls for* explanation because to anyone who knows the game it must appear to be incompatible with the model which we have learnt during the explanation of the game, and by reference to which we all explain and understand normal games.

However, the required explanation of White's 15th move is of a very different kind. What is needed now is not the acquisition of an explanatory model, but the removal of the real or apparent incompatibility between the player's move and the model of explanation he has already acquired. In such a case the perplexity can be removed only on the assumption that the incompatibility between the model and the game is merely apparent. As our model includes a presumed aim of both players, there are the following three possibilities: (a) White has made a mistake: he has overlooked the threat to his queen. In that case, the explanation is that White thought his move conducive to his end, but it was not. (b) Black has made a mistake: White set a trap for him. In that case, the explanation is that Black thought White's move was not conducive to White's end, but it was. (c) White is not pursuing the end which any chess player may be presumed to pursue: he is not trying to win his game. In that case, the explanation is that White has made a move which he knows is not conducive to the end of winning his game because, let us say, he wishes to please Black who is his boss.

Let us now set out the differences and similarities between the two types of understanding involved in these two kinds of explanation. I shall call the first kind "model"-understanding and explaining, respectively, because both involve the use of a model by reference to which understanding and explaining is effected. The second kind I shall call "unvexing," because the need for this type of explanation and understanding arises only when there is a perplexity arising out of the incompatibility of the model and the facts to be explained.

The first point is that unvexing presupposes model-understanding, but not vice versa. A person can neither have nor fail to have unvexing-understanding of White's fifteenth move at chess, if he does not already have model-understanding of chess. Obviously, if I don't know how to play chess, I shall fail to have model-understanding of White's fifteenth move. But I can neither fail to have nor, of course, can I have unvexing-understanding of it, for I cannot be perplexed by it. I merely fail to have model-understanding of this move as, indeed, of any other move of chess. On the other hand, I may well have model-understanding of chess without having unvexing understanding of every move. That is to say, I may well know how to play chess without understanding White's fifteenth move. A person cannot fail to have unvexing-understanding of the move unless he is vexed or perplexed by it, hence he cannot even fail to have unvexing-understanding unless he already has model-understanding. It is not true that one either understands or fails to understand. On certain occasions, one neither understands nor fails to understand.

The second point is that there are certain things which cannot call for unvexing-explanations. No one can for instance call for an unvexing-explanation of White's first move, which is Pawn to King's Four. For no one can be perplexed or vexed by this move. Either a person knows how to play chess or he does not. If he does, then he must understand this move, for if he does not understand it, he has not yet mastered the game. And if he does not know how to play chess, then he cannot yet have, or fail to have, unvexing-understanding, he cannot therefore need an unvexing-explanation. Intellectual problems do not arise out of ignorance, but out of insufficient knowledge. An ignoramus is puzzled by very little. Once a student can see problems, he is already well into the subject.

The third point is that model-understanding implies being able, without further thought, to have model-understanding of a good many other things, unvexing-understanding does not. A person who knows chess and therefore has model-understanding of it, must understand a good many chess moves, in fact all except those that call for unvexing-explanations. If he claims that he can understand White's first move, but no others, then he is either lying or deceiving himself or he really does not understand any move. On the other hand, a person who, after an unvexing-explanation, understands White's fifteenth move, need not be able, without further explanation, to understand Black's or any other further move which calls for unvexing-explanation.

What is true of explaining deliberate and highly stylized human behaviour such as playing a game of chess is also true of explaining natural phenomena. For what is characteristic of natural phenomena, that they recur in essentially the same way, that they are, so to speak, repeatable, is also true of chess games, as it is not of games of tennis or cricket. There is only one important difference: man himself has invented and laid down the rules of chess, as he has not invented or laid down the "rules or laws governing the behaviour of things." This difference between chess and phenomena is important, for it adds another way to the three already mentioned,[11] in which a perplexity can be removed by an unvexing-explanation, namely, by abandoning the original explanatory model. This is, of course, not possible in the case of games of chess, because the model for chess is not a "construction" on the basis of the already existing phenomena of chess, but an invention. The person who first thought up the model of chess could not have been mistaken. The person who first thought of a model explaining some phenomenon could have been mistaken.

Consider an example. We may think that the following phenomena belong together: the horizon seems to recede however far we walk towards it; we seem to be able to see further the higher the mountain we climb; the sun and moon seem every day to fall into the sea on one side but to come back from behind the mountains on the other side without being any the worse for it. We may explain these phenomena by two alternative models: (a) that the earth is a large disc; (b) that it is a large sphere. However, to a believer in the first theory there arises the following perplexity: how is it that when we travel long enough towards the horizon in any one direction, we do eventually come back to our starting point without ever coming to the edge of the earth? We may at first attempt to "save" the model by saying that there is only an apparent contradiction. We may say either that the model does not require us to come to an edge, for it may be possible only to walk round and round on the flat surface. Or we may say that the person must have walked over the edge without noticing it, or perhaps that the travellers are all lying. Alternatively, the fact that our model is "constructed" and not invented or laid down enables us to say, what we could not do in the case of chess, that the model is inadequate or unsuitable. We can choose another model which fits all the facts, for instance, that the earth is round. Of course, then we have to give an unvexing-explanation for why it *looks* flat, but we are able to do that.

We now can return to our original question, "Are scientific explanations true and full explanations or do they involve an infinite regress, leaving them for ever incomplete?"

Our distinction between model- and unvexing-explanations will help here. It is obvious that only those things which are perplexing *call for* and *can be given* unvexing-explanations. We have already seen that in disposing of one perplexity, we do not necessarily raise another. On the contrary, unvexing-explanations truly and completely explain what they set out to explain, namely, how something is possible which, on our explanatory model, seemed to be impossible. There can therefore be no infinite regress here. Unvexing-explanations are real and complete explanations.

Can there be an infinite regress, then, in the case of model-explanations? Take the following example. European children are puzzled by the fact that their antipodean counterparts do not drop into empty space. This perplexity can be removed by substituting for their explanatory model another one. The European children imagine that throughout space there is an all-pervasive force operating in the same direction as the force that pulls them to the ground. We must, in our revised model, substitute for this force another acting everywhere in the direction of the centre of the earth. Having thus removed their perplexity by giving them an adequate model, we can, however, go on to ask *why* there should be such a force as the force of gravity, why bodies should "naturally," in the absence of forces acting on them, behave in the way stated in Newton's laws. And we might be able to give such an explanation. We might for instance construct a model of space which would exhibit as derivable from it what in Newton's theory are "brute facts." Here we would have a case of the brute facts of one theory being explained within the framework of another, more general theory. And it is a sound methodological principle that we should continue to look for more and more general theories.

Note two points, however. The first is that we must distinguish, as we have seen, between *the possibility* and *the necessity* of giving an explanation. Particular occurrences can be explained by being exhibited as instances of regularities, and regularities can be explained by being exhibited as instances of more general regularities. Such explanations make things clearer. They organize the material before us. They introduce order where previously there was disorder. But absence of this sort of explanation (model-explanation) does not leave us with a puzzle or perplexity, an intellectual restlessness or cramp. The unexplained things are not unintelligible, incomprehensible, or irrational. Some things, on the other hand, call for, require, demand an explanation. As long as we are without such an explanation, we are perplexed, puzzled, intellectually perturbed. We need an unvexing-explanation.

Now, it must be admitted that we may be able to construct a more general theory, from which, let us say, Newton's theory can be derived. This would further clarify the phenomena of motion and would be intellectually satisfying. But failure to do so would not leave us with an intellectual cramp. The facts stated in Newton's theory do not require, or stand in need of, unvexing-explanations. They could do so only if we already had another theory or model with which Newton's theory was incompatible. They could not do so, by themselves, prior to the establishment of such another model.

The second point is that there is an objective limit to which such explanations tend, and beyond which they are pointless. There is a very good reason for wishing to explain a less general by a more general theory. Usually, such a unification goes hand in hand with greater precision in measuring the phenomena which both theories explain. Moreover, the more general theory, because of its greater generality, can explain a wider range of phenomena, including not only phenomena already explained by some other theories but also newly discovered phenomena, which the less general theory cannot explain. Now, the ideal limit to which such expansions of theories tend is an all-embracing theory which unifies all theories and explains all phenomena. Of course, such a limit can never be reached, since new phenomena are constantly discovered. Nevertheless, theories may be tending towards it. It will be remembered that the contention made against scientific theories was that there is no such limit because they involve an infinite regress. On that view, which I reject, there is no conceivable point at which scientific theories could be said to have explained the whole universe. On the view I am defending, there is such a limit, and it is the limit towards which scientific theories are actually tending. I claim that the nearer we come to this limit, the closer we are to a full and complete explanation of everything. For if we were to reach the limit, then though we could, of course, be left with a model which is itself unexplained and could be yet further explained by derivation from another model, there would be no need for, and no point in, such a further explanation. There would be no need for it, because any clearly defined model permitting us to expect the phenomena it is designed to explain offers full and complete explanations of these phenomena, however narrow the range. And while, at lower levels of generality, there is a good reason for providing more general models, since they further simplify, systematize, and organize the phenomena, this, which is the only reason for building more general theories, no longer applies once we reach the ideal limit of an all-embracing explanation.

It might be said that there is another reason for using different models: that they might enable us to discover new phenomena. Theories are not only instruments of explanation, but also of discovery. With this I agree, but it is irrelevant to my point: that *the needs of explanation* do not require us to go on for ever deriving one explanatory model from another.

It must be admitted, then, that in the case of model-explanations there is a regress, but it is neither vicious nor infinite. It is not vicious because, in order to explain a group of explicanda, a model-explanation *need* not itself be derived from another more general one. It gives a perfectly full and consistent explanation by itself. And the regress is not infinite, for there is a natural limit, an all-embracing model, which can explain all phenomena, beyond which it would be pointless to derive model-explanations from yet others.

What about our most serious question, "Why is there anything at all?" Sometimes, when we think about how one thing has developed out of another and that one out of a third, and so on back throughout all time, we are driven to ask the same question about the universe as a whole. We want to add up all things and refer to them by the name, "the world," and we want to know why the world exists and why there is not nothing instead. In such moments, the world seems to us a kind of bubble floating on an ocean of nothingness. Why should such flotsam be adrift in empty space? Surely, its emergence from the hyaline billows of nothingness is more mysterious even than Aphrodite's emergence from the sea. Wittgenstein expressed in these words the mystification we all feel: "Not *how* the world is, is the mystical, but *that* it is. The contemplation of the world *sub specie aeterni* is the contemplation of it as a limited whole. The feeling of the world as a limited whole is the mystical feeling."[12]

Professor J. J. C. Smart expresses his own mystification in these moving words:

"That anything should exist at all does seem to me a matter for the deepest awe. But whether other people feel this sort of awe, and whether they or I ought to is another question. I think we ought to. If so, the question arises: If 'Why should anything exist at all?' cannot be interpreted after the manner of the cosmological argument, that is, as an absurd request for the non-sensical postulation of a logically necessary being, what sort of question is it? What sort of question is this question 'Why should anything exist at all?' All I can say is that I do not yet know."[13]

It is undeniable that the magnitude and perhaps the very existence of the universe is awe-inspiring. It is probably true that it gives many people "the mystical feeling." It is also undeniable that our awe, our mystical feeling, aroused by contemplating the vastness of the world, is justified, in the same sense in which our fear is justified when we realize we are in danger. There is no more appropriate object for our awe or for the mystical feeling than the magnitude and perhaps the existence of the universe, just as there is no more appropriate object for our fear than a situation of personal peril. However, it does not follow from this that it is a good thing to cultivate, or indulge in, awe or mystical feelings, any more than it is necessarily a good thing to cultivate, or indulge in, fear in the presence of danger.

In any case, whether or not we ought to have or are justified in having a mystical feeling or a feeling of awe when contemplating the universe, having such a feeling is not the same as asking a meaningful question, although having it may well *incline us* to utter certain forms of words. Our question "Why is there anything at all?" may be no more than the expression of our feeling of awe or mystification, and not a meaningful question at all. Just as the feeling of fear may naturally but illegitimately give rise to the question "What sin have I committed?," so the feeling of awe or mystification may naturally but illegitimately lead to the question "Why is there anything at all?" What we have to discover, then, is whether this question makes sense or is meaningless.

Yes, of course, it will be said, it makes perfectly good sense. There is an undeniable fact and it calls for explanation. The fact is that the universe exists.

In the light of our experience, there can be no possible doubt that something or other exists, and the claim that the universe exists commits us to no more than that. And surely this calls for explanation, because the universe must have originated somehow. Everything has an origin and the universe is no exception. Since the universe is the totality of things, it must have originated out of nothing. If it had originated out of something, even something as small as one single hydrogen atom, what has so originated could not be the whole universe, but only the universe minus the atom. And then the atom itself would call for explanation, for it too must have had an origin, and it must be *an origin out of nothing*. And how can anything originate out of nothing? Surely that calls for explanation.

However, let us be quite clear what is to be explained. There are two facts here, not one. The first is that the universe exists, which is undeniable. The second is that the universe must have originated out of nothing, and that is not undeniable. It is true that, *if it has originated at all,* then it must have originated out of nothing, or else it is not the universe that has originated. But need it have originated? Could it not have existed for ever?[14] It might be argued that nothing exists for ever, that everything has originated out of something else. That may well be true, but it is perfectly compatible with the fact that the universe is everlasting. We may well be able to trace the origin of any thing to the time when, by some transformation, it has developed out of some other thing, and yet it may be the case that no thing has its origin in nothing, and the universe has existed for ever. For even if every *thing* has a beginning and an end, the total of mass and energy may well remain constant.

Moreover, the hypothesis that the universe originated out of nothing is, empirically speaking, completely empty. Suppose, for argument's sake, that the annihilation of an object without remainder is conceivable. It would still not be possible for any hypothetical observer to ascertain whether space was empty or not. Let us suppose that *within the range of observation of our observer* one object after another is annihilated without remainder and that only one is left. Our observer could not then tell whether in remote parts of the universe, beyond his range of observation, objects are coming into being or passing out of existence. What, moreover, are we to say of the observer himself? Is he to count for nothing? Must we not postulate him away as well, if the universe is to have arisen out of nothing?

Let us, however, ignore all these difficulties and assume that the universe really has originated out of nothing. Even that does not prove that the universe has not existed for ever. If the universe can conceivably develop out of nothing, then it can conceivably vanish without remainder. And it can arise out of nothing again, and subside into nothingness once more, and so on ad infinitum. Of course, "again" and "once more" are not quite the right words. The concept of time hardly applies to such universes. It does not make sense to ask whether one of them is earlier or later than, or perhaps simultaneous with, the other because we cannot ask whether they occupy the same or different spaces. Being separated from one another by "nothing," they are not separated from one another by "anything." We cannot therefore make any statement about their mutual spatio-temporal relations. It is impossible to distinguish between one long continuous universe and two universes separated by nothing. How, for instance, can we tell whether the universe including ourselves is not frequently annihilated and "again" reconstituted just as it was?

Let us now waive these difficulties as well. Let us suppose for a moment that we understand what is meant by saying that the universe originated out of nothing and that this has happened only once. Let us accept this as a fact. Does this fact call for explanation?

It does not call for an unvexing-explanation. That would be called for only if there were a perplexity due to the incompatibility of an accepted model with some fact. In our case, the fact to be explained is the origination of the universe out of nothing, hence there could not be such a perplexity, for we need not employ a model incompatible with this. If we had a model incompatible with our "fact," then that would be the wrong model and we would simply have to substitute another for it. The model we employ to explain the origin of the universe out of nothing could not be based on the similar origins of other things for, of course, there is nothing else with a similar origin.

All the same, it seems very surprising that something should have come out of nothing. It is contrary to the principle that every thing has an origin, that is, has developed out of something else. It must be admitted that there is this incompatibility. However, it does not arise because a well-established model does not square with an undeniable fact; it arises because a well-established model does not square with *an assumption* of which it is hard even to make sense and for which there is no evidence whatsoever. In fact, the only reason we have for making this assumption is a simple logical howler: that because every thing has an origin, the universe must have an origin, too, except, that, being the universe, it must have originated out of nothing. This is a howler, because it conceives of the universe as a big thing, whereas in fact it is the totality of things, that is, not a thing. That every thing has an origin does not entail that the totality of things has an origin. On the contrary, it strongly suggests that it has not. For to say that every thing has an origin implies that any given thing must have developed out of something else which in turn, being a thing, must have developed out of something else, and so forth. If we assume that every thing has an origin, we need not, indeed it is hard to see how we can, assume that the totality of things has an origin as well. There is therefore no perplexity, because we need not and should not assume that the universe has originated out of nothing.

If, however, in spite of all that has been said just now, someone still wishes to assume, contrary to all reason, that the universe has originated out of nothing, there would still be no perplexity, for then he would simply have to give up the principle which is incompatible with this assumption, namely, that no thing can originate out of nothing. After all, this principle *could* allow for exceptions. We have no proof that it does not. Again, there is no perplexity, because no incompatibility between our assumption and an inescapable principle.

But it might be asked, do we not need a model-explanation of our supposed fact? The answer is No. We do not need such an explanation, for there could not possibly be a model for this origin other than this origin itself. We cannot say that origination out of nothing is like birth, or emergence, or evolution, or anything else we know for it is not like anything we know. In all these cases, there is *something* out of which the new thing has originated.

To sum up. The question, "Why is there anything at all?" looks like a perfectly sensible question modelled on "Why does *this* exist?" or "How has *this* originated?" It looks like a question about the origin of a thing. However, it is not such a question, for the universe is not a thing, but the totality of things. There is therefore no reason to assume that the universe has an origin. The very assumption that it has is fraught with contradictions and absurdities. If, nevertheless, it were true that the universe has originated out of nothing, then this would not call either for an unvexing or a model-explanation. It would not call for the latter, because there could be no model of it taken from another part of our experience, since there is nothing analogous in our experience to origination out of nothing. It would not call for the former, because there can be no perplexity due to the incompatibility of a well-established model and an undeniable fact, since there is no undeniable fact and no well-established model. If, on the other hand, as is more probable, the universe has not originated at all, but is eternal, then the question why or how it has originated simply does not arise. There can then be no question about why anything at all exists, for it could not mean how or why the universe had originated, since ex hypothesi it has no origin. And what else could it mean?

Lastly, we must bear in mind that the hypothesis that the universe was made by God out of nothing only brings us back to the question who made God or how God originated. And if we do not find it repugnant to say that God is eternal, we cannot find it repugnant to say that the universe is eternal. The only difference is that we know for certain that the universe exists, while we have the greatest difficulty in even making sense of the claim that God exists.

To sum up. According to the argument examined, we must reject the scientific world picture because it is the outcome of scientific types of explanation which do not really and fully explain the world around us, but only tell us *how* things have come about, not *why,* and can give no answer to the ultimate question, why there is anything at all rather than nothing. Against this, I have argued that scientific explanations are real and full, just like the explanations of everyday life and of the traditional religions. They differ from those latter only in that they are more precise and more easily disprovable by the observation of facts.

My main points dealt with the question why scientific explanations were thought to be merely provisional and partial. The first main reason is the misunderstanding of the difference between teleological and causal explanations. It is first, and rightly, maintained that teleological explanations are answers to "Why?"-questions, while causal explanations are answers to "How?"-questions. It is further, and wrongly, maintained that, in order to obtain real and full explanations of anything, one must answer both "Why?" and "How?" questions. In other words, it is thought that all matters can and must be explained by both teleological and causal types of explanation. Causal explanations, it is believed, are merely provisional and partial, waiting to be completed by teleological explanations. Until a teleological explanation has been given, so the story goes, we have not *really* understood the explicandum. However, I have shown that both types are equally real and full explanations. The difference between them is merely that they are appropriate to different types of explicanda.

It should, moreover, be borne in mind that teleological explanations are not, in any sense, unscientific. They are rightly rejected in the natural sciences, not however because they are unscientific, but because no intelligences or purposes are found to be involved there. On the other hand, teleological explanations are very much in place in psychology, for we find intelligence and purpose involved in a good deal of human behaviour. It is not only not unscientific to give teleological explanations of deliberate human behaviour, but it would be quite unscientific to exclude them.

The second reason why scientific explanations are thought to be merely provisional and partial, is that they are believed to involve a vicious infinite regress. Two misconceptions have led to this important error. The first is the general misunderstanding of the nature of explanation, and in particular the failure to distinguish between the two types which I have called model- and unvexing-explanations, respectively. If one does not draw this distinction, it is natural to conclude that scientific explanations lead to a vicious infinite regress. For while it is true of those perplexing matters which are elucidated by unvexing-explanations that they are incomprehensible and cry out for explanation, it is not true that after an unvexing-explanation has been given, this itself is again capable, let alone in need of, a yet further explanation of the same kind. Conversely, while it is true that model-explanations of regularities can themselves be further explained by more general model-explanations, it is not true that, in the absence of such more general explanations, the less general are incomplete, hang in the air, so to speak, leaving the explicandum incomprehensible and crying out for explanation. The distinction between the two types of explanation shows us that an explicandum is either perplexing and incomprehensible, in which case an explanation of it *is necessary* for clarification and, when given, *complete,* or it is a regularity capable of being subsumed under a model, in which case a further explanation *is possible* and often profitable, but *not necessary* for clarification.

The second misconception responsible for the belief in a vicious infinite regress is the misrepresentation of scientific explanation *as essentially causal.* It has generally been held that, in a scientific explanation, the explicandum is the effect of some event, the cause, temporally prior to the explicandum. Combined with the principle of sufficient reason, (the principle that anything is in need of explanation which might conceivably have been different from what it is), this error generates the nightmare of determinism. Since any event might have been different from what it was, acceptance of this principle has the consequence that *every* event must have a reason or explanation. But if the reason is itself an event *prior in time,* then every reason must have a reason preceding it, and so the infinite regress of explanation is necessarily tied to the time scale stretching infinitely back into the endless past. It is, however, obvious from our account that science is not primarily concerned with the forging of such causal chains. The primary object of the natural sciences is not historical at all. Natural science claims to reveal, not the beginnings of things, but their underlying reality. It does not dig up the past, it digs down into the structure of things existing here and now. Some scientists do allow themselves to speculate, and rather precariously at that, about origins. But their hard work is done on the structure of what exists now. In particular those explanations which are themselves further explained are not explanations linking event to event in a gapless chain reaching back to creation day, but generalisations of theories tending towards a unified theory.

2. The Purpose of Man's Existence

Our conclusion in the previous section has been that science is in principle able to give complete and real explanations of every occurrence and thing in the universe. This has two important corollaries: (i) Acceptance of the scientific world picture cannot be *one's reason for* the belief that the universe is unintelligible and therefore meaningless, though coming to accept it, after having been taught the Christian world picture, may well have been, in the case of many individuals, *the only or the main cause* of their belief that the universe and human existence are meaningless. (ii) It is not in accordance with reason to reject this pessimistic belief on the grounds that scientific explanations are only provisional and incomplete and must be supplemented by religious ones.

In fact, it might be argued that the more clearly we understand the explanations given by science, the more we are driven to the conclusion that human life has no purpose and therefore no meaning. The science of astronomy teaches us that our earth was not specially created about 6,000 years ago, but evolved out of hot nebulae which previously had whirled aimlessly through space for countless ages. As they cooled, the sun and the planets formed. On one of these planets at a certain time the circumstances were propitious and life developed. But conditions will not remain favourable to life. When our solar system grows old, the sun will cool, our planet will be covered with ice, and all living creatures will eventually perish. Another theory has it that the sun will explode and that the heat generated will be so great that all organic life on earth will be destroyed. That is the comparatively short history and prospect of life on earth. Altogether it amounts to very little when compared with the endless history of the inanimate universe.

Biology teaches us that the species man was not specially created but is merely, in a long chain of evolutionary changes of forms of life, the last link, made in the likeness not of God but of nothing so much as an ape. The rest of the universe, whether animate or inanimate, instead of serving the ends of man, is at best indifferent, at worst savagely hostile. Evolution to whose operation the emergence of man is due is a ceaseless battle among members of different species, one species being gobbled up by another, only the fittest surviving. Far from being the gentlest and most highly moral, man is simply the creature best fitted to survive, the most efficient if not the most rapacious and insatiable killer. And in this unplanned, fortuitous, monstrous, savage world man is madly trying to snatch a few brief moments of joy, in the short intervals during which he is free from pain, sickness, persecution, war or famine until, finally, his life is snuffed out in death. Science has helped us to know and understand this world, but what purpose or meaning can it find in it?

Complaints such as these do not mean quite the same to everybody, but one thing, I think, they mean to most people: science shows life to be meaningless, because life is without purpose. The medieval world picture provided life with a purpose, hence medieval Christians could believe that life had a meaning. The scientific account of the world takes away life's purpose and with it its meaning.

There are, however, two quite different senses of "purpose." Which one is meant? Has science deprived human life of purpose in both senses? And if not, is it a harmless sense, in which human existence has been robbed of purpose? Could human existence still have meaning if it did not have a purpose in that sense?

What are the two senses? In the first and basic sense, purpose is normally attributed only to persons or their behaviour as in "Did you have a purpose in leaving the ignition on?" In the second sense, purpose is normally attributed only to things, as in "What is the purpose of that gadget you installed in the workshop?" The two uses are intimately connected. We cannot attribute a purpose to a thing without implying that someone did something, in the doing of which he had some purpose, namely, to bring about the thing with the purpose. Of course, *his* purpose is not identical with *its* purpose. In hiring labourers and engineers and buying materials and a site for a factory and the like, the entrepreneur's purpose, let us say, is to manufacture cars, but the purpose of cars is to serve as a means of transportation.

There are many things that a man may do, such as buying and selling, hiring labourers, ploughing, felling trees, and the like, which it is foolish, pointless, silly, perhaps crazy, to do if one has no purpose in doing them. A man who does these things without a purpose is engaging in inane, futile pursuits. Lives crammed full with such activities devoid of purpose are pointless, futile, worthless. Such lives may indeed be dismissed as meaningless. But it should also be perfectly clear that acceptance of the scientific world picture does not force us to regard our lives as being without a purpose in this sense. Science has not only not robbed us of any purpose which we had before, but it has furnished us with enormously greater power to achieve these purposes. Instead of praying for rain or a good harvest or offspring, we now use ice pellets, artificial manure, or artificial insemination.

By contrast, having or not having a purpose, in the other sense, is value neutral. We do not think more or less highly of a thing for having or not having a purpose. "Having a purpose," in this sense, confers no kudos, "being purposeless" carries no stigma. A row of trees growing near a farm may or may not have a purpose: it may or may not be a windbreak, may or may not have been planted or deliberately left standing there in order to prevent the wind from sweeping across the fields. We do not in any way disparage the trees if we say they have no purpose, but have just grown that way. They are as beautiful, made of as good wood, as valuable, as if they had a purpose. And, of course, they break the wind just as well. The same is true of living creatures. We do not disparage a dog when we say that it has no purpose, is not a sheep dog or a watch dog or a rabbiting dog, but just a dog that hangs around the house and is fed by us.

Man is in a different category, however. To attribute to a human being a purpose in that sense is not neutral, let alone complimentary: it is offensive. It is degrading for a man to be regarded as merely serving a purpose. If, at a garden party, I ask a man in livery, "What is your purpose?" I am insulting him. I might as well have asked, "What are you *for*?" Such questions reduce him to the level of a gadget, a domestic animal, or perhaps a slave. I imply that *we* allot to *him* the tasks, the goals, the aims which he is to pursue; that *his* wishes and desires and aspirations and purposes are to count for little or nothing. We are treating him, in Kant's phrase, merely as a means to our ends, not as an end in himself.

The Christian and the scientific world pictures do indeed differ fundamentally on this point. The latter robs man of a purpose in this sense. It sees him as a being with no purpose allotted to him by anyone but himself. It robs him of any goal, purpose, or destiny appointed for him by any outside agency. The Christian world picture, on the other hand, sees man as a creature, a divine artefact, something halfway between a robot (manufactured) and an animal (alive), a homunculus, or perhaps Frankenstein, made in God's laboratory, with a purpose or task assigned him by his Maker.

However, lack of purpose in this sense does not in any way detract from the meaningfulness of life. I suspect that many who reject the scientific outlook because it involves the loss of purpose of life, and therefore meaning, are guilty of a confusion between the two senses of "purpose" just distinguished. They confusedly think that if the scientific world picture is true, then their lives must be futile because that picture implies that man has no purpose given him from without. But this is muddled thinking, for, as has already been shown, pointlessness is implied only by purposelessness in the other sense, which is not at all implied by the scientific picture of the world. These people mistakenly conclude that there can be no purpose *in* life because there is no purpose *of* life; that *men* cannot themselves adopt and achieve purposes because *man*, unlike a robot or a watchdog, is not a creature with a purpose.[15]

However, not all people taking this view are guilty of the above confusion. Some really hanker after a purpose of life in this sense. To some people the greatest attraction of the medieval world picture is the belief in an omnipotent, omniscient, and all-good Father, the view of themselves as His children who worship Him, of their proper attitude to what befalls them as submission, humility, resignation in His will, and what is often described as the "creaturely feeling."[16] All these are attitudes and feelings appropriate to a being that stands to another in the same sort of relation, though of course on a higher plane, in which a helpless child stands to his progenitor. Many regard the scientific picture of the world as cold, unsympathetic, unhomely, frightening, because it does not provide for any appropriate object of this creaturely attitude. There is nothing and no one in the world, as science depicts it, in which we can have faith or trust, on whose guidance we can rely, to whom we can turn for consolation, whom we can worship or submit to—except other human beings. This may be felt as a keen disappointment, because it shows that the meaning of life cannot lie in submission to His will, in acceptance of whatever may come, and in worship. But it does not imply that life can have *no* meaning. It merely implies that it must have a different meaning from that which it was thought to have. Just as it is a great shock for a child to find that he must stand on his own feet, that his father and mother no longer provide for him, so a person who has lost his faith in God must reconcile himself to the idea that he has to stand on his own feet, alone in the world except for whatever friends he may succeed in making.

But is not this to miss the point of the Christian teaching? Surely, Christianity can tell us the meaning of life because it tells us the grand and noble end for which God has created the universe and man. No human life, however pointless it may seem, is meaningless because in being part of God's plan, every life is assured of significance.

This point is well taken. It brings to light a distinction of some importance: we call a person's life meaningful not only if it is worthwhile, but also if he has helped in the realization of some plan or purpose transcending his own concerns. A person who knows he must soon die a painful death can give significance to the remainder of his doomed life by, say, allowing certain experiments to be performed on him which will be useful in the fight against cancer. In a similar way, only on a much more elevated plane, every man, however humble or plagued by suffering, is guaranteed significance by the knowledge that he is participating in God's purpose.

What, then, on the Christian view, is the grand and noble end for which God has created the world and man in it? We can immediately dismiss that still popular opinion that the smallness of our intellect prevents us from stating meaningfully God's design in all its imposing grandeur.[17] This view cannot possibly be a satisfactory answer to our question about the purpose of life. It is, rather, a confession of the impossibility of giving one. If anyone thinks that this "answer" can remove the sting from the impression of meaninglessness and insignificance in our lives, he cannot have been stung very hard.

If, then, we turn to those who are willing to state God's purpose in so many words, we encounter two insuperable difficulties. The first is to find a purpose grand and noble enough to explain and justify the great amount of undeserved suffering in this world. We are inevitably filled by a sense of pathos when we read statements such as this: ". . . history is the scene of a divine purpose, in which the whole of history is included, and Jesus of Nazareth is the centre of that purpose, both as revelation and as achievement, as the fulfilment of all that was past, and the promise of all that was to come . . . If God is God, and if He made all these things, why did He do it? . . . God created a universe, bounded by the categories of time, space, matter, and causality, because He desired to enjoy for ever the society of a fellowship of finite and redeemed spirits which have made to His love the response of free and voluntary love and service."[18] Surely this cannot be right? Could a God be called omniscient, omnipotent, *and* all-good who, for the sake of satisfying his desire to be loved and served, imposes (or has to impose) on his creatures the amount of undeserved suffering we find in the world?

There is, however, a much more serious difficulty still: God's purpose in making the universe must be stated in terms of a dramatic story many of whose key incidents symbolize religious conceptions and practices which we no longer find morally acceptable: the imposition of a taboo on the fruits of a certain tree, the sin and guilt incurred by Adam and Eve by violating the taboo, the wrath of God,[19] the curse of Adam and Eve and all their progeny, the expulsion from Paradise, the Atonement by Christ's bloody sacrifice on the cross which makes available by way of the sacraments God's Grace by which alone men can be saved (thereby, incidentally, establishing the valuable power of priests to forgive sins and thus alone make possible a man's entry to heaven[20]) Judgment Day on which the sheep are separated from the goats and the latter condemned to eternal torment in hell-fire.

Obviously it is much more difficult to formulate a purpose for creating the universe and man that will justify the enormous amount of undeserved suffering which we find around us, if that story has to be fitted in as well. For now we have to explain not only why an omnipotent, omniscient, and all-good God should create such a universe and such a man, but also why, foreseeing every move of the feeble, weak-willed, ignorant, and covetous creature to be created, He should nevertheless have created him and, having done so, should be incensed and outraged by man's sin, and why He should deem it necessary to sacrifice His own son on the cross to atone for this sin which was, after all, only a disobedience of one of his commands, and why this atonement and consequent redemption could not have been followed by man's return to Paradise—particularly of those innocent children who had not yet sinned—and why, on Judgment Day, this merciful God should condemn some to eternal torment.[21] It is not surprising that in the face of these and other difficulties, we find, again and again, a return to the first view: that God's purpose cannot meaningfully be stated.

It will perhaps be objected that no Christian today believes in the dramatic history of the world as I have presented it. But this is not so. It is the official doctrine of the Roman Catholic, the Greek Orthodox, and a large section of the Anglican Church.[22] Nor does Protestantism substantially alter this picture. In fact, by insisting on "Justification by Faith Alone" and by rejecting the ritualistic, magical character of the medieval Catholic interpretation of certain elements in the Christian religion, such as indulgences, the sacraments, and prayer, while at the same time insisting on the necessity of grace, Protestantism undermined the moral element in medieval Christianity expressed in the Catholics' emphasis on personal merit.[23] Protestantism, by harking back to St. Augustine, who clearly realized the incompatibility of grace and personal merit,[24] opened the way for Calvin's doctrine of Predestination (the intellectual parent of that form of rigid determinism which is usually blamed on science) and Salvation or Condemnation from all eternity.[25] Since Roman Catholics, Lutherans, Calvinists, Presbyterians and Baptists officially subscribe to the views just outlined, one can justifiably claim that the overwhelming majority of professing Christians hold or ought to hold them.

It might still be objected that the best and most modern views are wholly different. I have not the necessary knowledge to pronounce on the accuracy of this claim. It may well be true that the best and most modern views are such as Professor Braithwaite's who maintains that Christianity is, roughly speaking, "morality plus stories," where the stories are intended merely to make the strict moral teaching both more easily understandable and more palatable.[26] Or it may be that one or the other of the modern views on the nature and importance of the dramatic story told in the sacred Scriptures is the best. My reply is that, even if it is true, it does not prove what I wish to disprove, that one can extract a sensible answer to our question, "What is the meaning of life?" from the kind of story subscribed to by the overwhelming majority of Christians, who would, moreover, reject any such modernist interpretation at least as indignantly as the scientific account. Moreover, though such views can perhaps avoid some of the worst absurdities of the traditional story, they are hardly in a much better position to state the purpose for which God has created the universe and man in it, because they cannot overcome the difficulty of finding a purpose grand and noble enough to justify the enormous amount of undeserved suffering in the world.

Let us, however, for argument's sake, waive all these objections. There remains one fundamental hurdle which no form of Christianity can overcome: the fact that it demands of man a morally repugnant attitude towards the universe. It is now very widely held[27] that the basic element of the Christian religion is an attitude of worship towards a being supremely worthy of being worshipped and that it is religious feelings and experiences which apprise their owner of such a being and which inspire in him the knowledge or the feeling of complete dependence, awe, worship, mystery, and self-abasement. There is, in other words, a bi-polarity (the famous "I-Thou relationship") in which the object, "the wholly-other," is exalted whereas the subject is abased to the limit. Rudolf Otto has called this the "creature-feeling"[28] and he quotes as an expression of it, Abraham's words when venturing to plead for the men of Sodom: "Behold now, I have taken upon me to speak unto the Lord, which am but dust and ashes" (Gen. XVIII.27). Christianity thus demands of men an attitude inconsistent with one of the presuppositions of morality: that man is not wholly dependent on something else, that man has free will, that man is in principle capable of responsibility. We have seen that the concept of grace is the Christian attempt to reconcile the claim of total dependence and the claim of individual responsibility (partial independence), and it is obvious that such attempts must fail. We may dismiss certain doctrines, such as the doctrine of original sin or the doctrine of eternal hellfire or the doctrine that there can be no salvation outside the Church as extravagant and peripheral, but we cannot reject the doctrine of total dependence without rejecting the characteristically Christian attitude as such.

3. The Meaning of Life

Perhaps some of you will have felt that I have been shirking the real problem. To many people the crux of the matter seems as follows. How can there be any meaning in our life if it ends in death? What meaning can there be in it that our inevitable death does not destroy? How can our existence be meaningful if there is no after-life in which perfect justice is meted out? How can life have any meaning if all it holds out to us are a few miserable earthly pleasures and even these to be enjoyed only rarely and for such a piteously short time?

I believe this is the point which exercises most people most deeply. Kirilov, in Dostoevsky's novel, *The Possessed,* claims, just before committing suicide, that as soon as we realize that there is no God, we cannot live any longer, we must put an end to our lives. One of the reasons which he gives is that when we discover that there is no paradise, we have nothing to live for.

". . . there was a day on earth, and in the middle of the earth were three crosses. One on the cross had such faith that He said to another, 'To-day thou shalt be with me in paradise.' The day came to an end, both died, and they went, but they found neither paradise nor resurrection. The saying did not come true. Listen: that man was the highest of all on earth . . . There has never been any one like Him before or since, and never will be . . . And if that is so, if the laws of Nature did not spare even *Him,* and made even him live in the midst of lies and die for a lie, then the whole planet is a lie and is based on a lie and a stupid mockery. So the very laws of the planet are a lie and a farce of the devil. What, then, is there to live for?"[29] And Tolstoy, too, was nearly driven to suicide when he came to doubt the existence of God and an after-life.[30] And this is true of many.

What, then, is it that inclines us to think that if life is to have a meaning, there would be an after-life? It is this. The Christian world view contains the following three propositions. The first is that since the Fall, God's curse of Adam and Eve, and the expulsion from Paradise, life on earth for mankind has not been worth while, but a vale of tears, one long chain of misery, suffering, unhappiness, and injustice. The second is that a perfect after-life is awaiting us after the death of the body. The third is that we can enter this perfect life only on certain conditions, among which is also the condition of enduring our earthly existence to its bitter end. In this way, our earthly existence which, in itself, would not (at least for many people if not all) be worth living, acquires meaning and significance: only if we endure it, can we gain admission to the realm of the blessed.

It might be doubted whether this view is still held today. However, there can be no doubt that even today we all imbibe a good deal of this view with our earliest education. In sermons, the contrast between the perfect life of the blessed and our life of sorrow and drudgery is frequently driven home and we hear it again and again that Christianity has a message of hope and consolation for all those "who are weary and heavy laden."[31]

It is not surprising, then, that when the implications of the scientific world picture begin to sink in, when we come to have doubts about the existence of God and another life, we are bitterly disappointed. For if there is no afterlife, then all we are left is our earthly life which we have come to regard as a necessary evil, the painful fee of admission to the land of eternal bliss. But if there is no eternal bliss to come and if this hell on earth is all, why hang on till the horrible end?

Our disappointment therefore arises out of these two propositions, that the earthly life is not worth living, and that there is another perfect life of eternal happiness and joy which we may enter upon if we satisfy certain conditions. We can regard our lives as meaningful, if we believe both. We cannot regard them as meaningful if we believe merely the first and not the second. It seems to me inevitable that people who are taught something of the history of science, will have serious doubts about the second. If they cannot overcome these, as many will be unable to do, then they must either accept the sad view that their life is meaningless or they must abandon the first proposition: that this earthly life is not worth living. They must find the meaning of their life in this earthly existence. But is this possible?

A moment's examination will show us that the Christian evaluation of our earthly life as worthless, which we accept in our moments of pessimism and dissatisfaction, is not one that we normally accept. Consider only the question of murder and suicide. On the Christian view, other things being equal, the most kindly thing to do would be for every one of us to kill as many of our friends and dear ones as still have the misfortune to be alive, and then to commit suicide without delay, for every moment spent in this life is wasted. On the Christian view, God has not made it that easy for us. He has forbidden us to hasten others or ourselves into the next life. Our bodies are his private property and must be allowed to wear themselves out in the way decided by Him, however painful and horrible that may be. We are, as it were, driving a burning car. There is only one way out, to jump clear and let it hurtle to destruction. But the owner of the car has forbidden it on pain of eternal tortures worse than burning. And so we do better to burn to death inside.

On this view, murder is a less serious wrong than suicide. For murder can always be confessed and repented and therefore forgiven, suicide cannot— unless we allow the ingenious way out chosen by the heroine of Graham Greene's play, The Living Room, who swallows a slow but deadly poison and, while awaiting its taking effect, repents having taken it. Murder, on the other hand, is not so serious because, in the first place, it need not rob the victim of anything but the last lap of his march in the vale of tears, and, in the second place, it can always be forgiven. Hamlet, it will be remembered, refrains from killing his uncle during the latter's prayers because, as a true Christian, he believes that killing his uncle at that point, when the latter has purified his soul by repentance, would merely be doing him a good turn, for murder at such a time would simply despatch him to undeserved and everlasting happiness.

These views strike us as odd, to say the least. They are the logical conse- quence of the official medieval evaluation of this our earthly existence. If this life is not worth living, then taking it is not robbing the person concerned of much. The only thing wrong with it is the damage to God's property, which is the same both in the case of murder and suicide. We do not take this view at all. Our view, on the contrary, is that murder is the most serious wrong because it consists in taking away from some one else against his will his most precious possession, his life. For this reason, when a person suffering from an incurable disease asks to be killed, the mercy killing of such a person is regarded as a much less serious crime than murder because, in such a case, the killer is not robbing the other of a good against his will. Suicide is not regarded as a real crime at all, for we take the view that a person can do with his own possessions what he likes.

However, from the fact that these are our normal opinions, we can infer nothing about their truth. After all, we could easily be mistaken. Whether life is or is not worthwhile, is a value judgment. Perhaps all this is merely a matter of opinion or taste. Perhaps no objective answer can be given. Fortunately, we need not enter deeply into these difficult and controversial questions. It is quite easy to show that the medieval evaluation of earthly life is based on a misguided procedure.

Let us remind ourselves briefly of how we arrive at our value judgments. When we determine the merits of students, meals, tennis players, bulls, or bathing belles, we do so on the basis of some criteria and some standard or norm. Criteria and standards notoriously vary from field to field and even from case to case. But that does not mean that we have *no* idea about what are the appropriate criteria or standards to use. It would not be fitting to apply the criteria for judging bulls to the judgment of students or bathing belles. They score on quite different points. And even where the same criteria are appro- priate as in the judgment of students enrolled in different schools and univer- sities, the standards will vary from one institution to another. Pupils who would only just pass in one, would perhaps obtain honours in another. The higher the standard applied, the lower the marks, that is, the merit conceded to the candidate.

The same procedure is applicable also in the evaluation of a life. We examine it on the basis of certain criteria and standards. The medieval Christian view uses the criteria of the ordinary man: a life is judged by what the person concerned can get out of it: the balance of happiness over unhappiness, pleasure over pain, bliss over suffering. Our earthly life is judged not worth while because it contains much unhappiness, pain, and suffering, little happiness, pleasure, and bliss. The next life is judged worth while because it provides eternal bliss and no suffering.

Armed with these criteria, we can compare the life of this man and that, and judge which is more worth while, which has a greater balance of bliss over suffering. But criteria alone enable us merely to make comparative judgments of value, not absolute ones. We can say which is more and which is less worth while, but we cannot say which is worth while and which is not. In order to determine the latter, we must introduce a standard. But what standard ought we to choose?

Ordinarily, the standard we employ is the average of the kind. We call a man and a tree tall if they are well above the average of their kind. We do not say that Jones is a short man because he is shorter than a tree. We do not judge a boy a bad student because his answer to a question in the Leaving Examination is much worse than that given in reply to the same question by a young man sitting for his finals for the Bachelor's degree.

The same principles must apply to judging lives. When we ask whether a given life was or was not worth while, then we must take into consideration the range of worthwhileness which ordinary lives normally cover. Our end poles of the scale must be the best possible and the worst possible life that one finds. A good and worthwhile life is one that is well above average. A bad one is one well below.

The Christian evaluation of earthly lives is misguided because it adopts a quite unjustifiably high standard. Christianity singles out the major shortcomings of our earthly existence: there is not enough happiness; there is too much suffering; the good and bad points are quite unequally and unfairly distributed; the underprivileged and underendowed do not get adequate compensation; it lasts only a short time. It then quite accurately depicts the perfect or ideal life as that which does not have any of these shortcomings. Its next step is to promise the believer that he will be able to enjoy this perfect life later on. And then it adopts as its standard of judgment the perfect life, dismissing as inadequate anything that falls short of it. Having dismissed earthly life as miserable, it further damns it by characterizing most of the pleasures of which earthly existence allows as bestial, gross, vile, and sinful, or alternatively as not really pleasurable.

This procedure is as illegitimate as if I were to refuse to call anything tall unless it is infinitely tall, or anything beautiful unless it is perfectly flawless, or any one strong unless he is omnipotent. Even if it were true that there is available to us an after-life which is flawless and perfect, it would still not be legitimate to judge earthly lives by this standard. We do not fail every candidate who is not an Einstein. And if we do not believe in an after-life, we must of course use ordinary earthly standards.

I have so far only spoken of the worthwhileness, only of what a person can get out of a life. There are other kinds of appraisal. Clearly, we evaluate people's lives not merely from the point of view of what they yield to the persons that lead them, but also from that of other men on whom these lives have impinged. We judge a life more significant if the person has contributed to the happiness of others, whether directly by what he did for others, or by the plans, discoveries, inventions, and work he performed. Many lives that hold little in the way of pleasure or happiness for its owner are highly significant and valuable, deserve admiration and respect on account of the contributions made.

It is now quite clear that death is simply irrelevant. If life can be worthwhile at all, then it can be so even though it be short. And if it is not worthwhile at all, then an eternity of it is simply a nightmare. It may be sad that we have to leave this beautiful world, but it is so only if and because it is beautiful. And it is no less beautiful for coming to an end. I rather suspect that an eternity of it might make us less appreciative, and in the end it would be tedious.

It will perhaps be objected now that I have not really demonstrated that life has a meaning, but merely that it can be worthwhile or have value. It must be admitted that there is a perfectly natural interpretation of the question, "What is the meaning of life?" on which my view actually proves that life has no meaning. I mean the interpretation discussed in section 2 of this lecture, where I attempted to show that, if we accept the explanations of natural science, we cannot believe that living organisms have appeared on earth in accordance with the deliberate plan of some intelligent being. Hence, on this view, life cannot be said to have a purpose, in the sense in which man-made things have a purpose. Hence it cannot be said to have a meaning or significance in that sense.

However, this conclusion is innocuous. People are disconcerted by the thought that *life as such* has no meaning in that sense only because they very naturally think it entails that no individual life can have meaning either. They naturally assume that *this* life or *that* can have meaning only if *life as such* has meaning. But it should by now be clear that your life and mine may or may not have meaning (in one sense) even if life as such has none (in the other). Of course, it follows from this that your life may have meaning while mine has not. The Christian view guarantees a meaning (in one sense) to every life, the scientific view does not (in any sense). By relating the question of the meaningfulness of life to the particular circumstances of an individual's existence, the scientific view leaves it an open question whether an individual's life has meaning or not. It is, however, clear that the latter is the important sense of "having a meaning." Christians, too, must feel that their life is wasted and meaningless if they have not achieved salvation. To know that even such lost lives have a meaning in another sense is no consolation to them. What matters is not that life should have a guaranteed meaning, whatever happens here or here-after, but that, by luck (Grace) or the right temperament and attitude (Faith) or a judicious life (Works) a person should make the most of his life.

"But here lies the rub," it will be said. "Surely, it makes all the difference whether there is an after-life. This is where morality comes in." It would be a mistake to believe that. Morality is not the meting out of punishment and reward. To be moral is to refrain from doing to others what, if they followed reason, they would not do to themselves, and to do for others what, if they followed reason, they would want to have done. It is, roughly speaking, to recognize that others, too, have a right to a worthwhile life. Being moral does not make one's own life worthwhile, it helps others to make theirs so.

Conclusion

I have tried to establish three points: (i) that scientific explanations render their explicanda as intelligible as pre-scientific explanations; they differ from the latter only in that, having testable implications and being more precisely formulated, their truth or falsity can be determined with a high degree of probability; (ii) that science does not rob human life of purpose, in the only sense that matters, but, on the contrary, renders many more of our purposes capable of realization; (iii) that common sense, the Christian world view, and the scientific approach agree on the criteria but differ on the standard to be employed in the evaluation of human lives; judging human lives by the standards of perfection, as Christians do, is unjustified; if we abandon this excessively high standard and replace it by an everyday one, we have no longer any reason for dismissing earthly existence as not worthwhile.

On the basis of these three points I have attempted to explain why so many people come to the conclusion that human existence is meaningless and to show that this conclusion is false. In my opinion, this pessimism rests on a combination of two beliefs, both partly true and partly false: the belief that the meaningfulness of life depends on the satisfaction of at least three conditions, and the belief that this universe satisfies none of them. The conditions are, first, that the universe is intelligible, second, that life has a purpose, and third, that all men's hopes and desires can ultimately be satisfied. It seemed to medieval Christians and it seems to many Christians today that Christianity offers a picture of the world which can meet these conditions. To many Christians and non-Christians alike it seems that the scientific world picture is incompatible with that of Christianity, therefore with the view that these three conditions are met, therefore with the view that life has a meaning. Hence they feel that they are confronted by the dilemma of accepting either a world picture incompatible with the discoveries of science or the view that life is meaningless.

I have attempted to show that the dilemma is unreal because life can be meaningful even if not all of these conditions are met. My main conclusion, therefore, is that acceptance of the scientific world picture provides no reason for saying that life is meaningless, but on the contrary every reason for saying that there are many lives which are meaningful and significant. My subsidiary conclusion is that one of the reasons frequently offered for retaining the Christian world picture, namely, that its acceptance gives us a guarantee of a meaning for human existence, is unsound. We can see that our lives can have a meaning even if we abandon it and adopt the scientific world picture instead. I have, moreover, mentioned several reasons for rejecting the Christian world picture: (i) the biblical explanations of the details of our universe are often simply false; (ii) the so-called explanations of the whole universe are incomprehensible or absurd; (iii) Christianity's low evaluation of earthly existence (which is the main cause of the belief in the meaninglessness of life) rests on the use of an unjustifiably high standard of judgment.

NOTES

1. Count Leo Tolstoy, "A Confession," reprinted in *A Confession, The Gospel in Brief, and What I Believe,* No. 229, The World's Classics (London: Geoffrey Cumberlege, 1940).

2. See e.g. Edwyn Bevan, *Christianity,* pp. 211-227. See also H. J. Paton, *The Modern Predicament* (London: George Allen and Unwin Ltd., 1955) pp. 103-116, 374.

3. See for instance, L. E. Elliott-Binns, *The Development of English Theology in the Later Nineteenth Century* (London: Longmans, Green & Co., 1952) pp. 30-33.

4. See "Phaedo" (*Five Dialogues* by Plato, Everyman's Library No. 456) para. 99, p. 189.

5. "On the Ultimate Origination of Things" (*The Philosophical Writings of Leibniz,* Everyman's Library No. 905) p. 32.

6. See "Monadology" (*The Philosophical Writings of Leibniz,* Everyman's Library No. 905) para. 32-38, pp. 8-10.

7. To borrrow the useful term coined by Professor D. A. T. Gasking of Melbourne University.

8. See e.g., J. J. C. Smart, "The Existence of God," reprinted in *New Essays in Philosophical Theology,* ed. by A. Flew and A. MacIntyre (London: S.C.M. Press, 1957) pp. 35-39.

9. That creation out of nothing is not a clarificatory notion becomes obvious when we learn that "in the philosophical sense" it does not imply creation at a particular time. The universe could be regarded as a creation out of nothing even if it had no beginning. See e.g. E. Gilson, *The Christian Philosophy of St. Thomas Aquinas* (London: Victor Gollancz Ltd. 1957) pp. 147-155 and E. L. Mascall, *Via Media* (London: Longmans, Green & Co., 1956) pp. 28 ff.

10. In what follows I have drawn heavily on the work of Ryle and Toulmin. See for instance G. Ryle, *The Concept of Mind* (London: Hutchinson's University Library, 1949) pp. 56-60 &c. and his article, "If, So, and Because," in *Philosophical Analysis* by Max Black, and S. E. Toulmin, *Introduction to the Philosophy of Science* (London: Hutchinson's University Library, 1953).

11. See above, p. 10, points (a)-(c).

12. L. Wittgenstein, *Tractatus Logico-Philosophicus* (London: Routledge & Kegan Paul Ltd., 1922), Sect. 6.44-6.45.

13. Op. cit. p. 46. See also Rudolf Otto, *The Idea of the Holy* (London: Geoffrey Cumberlege, 1952) esp. pp. 9-29.

14. Contemporary theologians would admit that it cannot be proved that the universe must have had a beginning. They would admit that we know it only through revelation. (See footnote No. 9.) I take it more or less for granted that Kant's attempted proof of the Thesis in his First Antinomy of Reason [Immanuel Kant's *Critique of Pure Reason,* trans. by Norman Kemp Smith (London: Macmillan and Co. Ltd., 1950) pp. 396-402] is invalid. It rests on a premise which is false: that the completion of the infinite series of succession of states, which must have preceded the present state if the world has had no beginning, is logically impossible. We can persuade ourselves to think that this infinite series is logically impossible if we insist that it is a series which must, literally, be *completed.* For the verb "to complete," as normally used, implies an activity which, in turn, implies

an agent who must have *begun* the activity at some time. If an infinite series is a whole that must be *completed* then, indeed, the world must have had a beginning. But that is precisely the question at issue. If we say, as Kant does at first, "that an eternity has elapsed," we do not feel the same impossibility. It is only when we take seriously the words "synthesis" and "completion," both of which suggest or imply "work" or "activity" and therefore "beginning," that it seems necessary that an infinity of successive states cannot have elapsed. [See also R. Crawshay-Williams, *Methods and Criteria of Reasoning* (London: Routledge & Kegan Paul, 1957) App. iv.]

15. See e.g. "Is Life Worth Living?" B.B.C. Talk by the Rev. John Sutherland Bonnell in *Asking Them Questions,* Third Series, ed. by R. S. Wright (London: Geoffrey Cumberlege, 1950).

16. See e.g. Rudolf Otto, *The Idea of the Holy,* pp. 9–11. See also C. A. Campbell, *On Selfhood and Godhood* (London: George Allen & Unwin Ltd., 1957) p. 246, and H. J. Paton, *The Modern Predicament,* pp. 69–71.

17. For a discussion of this issue, see the eighteenth century controversy between Deists and Theists, for instance, in Sir Leslie Stephen's *History of English Thought in the Eighteenth Century* (London: Smith, Elder & Co., 1902) pp. 112–119 and pp. 134–163. See also the attacks by Toland and Tindal on "the mysterious" in *Christianity not Mysterious* and *Christianity as Old as the Creation, or the Gospel a Republication of the Religion of Nature,* resp., parts of which are reprinted in Henry Bettenson's *Doctrines of the Christian Church,* pp. 426–431. For modern views maintaining that mysteriousness is an essential element in religion, see Rudolf Otto, *The Idea of the Holy,* esp. pp. 25–40, and most recently M. B. Foster, *Mystery and Philosophy* (London: S.C.M. Press, 1957) esp. Chs. IV. and VI. For the view that statements about God must be nonsensical or absurd, see e.g. H. J. Paton, op. cit. pp. 119–120, 367–369. See also "Theology and Falsification" in *New Essays in Philosophical Theology,* ed. by A. Flew and A. MacIntyre (London: S.C.M. Press, 1955) pp. 96–131; also N. McPherson, "Religion as the Inexpressible," ibid, esp. pp. 137–143.

18. Stephen Neill, *Christian Faith To-day* (London: Penguin Books, 1955) pp. 240–241.

19. It is difficult to feel the magnitude of this first sin unless one takes seriously the words, "Behold, the man has eaten of the fruit of the tree of knowledge of good and evil, and is become as one of us; and now, may he not put forth his hand, and take also of the tree of life, and eat, and live for ever?" Genesis iii, 22.

20. See in this connection the pastoral letter of 2nd February, 1905, by Johannes Katschtaler, Prince Bishop of Salzburg on the honour due to priests, contained in *Quellen zur Geschichte des Papsttums,* by Mirbt pp. 497–9, translated and reprinted in *The Protestant Tradition,* by J. S. Whale (Cambridge: University Press, 1955) pp. 259–262.

21. How impossible it is to make sense of this story has been demonstrated beyond any doubt by Tolstoy in his famous "Conclusion of A Criticism of Dogmatic Theology," reprinted in *A Confession, The Gospel in Brief, and What I Believe.*

22. See "The Nicene Creed," "The Tridentine Profession of Faith," "The Syllabus of Errors," reprinted in *Documents of the Christian Church,* pp. 34, 373 and 380 resp.

23. See e.g. J. S. Whale, *The Protestant Tradition,* Ch. IV., esp. pp. 48–56.

24. See ibid., pp. 61 ff.

25. See "The Confession of Augsburg" esp. Articles II., IV., XVIII., XIX., XX.; "Christianae Religionis Institutio," "The Westminster Confession of Faith," esp. Articles III., VI., IX., X., XI., XVI., XVII.; "The Baptist Confession of Faith," esp. Articles III., XXI., XXIII., reprinted in *Documents of the Christian Church,* pp. 294 ff., 298 ff., 344 ff., 349 ff.

26. See e.g. his *An Empiricist's View of the Nature of Religious Belief* (Eddington Memorial Lecture).

27. See e.g. the two series of Gifford Lectures most recently published: *The Modern Predicament* by H. J. Paton (London: George Allen & Unwin Ltd., 1955) pp. 69 ff. and *On Selfhood and Godhood* by C. A. Campbell (London: George Allen & Unwin Ltd., 1957) pp. 231–250.

28. Rudolf Otto, *The Idea of the Holy,* p. 9.

29. Fyodor Dostoyevsky, *The Devils* (London: The Penguin Classics, 1953) pp. 613–614.

30. Leo Tolstoy, *A Confession, The Gospel in Brief, and What I Believe,* The World's Classics, p. 24.

31. See for instance J. S. Whale, *Christian Doctrine,* pp. 171, 176–178, &c. See also Stephen Neill, *Christian Faith To-day,* p. 241.

1. What is Roszak's main objection to the scientific view of the world and its conceptions of knowledge, objectivity, and rationality? What does he mean by the distinctions between "augmentative" and "argumentative" knowledge? What do the distinctions between "quality" and "quantity" have to do with these distinctions? Why does Roszak appeal to Plato's ideal of knowledge in defending the "augmentative" view? (Reread the introduction to Part 5.)

2. According to Roszak, science is a "monster' because it renders life (and the universe) meaningless, and is thus immoral or amoral. Do you agree? What does Baier have to say about this charge?

3. How does Marx respond to Roszak's views? How does the "humanist tradition," appealed to by Roszak (and alluded to by Fromm: see introduction to Part 5), view modern science? How do the scientific ideals of objectivity, rationality, and knowledge turn science into a monster and adversely affect our culture and morals, on the views of Roszak and other "humanist" critics? What does the loss of meaning or purpose in the scientific view of the world have to do with their charges?

4. How, on Baier's view, can life be made "meaningful" without adopting the Christian world view, and without rejecting modern science? Do you agree with him?

5. Why, according to Baier's account, would the world appear "meaningless" for someone who rejected religious explanations of things? What is it about scientific explanations that are supposed to make life and the universe meaningless? (Here you may find it helpful to reread some of the articles in Part 2, e.g., Hospers.)

6. Baier argues (see his conclusions, (i)–(iii)) that scientific explanations make the world just as meaningful and intelligible as religious explanations and that life is, in a way, even more meaningful according to the scientific view of the universe than it is according to the Christian view (or at least the value placed upon [this?] life is more important). Why does he maintain these views? If he is right, would Roszak's critique of science be undermined? Why or why not?

7. Reread the Feyerabend article (Part 1). Why does he think science *is* a religion in our society? If he is right, would this detract from the plausibility of accounts, such as Roszak's and Baier's, which presuppose that science and religion are different and even conflicting world views?

8. What is Feyerabend's proposal regarding science education and the separation of science and the state? How will this "save society from science"? What do you think Roszak would say about this proposal?

9. How might Feyerabend attack Baier's views about science and religion? How would his idea that both science and religion are "ideologies" which deserve some credence—but not blind allegiance, as providing answers to all questions—bear upon Baier's point of view?

10. Baier's view assumes that morality can be made sense of without a religious foundation, and also that value judgments cannot be reduced to scientific terms, although they are compatible with them. Reread Part 5, especially Hempel, and briefly say whether Baier's (implicit) remarks about the nature of morality avoid the objections to the "objectivist" and "subjectivist" view of morality discussed in the introduction to Part 5.

11. Some authors, e.g., Roszak and Feyerabend (see Part 1), claim that science is a religion. How do they support this claim? To what extent is it true?

12. Can you think of any reasons to support the claim that science and technology are inherently evil, dehumanizing, alienating, manipulative, and so on? Is there any reason to support the contention that, since most of our problems are caused by science and technology, the only solution to them must come from (more) science and technology? (Is this like arguing that Jack the Ripper is the best person to teach women how to defend themselves against a mad rapist-murderer?)

13. Compare and contrast the technological imperative, *whatever can be done should be done,* with the ethical maxim, *whatever should be done can be done.* Provide counterexamples and examples in support of each of these maxims, and explain why and how they differ.

14. Feyerabend argues for a separation of science and education, analogous to the separation between church, state, and religion. Reread his article (Part 1) and briefly evaluate his arguments for this conclusion.

15. Some critics of science see a growing rift between democracy and the needs of technology (e.g., expert decision-makers). Give some concrete instances of this tension, and briefly consider the question whether democracy can survive the growing need for centralized expert planning in a technological society.

16. Is it naive to suppose that "critical" scientists and an educated public can control the abuses of science and technology through the democratic processes? Give some examples.

17. Why do so many critics of science and technology invoke phrases such as "the right to play God" in criticizing genetic engineering and behavior control? Why is it a criticism to accuse scientists of playing God?

18. How does the technological imperative—with its emphasis on technique and efficiency—affect our culture and personal outlook? For example, think of such things as manuals of technique for love-making, raising children, and so forth, our emphasis on efficiency, cost, and values in making decisions, and the role of expediency in politics and personal life. Are we dominated by these values, and how might we "work through" this sort of approach to personal and cultural situations?

19. Do you think that countercultural movements provide an adequate and realistic alternative to a technological culture? Give some examples. What are some of the values of such a culture, and what sort of changes in our lives might they bring about?

20. In attempting to weigh the overall effects of science on culture, why do so many critics either neglect or play down the positive benefits of, say, medicine, and the beneficial products of applied science and technology (electricity, plumbing, and so on)? And why do they neglect or play down the fact that human beings are, in part, technological beings, at least to the extent that life as we know it, and society as we know it, never has been, and never will be, possible without technology? How might someone like Roszak answer these questions?

SELECTED BIBLIOGRAPHY

[1] Barbour, I. *Issues in Science and Religion*. New York: Harper & Row, 1971. An impressive scholarly and analytic account of a variety of fundamental issues about science, the philosophy of science, religion, and their interrelations.

[2] Heidegger, M. *The Question Concerning Technology and Other Essays*. New York: Harper & Row, 1978. The title essay and "Science and Reflection" are indispensable readings for anyone who wishes to confront the basic meaning of science and technology. Heidegger is an important but difficult thinker; beginners might want to look at [3] first.

[3] Leiss, W. *The Domination of Nature*. Boston: Beacon Press, 1974. A lucid account, based upon insights from Husserl, Heidegger, and Marcuse, of the ideal of science and its impact on contemporary culture through the idea that knowledge requires domination over nature and human beings. Especially good on historical developments.

[4] Matson, F., *The Broken Image*. New York: Anchor-Doubleday, 1965. An insightful historical account of the interrelations between the physical and biological sciences and the self-images of the social and behavioral sciences. Attempts to show that developments in twentieth-century physics and biology are more consonant with humanistic approaches to the study of human behavior.

[5] Mitcham, C., and McKay, R., eds. *Philosophy and Technology*. New York: Free Press, 1972. A useful collection of essays, with superb bibliography, on many aspects of technology, culture, pure and applied science.

[6] Passmore, J. *Science and Its Critics*. Rutgers: State University Press, 1978. A lucid and intelligent discussion, by a philosopher and scholar of international note, of countercultural attacks on science. A useful antidote to [8].

[7] Polanyi, M. *Science, Faith and Society*. Chicago: Phoenix Books, 1964. A stimulating attempt at defending a liberal concept of the open society and freedom of inquiry with an account of science which puts morality at the center of things.

[8] Roszak, T. *Where the Wasteland Ends*. New York: Anchor–Doubleday, 1972. The most elaborate defense of the idea that science's impact on our culture is pervasive and destructive, together with a detailed proposal

for an alternative based upon the insights of the romantic poets and the humanists of an earlier day. Provocative reading.

[9] Truitt, W., and Solomon, T. G. eds. *Science, Technology and Freedom.* Boston: Houghton Mifflin, 1974. A useful collection of essays on some basic issues, including many of those surveyed in Parts 5 and 6 of this book.

[10] Whitehead, A. N. *Science and the Modern World.* New York: Macmillan, 1925. A seminal work that has long been a classic discussion of the origins of modern science and its impact on philosophy, religion, and culture. Still one of the best places to begin reading on these topics.

[11] Ziman, J. *The Force of Knowledge.* New York: Cambridge University Press, 1978. A very instructive and enjoyable book by a noted physicist that provides many helpful examples and useful information about the origins and nature of science and technology. Designed to show that any blanket claims about pure and applied science, the evils of technology, and so on, are not to be taken seriously.

PAPERBACKS AVAILABLE FROM PROMETHEUS BOOKS

CRITIQUES OF THE PARANORMAL

____ESP & PARAPSYCHOLOGY: A CRITICAL RE-EVALUATION C.E.M. Hansel	$7.95
____EXTRA-TERRESTRIAL INTELLIGENCE James L. Christian, editor	$6.95
____OBJECTIONS TO ASTROLOGY L. Jerome & B. Bok	3.95
____THE PSYCHOLOGY OF THE PSYCHIC David Marks & Richard Kammann	7.95
____PHILOSOPHY & PARAPSYCHOLOGY J. Ludwig, editor	8.95

HUMANISM

____ETHICS WITHOUT GOD K. Nielsen	4.95
____HUMANIST ALTERNATIVE Paul Kurtz, editor	4.95
____HUMANIST ETHICS Morris Storer, editor	8.95
____HUMANIST FUNERAL SERVICE Corliss Lamont	2.95
____HUMANIST MANIFESTOS I & II	1.95
____HUMANIST WEDDING SERVICE Corliss Lamont	1.95
____HUMANISTIC PSYCHOLOGY I. David Welch, George Tate, Fred Richards, editors	8.95
____MORAL PROBLEMS IN CONTEMPORARY SOCIETY Paul Kurtz, editor	5.95
____VOICE IN THE WILDERNESS Corliss Lamont	4.95
____RABBI AND MINISTER: THE FRIENDSHIP OF STEPHEN S. WISE AND JOHN HAYNES HOLMES Carl Hermann Voss	6.95

PHILOSOPHY & ETHICS

____ART OF DECEPTION Nicholas Capaldi	5.95
____BENEFICENT EUTHANASIA M. Kohl, editor	7.95
____ESTHETICS CONTEMPORARY Richard Kostelanetz, editor	9.95
____EXUBERANCE: A PHILOSOPHY OF HAPPINESS Paul Kurtz	3.00
____FULLNESS OF LIFE Paul Kurtz	5.95
____FREEDOM OF CHOICE AFFIRMED Corliss Lamont	4.95
____HUMANHOOD: ESSAYS IN BIOMEDICAL ETHICS Joseph Fletcher	6.95
____JOURNEYS THROUGH PHILOSOPHY N. Capaldi & L. Navia, editors	10.95
____MORAL EDUCATION IN THEORY & PRACTICE Robert Hall & John Davis	7.95
____PHILOSOPHY: AN INTRODUCTION Antony Flew	5.95
____THINKING STRAIGHT Antony Flew	4.95
____WORLDS OF PLATO & ARISTOTLE J.B. Wilbur & H.J. Allen, editors	5.95
____WORLDS OF THE EARLY GREEK PHILOSOPHERS J.B. Wilbur & H.J. Allen, editors	5.95
____PHILOSOPHY: AN INTRODUCTION Antony Flew	6.95
____INTRODUCTORY READINGS IN THE PHILOSOPHY OF SCIENCE E.D. Klemke, Robert Hollinger, A. David Kline, editors	10.95

SEXOLOGY

____THE FRONTIERS OF SEX RESEARCH *Vern Bullough, editor*   6.95

____NEW BILL OF SEXUAL RIGHTS & RESPONSIBILITIES *Lester Kirkendall*   1.95

____NEW SEXUAL REVOLUTION *Lester Kirkendall, editor*   4.95

____PHILOSOPHY & SEX *Robert Baker & Fred Elliston, editors*   6.95

____SEX WITHOUT LOVE: A PHILOSOPHICAL EXPLORATION *Russell Vannoy*   7.95

SKEPTICS BOOKSHELF

____CLASSICS OF FREE THOUGHT *Paul Blanshard, editor*   5.95

____CRITIQUES OF GOD *Peter Angeles, editor*   6.95

____WHAT ABOUT GODS? (for children) *Chris Brockman*   3.95

SOCIAL ISSUES

____AGE OF AGING: A READER IN SOCIAL GERONTOLOGY
Abraham Monk, editor   8.95

____REVERSE DISCRIMINATION *Barry Gross, editor*   7.95

The books listed above can be obtained from your book dealer
or directly from Prometheus Books.
Please check off the appropriate books.
Remittance must accompany all orders from individuals.
Please include $1.00 postage and handling for each book.
(N.Y. State Residents add 7% sales tax)

Send to _____
(Please type or print clearly)

Address _____

City _____ State_____Zip_____

Amount Enclosed_____

⅙ Prometheus Books
Box 55 Kensington Station
Buffalo, New York 14215